Leaps and Bounds toward Math Understanding 7/8

Senior Author
Marian Small

Authors
Kathy Kubota-Zarivnij
Amy Lin

Advisory Panel
Katharine Borgen, BC
Cathy Campbell, AB
Richard Donnelly, SK
Joyce Tonner, ON

Reviewers

Debora Benedict
School District 44
North Vancouver, BC

Tanya Cook
School District 34
Abbotsford, BC

Stefanie DeAngelis
Peel District School Board, ON

Kerry Dwyer-Mitchell
Hamilton-Wentworth Catholic
District School Board, ON

Marie Elliot
Kawartha Pine Ridge
District School Board, ON

Carole Fullerton
Mind-Full Consulting
Vancouver, BC

Robin Harris
Halifax Regional School Board, NS

Murray Jackson
Upper Canada District
School Board, ON

Deanna Lightbody
School District 35
Langley, BC

Deb McLean
District School Board
of Niagara, ON

Michelle Middleton
School District 34
Abbotsford, BC

Scott Moore
Thames Valley District
School Board, ON

Mike Pue
School District 35
Langley, BC

Leslie Stoffberg
School District 43
Coquitlam, BC

Scott Young
District School Board
of Niagara, ON

NELSON EDUCATION

NELSON EDUCATION

Leaps and Bounds Toward Math Understanding 7/8
Teacher's Resource

Editorial Director
Linda Allison

Publisher, Mathematics
Colin Garnham

Managing Editor, Development
Erynn Marcus

Product Manager
Linda Krepinsky

Program Manager
Mary Reeve

Project Manager
Alisa Yampolsky

Developmental Editors
Shirley Barrett
Megan Robinson
Gurmeet K. Sodhi-Bains
Jackie Williams

Editorial Assistant
Claire Sheldon

Senior Content Production Manager
Sujata Singh

Content Production Editor
Joe Zingrone

Copyeditor
Gerry Jenkison

Proofreaders
Judy Sturrup
Linda Szostak

Indexer
Marilyn Augst

Senior Production Coordinator
Sharon Latta Paterson

Production Manager
Helen Jager Locsin

Design Director
Ken Phipps

Interior Design
Courtney Hellam
Visutronx

Cover Design
Jennifer Leung

Cover Photo
Shutterstock; Getty Images

Asset Coordinators
Suzanne Peden
Visutronx

Illustrators
Dave McKay
Brian McLachlan
Visutronx

Compositor
MPS Limited

Printer
Transcontinental Printing

COPYRIGHT © 2012 by
Nelson Education Ltd.

ISBN-13: 978-0-17-635152-6
ISBN-10: 0-17-635152-3

Printed and bound in Canada
1 2 3 4 15 14 13 12

For more information contact
Nelson Education Ltd.,
1120 Birchmount Road, Toronto,
Ontario, M1K 5G4. Or you can
visit our Internet site at
http://www.nelson.com.

ALL RIGHTS RESERVED. No part of this work covered by the copyright herein, except for any reproducible pages included in this work, may be reproduced, transcribed, or used in any form or by any means—graphic, electronic, or mechanical, including photocopying, recording, taping, Web distribution, or information storage and retrieval systems—without the written permission of the publisher.

For permission to use material from this text or product, submit a request online at www.cengage.com/permissions. Further questions about permissions can be emailed to permissionrequest@cengage.com.

Every effort has been made to trace ownership of all copyrighted material and to secure permission from copyright holders. In the event of any question arising as to the use of any material, we will be pleased to make the necessary corrections in future printings.

Reproduction of BLMs is permitted for classroom/instruction purposes only and only to the purchaser of this product.

Note:
Solutions for the Student Resource questions can be found at www.nelson.com/leapsandbounds.
User name: leapsandbounds78
Password: SRsolutions78

Contents

Program Overview

Teaching Notes

Strand: Number

 Copyright © 2012 by Nelson Education Ltd.

Strand: Measurement

Copyright © 2012 by Nelson Education Ltd.

Copyright © 2012 by Nelson Education Ltd.

Blackline Masters

 Copyright © 2012 by Nelson Education Ltd.

What Is *Leaps and Bounds*?

Leaps and Bounds Toward Math Understanding is a supplementary resource for students struggling in mathematics, Grades 3 to 8. Often these students are not struggling in other areas. This resource is designed to assist teachers in providing precise, targeted remediation for these students—as a whole class, in small groups, or individually. With *Leaps and Bounds*, teachers can help students better understand the prerequisite math so they can be successful in meeting the curriculum requirements for their grade. *Leaps and Bounds* materials determine gaps in understanding and then provide tailored interventions that target the missing knowledge rather than simply repeating all curriculum expectations and outcomes from previous grades.

For each topic in each strand, a short diagnostic assessment provides a snapshot of what students understand and what they do not. Each diagnostic tool leads to tailored intervention materials that provide differentiated instruction for each student or group of students. These materials are organized into pathways with open-ended and guided interventions. Both approaches to intervention are developmentally appropriate and based on knowledge gaps identified through the diagnostic tools in conjunction with the other assessment processes used in the classroom.

> There is a belief that students, especially students struggling in mathematics, want and need you to model each new concept or skill. However, what we are learning is that students can figure out many mathematical ideas on their own and in their own way. Some can also benefit from clearly explained, conceptually based development of mathematical ideas. *Leaps and Bounds* recognizes that many students will not be able to make a substantial leap if the learning is overly structured; they become too dependent on someone else doing too much of the thinking for them. If we want leaps forward, and not just baby steps, we need to provide accessible, conceptually based interventions with lots of visual support.
>
> **—Marian Small**

Who Is *Leaps and Bounds* Designed For?

Leaps and Bounds is designed for students experiencing difficulty in mathematics in situations such as the following:

- an age-appropriate classroom program
- an individualized educational program either in a regular classroom or in a separate instructional setting
- a formal after-school tutoring program
- additional help sessions

The resource may be used by a classroom teacher, special education teacher, or tutor.

Copyright © 2012 by Nelson Education Ltd.

A Research Foundation

Leaps and Bounds Toward Math Understanding addresses the need for differentiated instruction based on how students learn. It draws on what we know about how students learn by sequencing the content in a developmentally appropriate way and providing alternatives in approach.

Leaps and Bounds has a solid research foundation that reflects
- developmental learning of mathematics, as determined from the *PRIME* research
- common areas of difficulty in mathematics and the best instructional practices for addressing these areas of difficulty
- current research about how to address different learning styles as well as alternative strategies for learning a mathematical concept

Developmental Learning in Mathematics

Marian Small and colleagues at the University of Toronto conducted Canadian research from 2002 to 2004 on the developmental learning of mathematics with 12 000 students from kindergarten to Grade 7 in seven provinces. The data collected were the foundation of developmental maps published in *PRIME* (*Professional Resources and Instruction for Math Educators,* Nelson Education Ltd., 2005–2010). The maps for all 5 strands describe stages, or phases, that students travel through as they learn mathematics and indicate key behaviours that students exhibit at each phase. The maps also reflect the curriculum used in elementary schools across Canada.

Developmental Learning and Leaps and Bounds

The *PRIME* research and developmental maps were invaluable in the creation of *Leaps and Bounds*. They formed the foundation for the creation of carefully sequenced interventions. For example, development in decimal operations goes from adding and subtracting decimals to multiplying and dividing decimals by whole numbers to multiplying and dividing by decimals. So, in *Leaps and Bounds 7/8*, four intervention pathways reflect this conceptual development:
Pathway 1: Dividing Whole Numbers by Decimals
Pathway 2: Dividing Decimals by Whole Numbers
Pathway 3: Multiplying with Decimals
Pathway 4: Adding and Subtracting Decimals

The pathways in all topics re-teach critical concepts and allow students to progress in their understanding in a developmentally appropriate way. A diagnostic tool that highlights common difficulties is provided to determine which pathway is appropriate for an individual student. Although a student who is significantly behind might be assigned Pathway 3 or 4, it is not always necessary to then complete Pathway 2 and then Pathway 1. After the student has completed Pathway 3 or 4, appropriate questions from the diagnostic tool can be used to determine whether further intervention is required.

 Copyright © 2012 by Nelson Education Ltd.

Common Areas of Difficulty in Mathematics

Most teachers know from experience where students tend to have difficulty in mathematics. There is also ample research literature on this (see the References at www.nelson.com/leapsandbounds). In addition, one of the interesting by-products of the *PRIME* research on how students learn mathematics was that common areas of difficulty surfaced.

An understanding of developmental learning in math can be key to remediating these common difficulties, since some arise from students being introduced to concepts before they are ready or because the concepts are introduced in a developmentally inappropriate way. We know from the research that specific strategies can be used to target some of the difficulties.

Common Areas of Difficulty and Leaps and Bounds

Simply creating pathways that reflect what we know about how students develop their understanding of mathematics will go a long way to address common difficulties. However, *Leaps and Bounds* also uses what we know about where and why students struggle in mathematics and how to help them as a basis for creating intervention pathways. So, the pathways not only allow students an opportunity to re-examine concepts in a conceptually meaningful way but also include tasks and questions that target common areas of difficulty.

As well, the *Leaps and Bounds* Teacher's Resource provides a list of the common areas of difficulty in every topic overview. For example:

Why might students struggle with decimal operations?

Students might struggle with decimal operations for any of the following reasons:

- They might have difficulty with whole number calculations, which then affects their performance with decimals.
- They might not have a strong enough understanding of thousandths and, therefore, might have difficulty computing with them.
- They might struggle with adding or subtracting decimals with different numbers of digits (e.g., they might line up $1.34 + 12.3$ by the 3 digits instead of by the decimal points and digits of equivalent place value).
- They might struggle when the number of tenths, hundredths, or thousandths in the minuend is fewer than the number of tenths, hundredths, or thousandths in the subtrahend (e.g., they calculate $0.327 - 0.269$ incorrectly as 0.142).
- They might misplace the decimal point in the answer because they do not estimate and/or they might have limited number sense.
- They might struggle with remainders (e.g., when dividing 4.1 by 3, they get 1.3 and do not know what to do with the 2 tenths remainder).
- They might not recognize that division can mean counting the number of groups in an amount and not just sharing. This would make it difficult to understand dividing a whole number by a decimal.

In the teaching notes for each intervention, this Teacher's Resource includes key behaviours to look for and key questions to ask as students work, which focus on the critical concepts and the common areas of difficulty.

Current Research on Supporting Struggling Students

Current research on supporting struggling students suggests the need to incorporate the following in any remediation program:

- differentiated instruction
- visual representation
- meaningful practice
- scaffolding
- math discussion

Differentiated Instruction and Leaps and Bounds

To differentiate instruction for any student or group of students, both the content and the strategies used to teach the content can be individualized. *Leaps and Bounds* does both.

Individualizing Content

As described earlier, each topic in each strand provides multiple intervention pathways for students to follow, depending on individual needs. A diagnostic tool is provided to help determine which pathway is most suitable for the student or group of students.

Individualizing Instructional Approaches

Having determined which pathway is most suitable, the teacher has another choice to make, that is, whether to use a more open approach (the open-ended intervention) or a more guided approach (the guided intervention). Thus, the instructional approach can be individualized or differentiated. The 2 types of intervention cover the same content but in different ways:

- The open-ended interventions typically provide a brief introduction with minimal instruction, often in a context, followed by a problem to solve or a task that has multiple possible solutions. It allows students who prefer exploring in their own way more opportunity to do so.
- The guided interventions begin with an instructional section, followed by **Try These** questions that are sequenced to guide the student through the content in a more deliberate way. This suits students who prefer more direction, although a variety of strategies for doing the math are still explored.

The teacher can choose the intervention that is more suitable for the student's needs or style and most appropriate for his or her specific learning situation. Teachers might also use both interventions, in either order.

Visual Representations and Leaps and Bounds

Visual models make it easier to learn as they help explain and make sense of the abstract concepts in math. They also provide images that students can call upon later to help them use visualization when solving mathematical problems and explain their thinking.

 Copyright © 2012 by Nelson Education Ltd.

In *Leaps and Bounds*, a variety of visual representations are used to help students "see," understand, and remember the math. Visual representations such as thousandths grids, number lines, place value charts, labelled diagrams, and illustrations are used to represent abstract concepts, as well as to engage students. Students then incorporate these models as they solve problems and explain their thinking.

Meaningful Practice and Leaps and Bounds

Meaningful practice is a balance of conceptual and procedural work that does not lose sight of the big ideas of mathematics. In *Leaps and Bounds*, the **Try These** questions in the guided interventions provide opportunities for students to apply, extend, and practise in a meaningful way what they have learned in the instructional section. These questions are carefully sequenced (from straightforward to more complex and challenging) and include a variety of visual representations and contexts (both real world and mathematical contexts) so that students can more easily generalize their learning.

Scaffolding and Leaps and Bounds

Scaffolding is used throughout *Leaps and Bounds* to prepare students and help them better understand what they are asked to do. For example,

- The "Before Using the Intervention" activity in the Teacher's Resource is a series of questions for each intervention that teachers can use to prepare students for what they will be doing in the Student Resource.
- Marginal definition boxes and Remember boxes are used in the Student Resource to remind students of key terms and concepts.
- Pathways within each topic are designed to be parallel. So, a student who is successful with Pathway 2 will be prepared to work on Pathway 1 if the teacher feels that the student needs more intervention.
- The **Try These** questions of the guided interventions are organized so the questions evolve from simple to more challenging.
- Questions or problems are broken down into logical and sequential parts (either using parts a), b), and so on, or using bullets) that help students move toward the solution in steps.

Math Discussion and Leaps and Bounds

Because *Leaps and Bounds* is a supplementary remediation resource, most of the mathematical discussions will occur among students working in pairs or small groups. There will also be valuable discussions between teacher and student. The open interventions in *Leaps and Bounds* are designed to be open enough to make discussion not only inevitable but valuable. The guided interventions also include some open questions that invite discussion about the bigger ideas behind the lesson. As well, in the Teacher's Resource for each intervention, a series of Consolidating and Reflecting questions are provided, along with sample responses. These are designed to prompt a rich discussion about what students have been doing.

Copyright © 2012 by Nelson Education Ltd.

How to Use *Leaps and Bounds*

Leaps and Bounds Toward Math Understanding provides interventions for various topics in each of the 5 strands of mathematics. Each intervention can be used in a variety of instructional groupings—whole class, small group, or individual. The plan below outlines how a classroom teacher might use *Leaps and Bounds* in conjunction with core math resources such as *Nelson Mathematics* or *Nelson Math Focus.*

Step 1 Diagnostic Assessment

When you start a new topic in the core resource, use the review questions at the beginning of a chapter (e.g., Getting Started) to identify students who may be struggling. Have these students complete the diagnostic tool for that topic in the *Leaps and Bounds* Teacher's Resource to get a more detailed picture of their understanding. You may have students complete the questions independently in writing, or you may want to read the questions and have students respond orally.

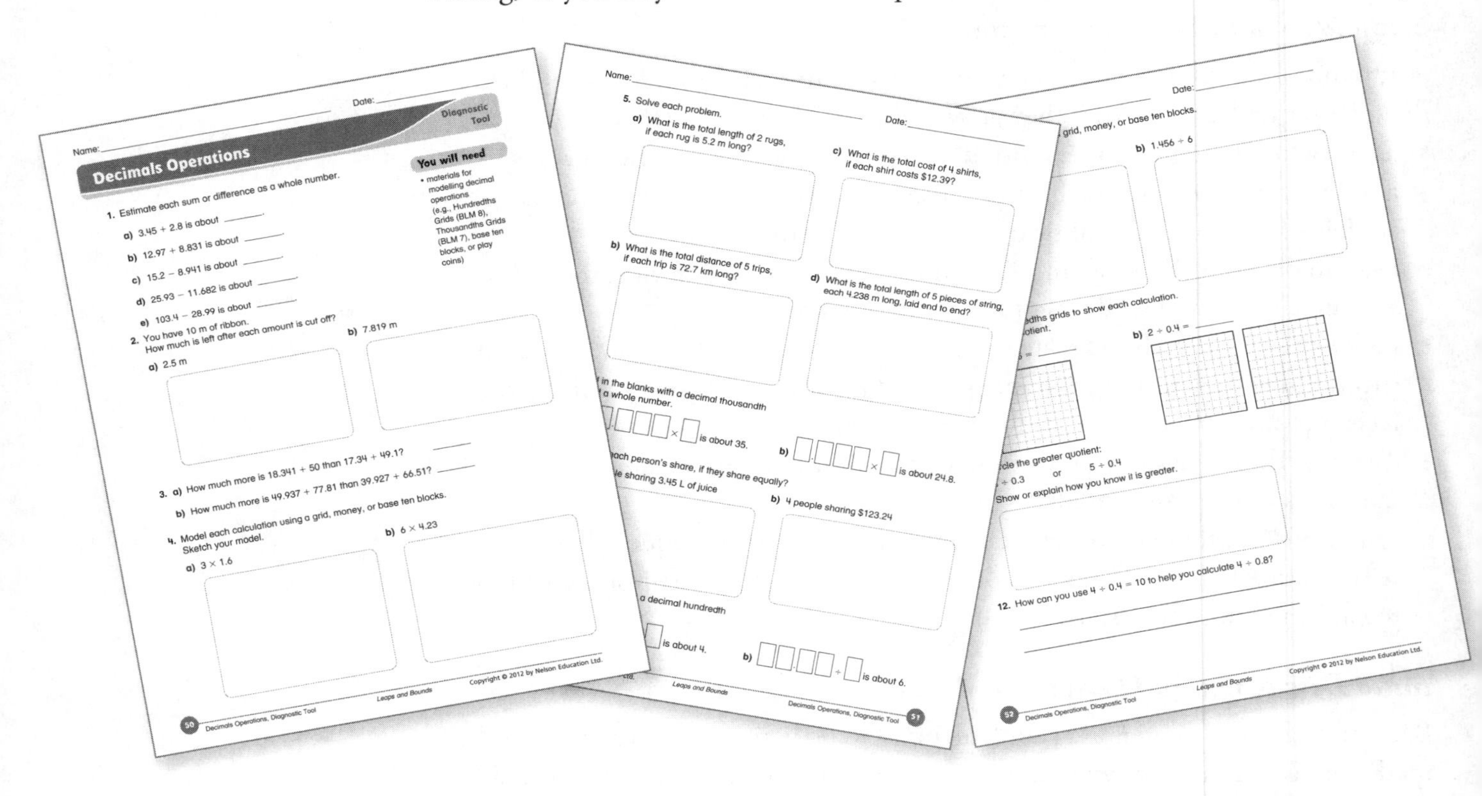

Name: Date:

Diagnostic Tool

Decimals Operations

You will need
- materials for modelling decimal operations (e.g., Hundredths Grids (BLM 8), Thousandths Grids (BLM 7), base ten blocks, or play coins)

1. Estimate each sum or difference as a whole number.
   a) 3.45 + 2.8 is about ____
   b) 12.97 + 8.831 is about ____
   c) 15.2 − 8.941 is about ____
   d) 25.93 − 11.682 is about ____
   e) 103.4 − 28.99 is about ____
2. You have 10 m of ribbon. How much is left after each amount is cut off?
   a) 2.5 m
   b) 7.819 m
3. a) How much more is 18.341 + 50 than 17.34 + 49.1? ____
   b) How much more is 49.937 + 77.81 than 39.927 + 66.51? ____
4. Model each calculation using a grid, money, or base ten blocks. Sketch your model.
   a) 3 × 1.6
   b) 6 × 4.23

50 Decimals Operations, Diagnostic Tool Leaps and Bounds Copyright © 2012 by Nelson Education Ltd.

Name: Date:

5. Solve each problem.
   a) What is the total length of 2 rugs, if each rug is 5.2 m long?
   b) What is the total distance of 5 trips, if each trip is 72.7 km long?
   c) What is the total cost of 4 shirts, if each shirt costs $12.39?
   d) What is the total length of 5 pieces of string, each 4.238 m long, laid end to end?

...l in the blanks with a decimal thousandth ...a whole number.
   ... × ☐ is about 35.
   b) ☐.☐☐☐ × ☐ is about 24.8.

...ach person's share, if they share equally?
   ...e sharing 3.45 L of juice
   b) 4 people sharing $123.24

... a decimal hundredth
   ... is about 4.
   b) ☐☐.☐☐ ÷ ☐ is about 6.

Leaps and Bounds Decimals Operations, Diagnostic Tool 51

Date:

... grid, money, or base ten blocks.
   b) 1.456 ÷ 6

...dths grids to show each calculation.
...otient.
   b) 2 ÷ 0.4 = ____

...cle the greater quotient:
   ... ÷ 0.3 or 5 ÷ 0.4
   Show or explain how you know it is greater.

12. How can you use 4 ÷ 0.4 = 10 to help you calculate 4 ÷ 0.8?

52 Decimals Operations, Diagnostic Tool Leaps and Bounds Copyright © 2012 by Nelson Education Ltd.

 Copyright © 2012 by Nelson Education Ltd.

Step 2 Pathway

Use the results of the diagnostic tool to choose an appropriate pathway for each student. Solutions are provided in this Teacher's Resource with a key to pathways.

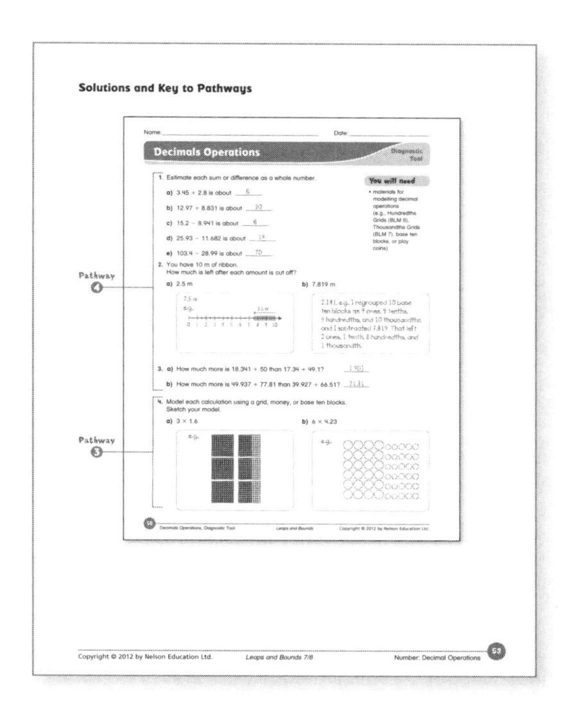
Solutions and Key to Pathways

Name: Date:

Decimals Operations Diagnostic Tool

1. Estimate each sum or difference as a whole number.

a) 3.45 + 2.8 is about 6

b) 12.97 + 8.831 is about 22

c) 15.2 − 8.941 is about 6

d) 25.93 − 11.682 is about 14

e) 103.4 − 28.99 is about 75

2. You have 10 m of ribbon. How much is left after each amount is cut off?

a) 2.5 m b) 7.819 m

3. a) How much more is 18.341 + 50 than 17.34 + 49.1?

b) How much more is 49.937 + 77.81 than 39.927 + 66.51?

4. Model each calculation using a grid, money, or base ten blocks. Sketch your model.

a) 3 × 1.6 b) 6 × 4.23

You will need
- materials for modelling decimal operations (e.g., Hundredths Grids (BLM 8), Thousandths Grids (BLM 7), base ten blocks, or play coins)

Pathway 4

Pathway 3

Copyright © 2012 by Nelson Education Ltd. *Leaps and Bounds 7/8* Number: Decimal Operations 53

Diagnostic Tool Results	Intervention Pathway
If students struggle with Questions 10 to 12	use Pathway 1: Dividing Whole Numbers by Decimals *Teacher's Resource pages 56–57* *Student Resource pages 56–61*
If students struggle with Questions 7 to 9	use Pathway 2: Dividing Decimals by Whole Numbers *Teacher's Resource pages 58–59* *Student Resource pages 62–67*
If students struggle with Questions 4 to 6	use Pathway 3: Multiplying with Decimals *Teacher's Resource pages 60–61* *Student Resource pages 68–73*
If students struggle with Questions 1 to 3	use Pathway 4: Adding and Subtracting Decimals *Teacher's Resource pages 62–63* *Student Resource pages 74–79*

Step 3 Intervention

Choose either the open-ended intervention or the guided intervention for the pathway. The remainder of the class could be working on an exploration lesson from the core resource.

Use the "Before Using the Intervention" section of the Teacher's Resource to get students ready. Then introduce the Student Resource pages. Read through the open-ended tasks or work through the guided instruction with students. While students then continue their work on either the open-ended or guided task in pairs, you could return to the core resource with the rest of the class.

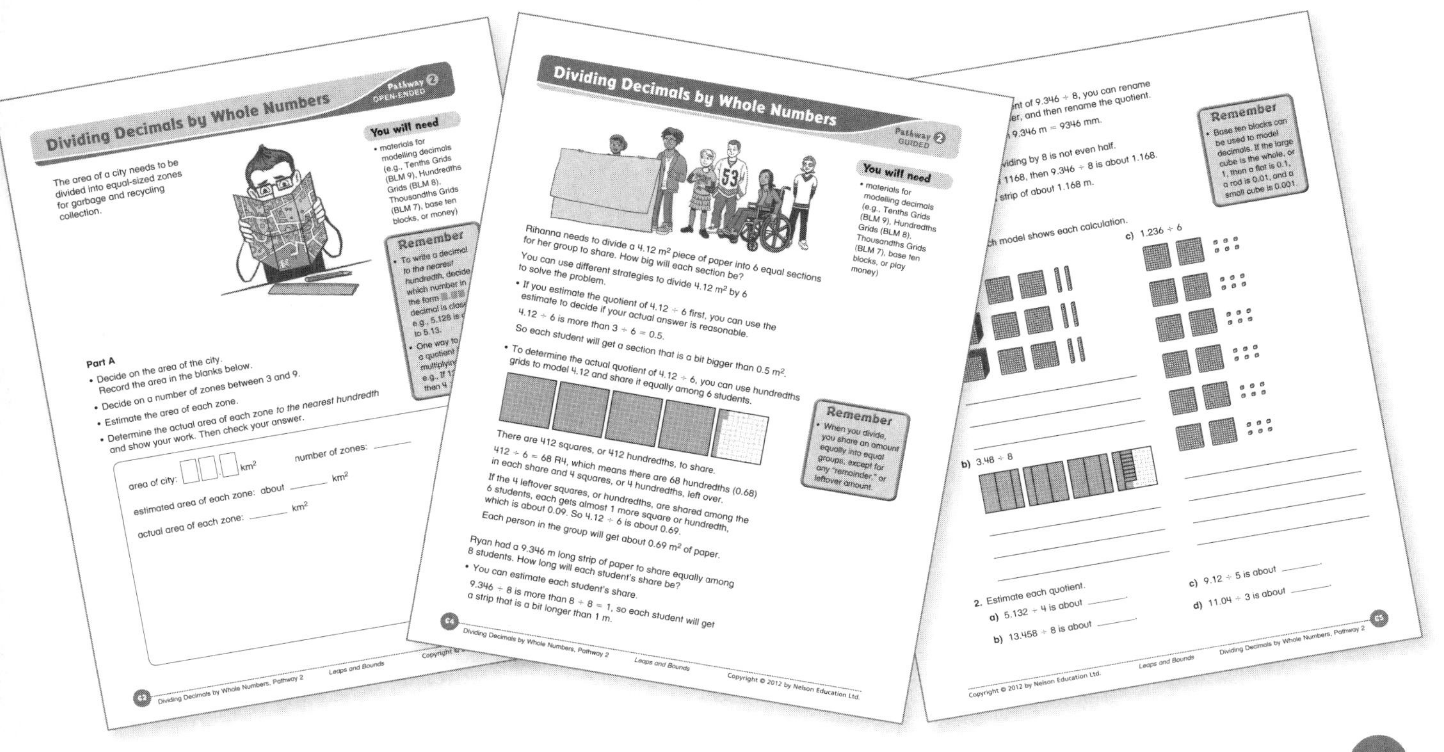
Dividing Decimals by Whole Numbers Pathway 2 OPEN-ENDED

The area of a city needs to be divided into equal-sized zones for garbage and recycling collection.

Part A
- Decide on the area of the city. Record the area in the blanks below.
- Decide on a number of zones between 3 and 9.
- Estimate the area of each zone.
- Determine the actual area of each zone to the nearest hundredth and show your work. Then check your answer.

area of city: km² number of zones:

estimated area of each zone: about km²

actual area of each zone: km²

Dividing Decimals by Whole Numbers Pathway 2 GUIDED

Rihanna needs to divide a 4.12 m² piece of paper into 6 equal sections for her group to share. How big will each section be?

You can use different strategies to divide 4.12 m² by 6 to solve the problem.

- If you estimate the quotient of 4.12 ÷ 6 first, you can use the estimate to decide if your actual answer is reasonable.
4.12 ÷ 6 is more than 3 ÷ 6 = 0.5.
So each student will get a section that is a bit bigger than 0.5 m².
- To determine the actual quotient of 4.12 ÷ 6, you can use hundredths grids to model 4.12 and share it equally among 6 students.

There are 412 squares, or 412 hundredths, to share.
412 ÷ 6 = 68 R4, which means there are 68 hundredths (0.68) in each share and 4 squares, or 4 hundredths, left over.
If the 4 leftover squares, or hundredths, are shared among the 6 students, each gets almost 1 more square or hundredth, which is about 0.09. So 4.12 ÷ 6 is about 0.69.
Each person in the group will get about 0.69 m² of paper.

Ryan had a 9.346 m long strip of paper to share equally among 8 students. How long will each student's share be?
- You can estimate each student's share.
9.346 ÷ 8 is more than 8 ÷ 8 = 1, so each student will get a strip that is a bit longer than 1 m.

Remember
- When you divide, you share an amount equally into equal groups, except for any "remainder," or leftover amount.

b) 3.48 ÷ 8

c) 1.236 ÷ 6

2. Estimate each quotient.
a) 5.132 ÷ 4 is about
b) 13.458 ÷ 8 is about
c) 9.12 ÷ 5 is about
d) 11.04 ÷ 3 is about

Step 4 Consolidation

When students have completed their work with the Student Resource pages, bring out and consolidate the key ideas from the intervention. Use the questions in the "Consolidating and Reflecting" section of the Teacher's Resource, adapted with examples from students' work.

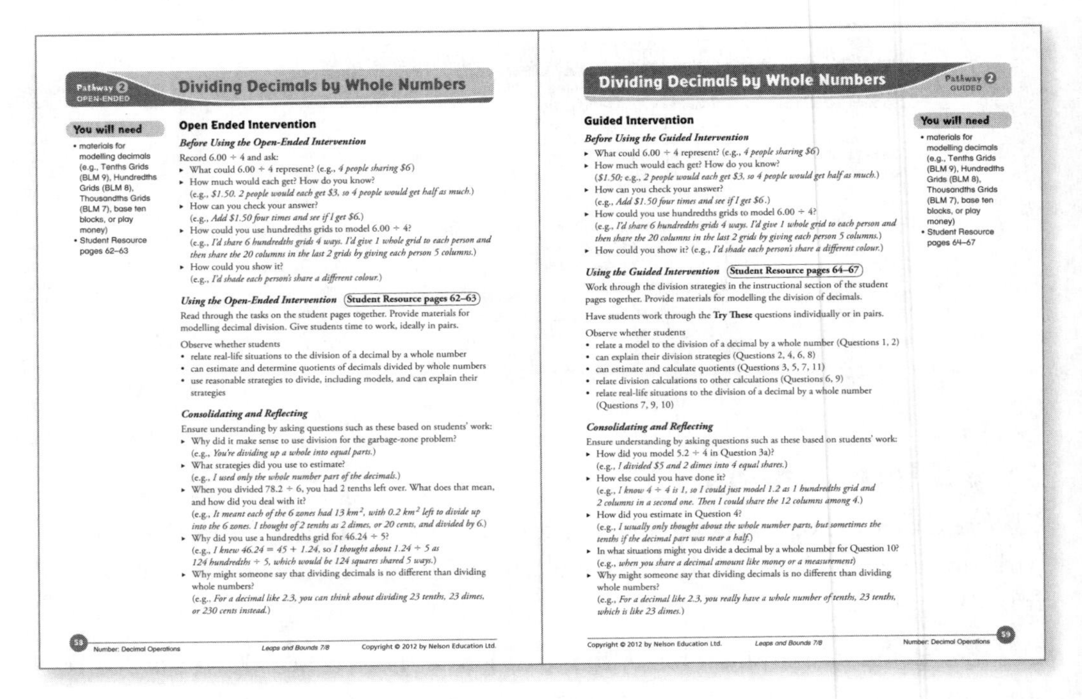

Pathway 2 OPEN-ENDED

Dividing Decimals by Whole Numbers

You will need

- materials for modelling decimals (e.g., Tenths Grids (BLM 9), Hundredths Grids (BLM 8), Thousandths Grids (BLM 7), base ten blocks, or play money)
- Student Resource pages 62–63

Open Ended Intervention

Before Using the Open-Ended Intervention

Record 6.00 ÷ 4 and ask:

- What could 6.00 ÷ 4 represent? (e.g., *4 people sharing $6*)
- How much would each get? How do you know? (e.g., *$1.50. 2 people would each get $3, so 4 people would get half as much.*)
- How can you check your answer? (e.g., *Add $1.50 four times and see if I get $6.*)
- How could you use hundredths grids to model 6.00 ÷ 4? (e.g., *I'd share 6 hundredths grids 4 ways. I'd give 1 whole grid to each person and then share the 20 columns in the last 2 grids by giving each person 5 columns.*)
- How could you show it? (e.g., *I'd shade each person's share a different colour.*)

Using the Open-Ended Intervention **(Student Resource pages 62–63)**

Read through the tasks on the student pages together. Provide materials for modelling decimal division. Give students time to work, ideally in pairs.

Observe whether students

- relate real-life situations to the division of a decimal by a whole number
- can estimate and determine quotients of decimals divided by whole numbers
- use reasonable strategies to divide, including models, and can explain their strategies

Consolidating and Reflecting

Ensure understanding by asking questions such as these based on students' work:

- Why did it make sense to use division for the garbage-zone problem? (e.g., *You're dividing up a whole into equal parts.*)
- What strategies did you use to estimate? (e.g., *I used only the whole number part of the decimals.*)
- When you divided 78.2 ÷ 6, you had 2 tenths left over. What does that mean, and how did you deal with it? (e.g., *It meant each of the 6 zones had 13 km^2, with 0.2 km^2 left to divide up into the 6 zones. I thought of 2 tenths as 2 dimes, or 20 cents, and divided by 6.*)
- Why did you use a hundredths grid for 46.24 ÷ 5? (e.g., *I knew 46.24 = 45 + 1.24,* so *I thought about 1.24 ÷ 5 as 124 hundredths ÷ 5, which would be 124 squares shared 5 ways.*)
- Why might someone say that dividing decimals is no different than dividing whole numbers? (e.g., *For a decimal like 2.3, you can think about dividing 23 tenths, 23 dimes, or 230 cents instead.*)

58 Number: Decimal Operations — *Leaps and Bounds 7/8* — Copyright © 2012 by Nelson Education Ltd.

Dividing Decimals by Whole Numbers

Pathway 2 GUIDED

You will need

- materials for modelling decimals (e.g., Tenths Grids (BLM 9), Hundredths Grids (BLM 8), Thousandths Grids (BLM 7), base ten blocks, or play money)
- Student Resource pages 64–67

Guided Intervention

Before Using the Guided Intervention

- What could 6.00 ÷ 4 represent? (e.g., *4 people sharing $6*)
- How much would each get? How do you know? (*$1.50;* e.g., *2 people would each get $3, so 4 people would get half as much.*)
- How can you check your answer? (e.g., *Add $1.50 four times and see if I get $6.*)
- How could you use hundredths grids to model 6.00 ÷ 4? (e.g., *I'd share 6 hundredths grids 4 ways. I'd give 1 whole grid to each person and then share the 20 columns in the last 2 grids by giving each person 5 columns.*)
- How could you show it? (e.g., *I'd shade each person's share a different colour.*)

Using the Guided Intervention **(Student Resource pages 64–67)**

Work through the division strategies in the instructional section of the student pages together. Provide materials for modelling the division of decimals.

Have students work through the **Try These** questions individually or in pairs.

Observe whether students

- relate a model to the division of a decimal by a whole number (Questions 1, 2)
- can explain their division strategies (Questions 2, 4, 6, 8)
- can estimate and calculate quotients (Questions 3, 5, 7, 11)
- relate division calculations to other calculations (Questions 6, 9)
- relate real-life situations to the division of a decimal by a whole number (Questions 7, 9, 10)

Consolidating and Reflecting

Ensure understanding by asking questions such as these based on students' work:

- How did you model 5.2 ÷ 4 in Question 3a)? (e.g., *I divided $5 and 2 dimes into 4 equal shares.*)
- How else could you have done it? (e.g., *I know 4 ÷ 4 is 1, so I could just model 1.2 as 1 hundredths grid and 2 columns in a second one. Then I could share the 12 columns among 4.*)
- How did you estimate in Question 4? (e.g., *I usually only thought about the whole number parts, but sometimes the tenths if the decimal part was near a half.*)
- In what situations might you divide a decimal by a whole number for Question 10? (e.g., *when you share a decimal amount like money or a measurement*)
- Why might someone say that dividing decimals is no different than dividing whole numbers? (e.g., *For a decimal like 2.3, you really have a whole number of tenths, 23 tenths, which is like 23 dimes.*)

Copyright © 2012 by Nelson Education Ltd. — *Leaps and Bounds 7/8* — Number: Decimal Operations 59

Copyright © 2012 by Nelson Education Ltd.

Frequently Asked Questions

How were *Leaps and Bounds* topics chosen?

Leaps and Bounds 7/8 includes materials for intervention in all 5 strands of mathematics. The topics for each strand were selected based on 2 factors. The first involved an analysis of Grade 5 to 8 curriculums across the country. No topics new to Grade 8 were included since a Grade 8 student could not yet be significantly behind in those topics. The second factor was an examination of the research literature—what we know about what aspects of each topic students struggle with. Prerequisite concepts and skills needed for success in Grade 7 or 8 mathematics were determined from curriculum documents, *PRIME* developmental maps (Marian Small, Nelson Education Ltd., 2005–2010), and other educational research. The prerequisites determine the topics, which are in turn divided into levels, called pathways.

How were the pathways determined?

To determine the pathways for each topic, an analysis of the prerequisites was undertaken. Attention was given to the curriculum outcomes and expectations, but the decision about which pathways to choose was based on a deep analysis of the mathematics. We pinpoint the missing knowledge and provide an intervention to fill that gap. By being this precise, we ensure students do not waste valuable learning time on unnecessary material.

Some of the pathways for a topic vary in the complexity addressed. For example, one pathway might focus on much smaller numbers than another within the same overall topic. But the pathways are treated in a parallel manner so that students at different levels of readiness can be taught the same fundamental concept at the same time.

Generally, Pathway 3 knowledge is prerequisite to Pathway 2 knowledge and Pathway 2 knowledge is prerequisite to Pathway 1 knowledge. However, there are exceptions. Sometimes there are only 2 pathways to cover all necessary prerequisites. At other times, there might be pathways where one is not a prerequisite to the other: both contribute critical knowledge for the student to meet success in the regular grade-level outcomes or expectations.

Each pathway has 2 types of structure to choose from: open-ended or guided.

Why open-ended and guided interventions?

Two types of structure are provided so that you may choose the intervention that is most suitable for the student's needs or style and most appropriate for the specific classroom learning situation. Either the guided intervention or the open-ended intervention or both can be used. If both are used, either order is acceptable.

Why is Pathway 1 for the stronger student and not the weaker student?

Pathway 1 contains content closest to grade level. Having this pathway appear first in the sequence increases the chance that no student is required to cover more material than necessary. A student farther behind would start with one of the later pathways.

The nature and extent of a student's area of struggle in a topic is determined through use of the diagnostic tool in addition to other classroom assessments and observations. Therefore, each student will engage in only the appropriate pathway activities rather than work through all of them.

Must a student do all pathways consecutively?

A student who completes Pathway 2 or 3 may or may not still require the support of the materials in higher-level pathways. If the underlying problem is resolved, the student can bypass pathways between the one successfully completed and the normal grade-level work. To decide if work in additional pathways is required, the relevant items on the diagnostic for the higher pathway(s) can be re-administered; if students still struggle, the additional pathway(s) might be beneficial.

Alternatively, you might decide to ask students to complete a limited version of the open-ended intervention or one or two items in the guided intervention in the higher pathway to determine if more work in that pathway is needed.

To find out whether the student's work is at an acceptable level once he or she has completed a pathway, you can re-administer the diagnostic tool or use other assessment items relevant for that topic.

How were the diagnostic questions developed?

The diagnostic questions were created to address what we know from the literature about common misconceptions as well as what we know from the *PRIME* research about developmental growth. For example, in the decimal operations topics, the diagnostic tool includes questions on operating with decimals with different numbers of digits after the decimal place, since that is often what confuses students. There is also a focus on how a smaller decimal divisor affects a quotient.

The questions in the diagnostic tool are broken down so that precise misconceptions can be identified. At the same time, the diagnostic sections are short enough to be practical and not overwhelming.

Copyright © 2012 by Nelson Education Ltd.

Components

Leaps and Bounds is designed to address all 5 math strands by diagnosing areas of weakness and filling in knowledge gaps to allow students to meet success at their current grade level. It provides instructional support to allow students who are struggling to grow—by leaps and bounds.

The 3 format options for the Teacher's Resource are
- a bound softcover book
- a digital version online with interactive whiteboard files
- a digital version on DVD with interactive whiteboard files

The 3 format options for the Student Resource are
- a bound softcover workbook
- blackline masters
- a CD-ROM with student pages in MS Word or PDF

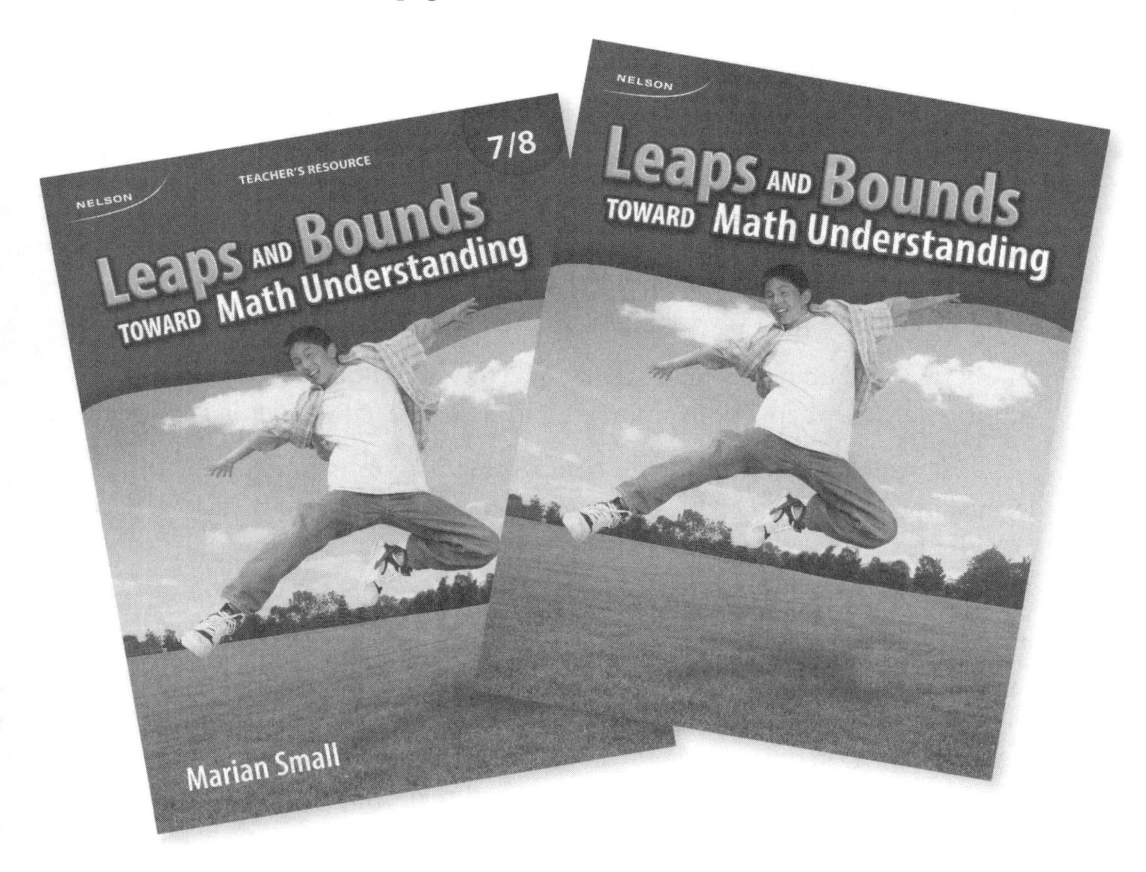

The pedagogy of *Leaps and Bounds* is based on the teacher diagnosing a student's areas of weakness and then providing instruction and guidance to move that student forward, with a focus on conceptual understanding. *Leaps and Bounds* supports multiple strategies to meet individual needs. The Teacher's Resource provides the diagnostic tools, keys for choosing appropriate pathways, and notes on selecting and using the relevant sections of the Student Resource.

Strand: Number

Number Strand Overview

How were the number topics chosen?

This resource provides materials for assisting students with 10 number topics. These topics were drawn from the curriculum outcomes from across the country for Grades 5 to 7. Topics are divided into distinct levels—called *pathways*—that address gaps in students' prerequisite skills and knowledge.

How were the pathways determined?

At the Grade 7/8 level, the focus moves from whole numbers to decimals, fractions, integers, and proportional reasoning. However, there remains some attention to whole numbers in these materials, specifically, representing and comparing large whole numbers, divisibility, and whole number operations.

There are 3 pathways related to using large numbers. Two involve interpreting numbers in the hundred thousands and numbers greater than 1 million written in standard form; the third helps students see how to use decimals with million and billion benchmarks to describe large whole numbers, for example, 3.2 million.

Work on multiplication of whole numbers goes as far back as multiplying 2 two-digit numbers and multiplying one-digit numbers by multi-digit numbers. A separate pathway focuses on dividing multi-digit numbers by one-digit numbers. Another pathway addresses order of operations with whole numbers.

Another whole numbers topic focuses on factors and multiples of relatively simple whole numbers, prime numbers and perfect squares, and divisibility rules.

Decimals are addressed in 3 topics: representing and comparing decimals, decimal operations, and relating real-life problem situations to the use of decimal operations.

The simplest pathway in the topic of representing and comparing decimals involves multiplying decimal tenths and hundredths by powers of 10; the focus is on using place value concepts and explanations. In representing decimals, the materials go back to representing decimal thousandths, but not specifically back to hundredths or tenths. *Leaps and Bounds 5/6* may be used for students needing earlier work. The pathways progress to comparing decimal thousandths and then understanding decimals less than 1 thousandth. Not all pathways are required in all curriculums.

Copyright © 2012 by Nelson Education Ltd.

The topic on decimal operations includes adding and subtracting decimals, primarily—but not exclusively—decimal thousandths. There are also pathways involving multiplying decimals by whole numbers, dividing decimals by whole numbers, and dividing whole numbers by simple decimals. Not all pathways are required in all curriculums.

A separate topic helps students recognize the operation(s) suggested by a problem situation. The pathways focus on the different meanings of subtraction (e.g., takeaway or comparison), multiplication (e.g., equal groups or area), and division (e.g., how many groups or how many in each group) in situations involving decimal numbers. There is no separate pathway for addition, since recognizing addition situations is rarely difficult for students.

Two topics focus on fractions. One emphasizes fraction comparisons and the other emphasizes fraction operations. The pathways involving fraction comparisons go back to showing equivalence and then move to the comparison of proper fractions, improper fractions, and mixed numbers. Multiplication and division of fractions are not fully addressed, since they are generally Grade 8 topics, but repeated addition and repeated subtraction of fractions are addressed, as are adding and subtracting fractions and mixed numbers.

The proportional reasoning topic highlights ratios, rates, and percents. There is a pathway for each topic.

The integers topic covers representing and comparing integers and also includes separate pathways for adding integers and subtracting integers.

What number topics have been omitted?

Basic multiplication facts and strategies, mental math with whole numbers, and addition and subtraction of whole numbers are dealt with in both *Leaps and Bounds 3/4* and *5/6*, so they were not repeated here as pathways.

Representing and comparing decimal tenths and hundredths is embedded in the pathways dealing with thousandths. If students need even more work with simpler decimals, they could be directed to the *Leaps and Bounds 5/6* material.

Materials

The materials for assisting students struggling with number topics are likely in the classroom or easily accessible. Blackline masters are provided at the back of this resource. Many of these materials are optional and so are listed only in the Teacher's Resource.

- base ten blocks
- play coins
- metre sticks
- measuring tapes
- counters
- square tiles
- containers, spoons, and millilitre and litre measures (optional)
- calculators
- pattern blocks
- fraction materials

BLM 1: Millions and Billions
BLM 2: Place Value Charts (to Billions)
BLM 3: Place Value Charts (to Hundred Thousands)
BLM 4: Place Value Charts (to Millionths)
BLM 5: Place Value Charts (to Ten Thousandths)
BLM 6: Place Value Charts (to Thousandths)
BLM 7: Thousandths Grids
BLM 8: Hundredths Grids
BLM 9: Tenths Grids
BLM 10: Fraction Strips
BLM 11: 2 cm Grid Paper
BLM 12: 1 cm Grid Paper
BLM 18: Fraction Circles/Spinners
BLM 19–24: Pattern Blocks
BLM 25: Number Lines

Number Topics and Pathways

Topics and pathways in this strand are shown below.
Each pathway has an open-ended intervention and a guided intervention.

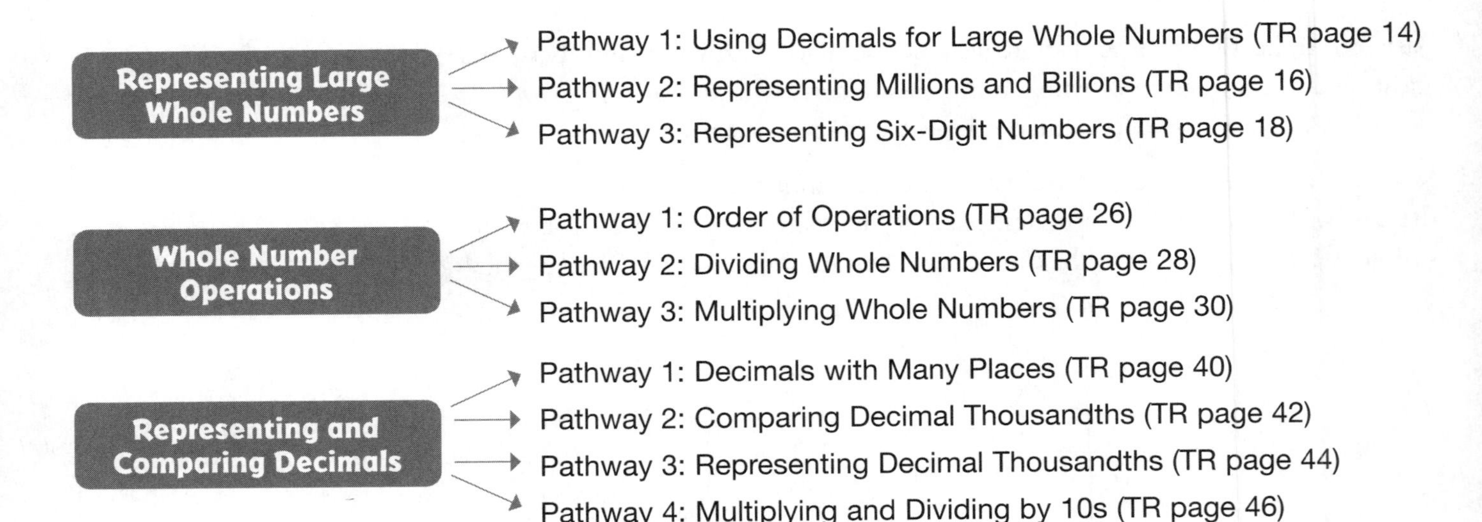

 Copyright © 2012 by Nelson Education Ltd.

Decimal Operations

- Pathway 1: Dividing Whole Numbers by Decimals (TR page 56)
- Pathway 2: Dividing Decimals by Whole Numbers (TR page 58)
- Pathway 3: Multiplying with Decimals (TR page 60)
- Pathway 4: Adding and Subtracting Decimals (TR page 62)

Relating Situations to Operations

- Pathway 1: Recognizing Division Situations (TR page 70)
- Pathway 2: Recognizing Multiplication Situations (TR page 72)
- Pathway 3: Recognizing Subtraction Situations (TR page 74)

Comparing Fractions

- Pathway 1: Fractions and Mixed Numbers (TR page 82)
- Pathway 2: Proper Fractions (TR page 84)
- Pathway 3: Equivalent Fractions (TR page 86)

Fraction Operations

- Pathway 1: Repeated Addition of Fractions (TR page 98)
- Pathway 2: Adding and Subtracting Mixed Numbers (TR page 100)
- Pathway 3: Subtracting Fractions (TR page 102)
- Pathway 4: Adding Fractions (TR page 104)

Rates, Percents, and Ratios

- Pathway 1: Using Rates (TR page 112)
- Pathway 2: Using Percents (TR page 114)
- Pathway 3: Using Ratios (TR page 116)

Multiplicative Relationships

- Pathway 1: Divisibility Rules (TR page 124)
- Pathway 2: Prime Numbers and Perfect Squares (TR page 126)
- Pathway 3: Factors and Multiples (TR page 128)

Integers

- Pathway 1: Subtracting Integers (TR page 136)
- Pathway 2: Adding Integers (TR page 138)
- Pathway 3: Representing and Comparing Integers (TR page 140)

Copyright © 2012 by Nelson Education Ltd.

Strand: Number

Representing Large Whole Numbers

Planning For This Topic

Materials for assisting students with representing, comparing, and renaming large numbers consist of a diagnostic tool and 3 intervention pathways. Pathway 1 focuses on the use of decimals to represent large whole numbers (e.g., 3 300 000 as 3.3 million). Pathway 2 focuses on standard form for numbers greater than 1 million. Pathway 3 focuses on standard form for six-digit numbers.

Each pathway has an open-ended intervention and a guided intervention. Choose the type of intervention more suitable for your students' needs and your particular circumstances.

Curriculum Connections

Grades 5 to 8 curriculum connections for this topic are provided online. See www.nelson.com/leapsandbounds. The Ontario curriculum does not specifically mention place value for numbers beyond 1 million, so Pathways 1 and 2 may be viewed as optional for these students. The WNCP curriculum extends whole number representations to 1 million in Grade 5 and greater than 1 million in Grade 6. Neither curriculum specifically states that students need to interpret and use large whole numbers written using decimals (Pathway 1). However, these activities are helpful for students in everyday life and in work they do with such numbers in science.

Why might students struggle with large numbers?

Students might struggle with large numbers for any of the following reasons:

- Students might have little, if any, concrete experience and limited life experience with large numbers, whether in standard form or in decimal form (e.g., 1.1 million).
- Students might not know where certain large whole numbers might appear in context (e.g., the number of people in India is in the billions place).
- Written conventions for numbers are based on place value and not on the way the number word sounds (e.g., "one million three hundred" is not 1 000 000 300).
- Students might not understand the periodic nature of the place value system, so it is not obvious why you read 123 400 as *123 thousand 400*.
- They might not know where to put the spaces in large whole numbers; that is, they do not understand the role of the periods in reading and writing numerals (e.g., they might write *forty-seven thousand twelve* as 47 000 12).
- It is not intuitively obvious why the value of a digit changes depending on its place in a numeral (e.g., why the value of the 3 in 302 415 is different from the value of 3 in 203 415).

Professional Learning Connections

PRIME: Number and Operations, Background and Strategies (Nelson Education, 2005), pages 63–68

Making Math Meaningful to Canadian Students K–8 (Nelson Education Ltd., 2008), pages 137–144, 146–147

Big Ideas from Dr. Small, Grades 4–8 (Nelson Education Ltd., 2010), pages 15–19

More Good Questions (dist. by Nelson Education Ltd., 2010), page 67

 Copyright © 2012 by Nelson Education Ltd.

- Some students struggle with representations of whole numbers other than standard form (e.g., they might find it difficult to write or think of 3 200 000 as 3200 thousands).
- Students might be unfamiliar with the use of large number units, such as millions or billions, in combination with decimal values.
- Students who are not generally comfortable with decimals may be reluctant to use decimals to represent whole numbers.

Diagnostic Tool: Representing Large Whole Numbers

Use the diagnostic tool to determine the most suitable intervention pathway for representing large whole numbers. Provide Diagnostic Tool: Representing Large Whole Numbers, Teacher's Resource pages 8 to 10, and have students complete it in writing or orally. If students need more place value charts, provide Place Value Charts (to Billions) (BLM 2).

See solutions on Teacher's Resource pages 11 to 13.

Intervention Pathways

These intervention pathways help students work with various ways of representing large whole numbers. This prepares them for working flexibly with large whole numbers of any size.

There are 3 pathways:

- Pathway 1: Using Decimals for Large Whole Numbers
- Pathway 2: Representing Millions and Billions
- Pathway 3: Representing Six-Digit Numbers

Use the chart below (or the Key to Pathways on Teacher's Resource pages 11 to 13) to determine which pathway is most suitable for each student or group of students.

Diagnostic Tool Results	Intervention Pathway
If students struggle with Questions 1e–f, 2e–f, 3c, 4e–f, 5e–f, 6e–f, 7e–f, 8c	use Pathway 1: Using Decimals for Large Whole Numbers *Teacher Resource pages 14–15* *Student Resource pages 1–5*
If students struggle with Questions 1c–d, 2c–d, 3b, 4c–d, 5c–d, 6c–d, 7c–d, 8b	use Pathway 2: Representing Millions and Billions *Teacher Resource pages 16–17* *Student Resource pages 6–11*
If students struggle with Questions 1a–b, 2a–b, 3a, 4a–b, 5a–b, 6a–b, 7a–b, 8a	use Pathway 3: Representing Six-Digit Numbers *Teacher Resource pages 18–19* *Student Resource pages 12–15*

Copyright © 2012 by Nelson Education Ltd.

Name: ____________________ Date: ____________________

Representing Large Whole Numbers

Diagnostic Tool

You can use a place value chart like this to help you answer Questions 1 to 6.

Billions			Millions			Thousands			Ones		
H	T	O	H	T	O	H	T	O	H	T	O

1. What do the digits 4 and 7 represent in each number?

a) **4**2**7** 213 ________ ________

b) 1**4**3 **7**28 ________ ________

c) **4** 2**7**3 689 ________ ________

d) 2**4**1 1**7**6 538 916 ________ ________

e) **4.7** million ________ ________

f) 2.**47** billion ________ ________

2. Write each number in standard form.

a) 4 hundred thousand + 24 ________

b) 215 thousand + 46 ________

c) 3 million + 40 thousand + 100 ________

d) 12 billion + 17 thousand + 10 ________

e) 4.2 million ________

f) 3.71 billion ________

3. How are the values of the two 6s in each number different?

a) **6**00 0**6**0 ________

b) **6**0 00**6** 030 ________

c) **6**.2**6** million ________

Copyright © 2012 by Nelson Education Ltd.

Name:____________________ Date:____________________

4. Create 2 numbers to match each description.

a) 6 in the hundred thousands place and 2 in the hundreds place

__________ __________

b) 6 in the ten thousands place and 2 in the tens place

__________ __________

c) 4 in the ten millions place and 5 in the thousands place

__________ __________

d) 1 in the billions place and 2 in the hundred millions place

__________ __________

e) about 1 hundred thousand more than 2.4 million

__________ __________

f) about 100 million less than 4.62 billion

__________ __________

5. Fill in the blank with a number or a place value word.

a) 3 hundred thousand = ______ thousand

b) 40 ten thousand = 4 __________

c) 2 million = ______ thousand

d) 30 billion = 3000 __________

e) 4.3 billion = ______ million

f) 2.1 million = 2100 __________

6. Estimate.

a) 461 589 is about ______ hundred thousand.

b) 304 127 is about ______ thousand.

c) 12 089 157 is about ______ million.

d) 11 314 578 138 is about ______ million.

e) 0.7 million is about ______ hundred thousand.

f) 1.77 billion is about ______ million.

Copyright © 2012 by Nelson Education Ltd.

Name: ______________________________ Date: ______________

You can use this place value chart to help you answer Question 7.

Billions			Millions			Thousands			Ones		
H	T	O	H	T	O	H	T	O	H	T	O

7. Order each set of numbers from least to greatest.

a) 304 209 940 302 94 302

______________, ______________, ______________

b) 38 158 624 135 622 143

______________, ______________, ______________

c) 3 121 043 31 043 699 3 112 043

______________, ______________, ______________

d) 4 002 003 700 384 674 121 583 912 934

______________, ______________, ______________

e) 3.4 million 1.3 billion 9.9 million

______________, ______________, ______________

f) 8.03 million 0.83 billion 0.5 million

______________, ______________, ______________

8. Match each number to what you think it most likely describes.

312 678	population of China
1 319 175 334	population of Canada
34.3 million	cost of a house in dollars

Copyright © 2012 by Nelson Education Ltd.

Solutions and Key to Pathways

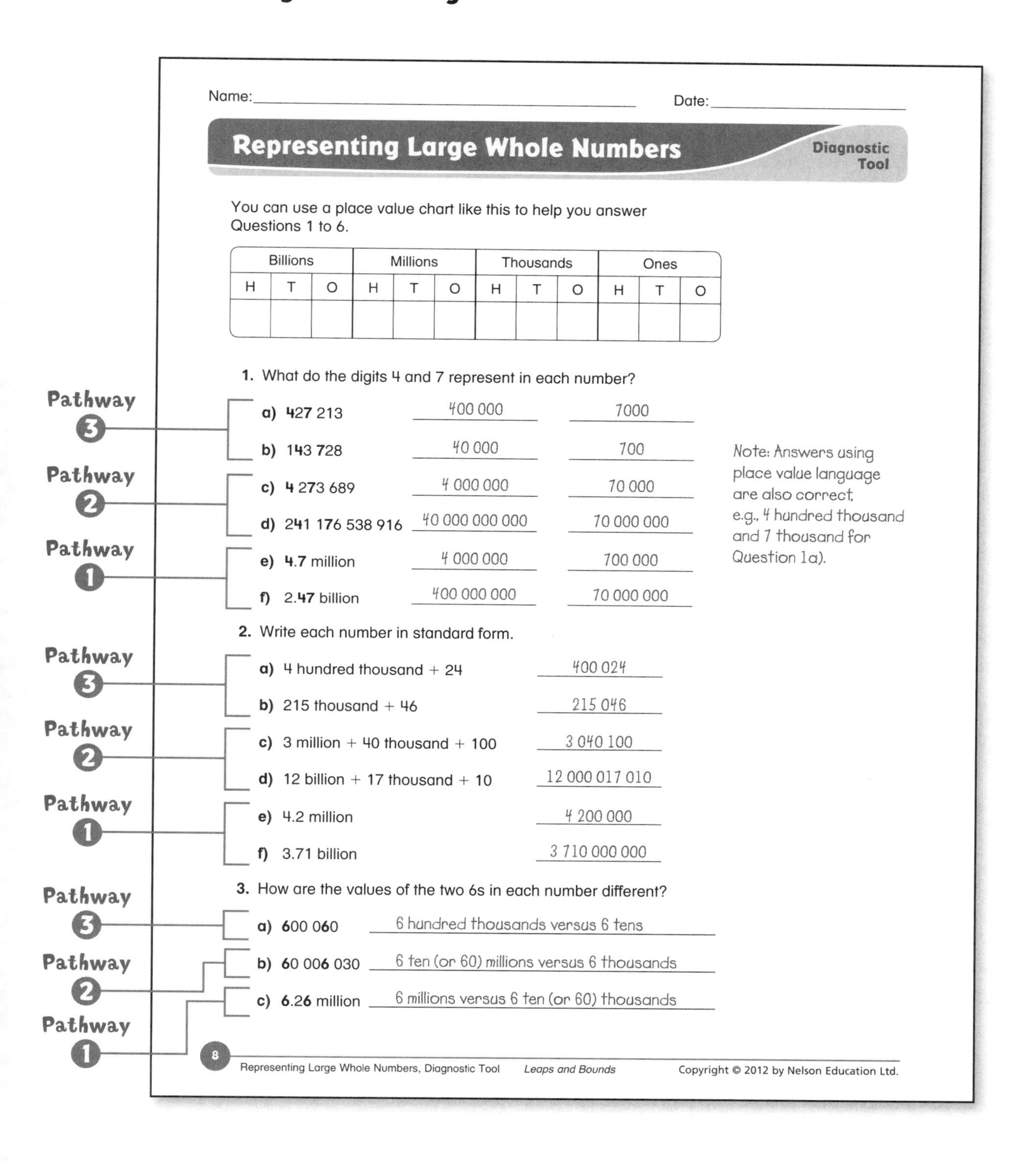

Name: ______________________ Date: ______________

Representing Large Whole Numbers

Diagnostic Tool

You can use a place value chart like this to help you answer Questions 1 to 6.

Billions			Millions			Thousands			Ones		
H	T	O	H	T	O	H	T	O	H	T	O

1. What do the digits 4 and 7 represent in each number?

Pathway 3
- **a)** **4**2**7** 213 — 400 000 — 7000
- **b)** 1**4**3 **7**28 — 40 000 — 700

Pathway 2
- **c)** **4** 2**7**3 689 — 4 000 000 — 70 000
- **d)** 2**4**1 1**7**6 538 916 — 40 000 000 000 — 70 000 000

Pathway 1
- **e)** **4.7** million — 4 000 000 — 700 000
- **f)** 2.**47** billion — 400 000 000 — 70 000 000

Note: Answers using place value language are also correct, e.g., 4 hundred thousand and 7 thousand for Question 1a).

2. Write each number in standard form.

Pathway 3
- **a)** 4 hundred thousand + 24 — 400 024
- **b)** 215 thousand + 46 — 215 046

Pathway 2
- **c)** 3 million + 40 thousand + 100 — 3 040 100
- **d)** 12 billion + 17 thousand + 10 — 12 000 017 010

Pathway 1
- **e)** 4.2 million — 4 200 000
- **f)** 3.71 billion — 3 710 000 000

3. How are the values of the two 6s in each number different?

- Pathway 3: **a)** **6**00 0**6**0 — 6 hundred thousands versus 6 tens
- Pathway 2: **b)** **6**0 00**6** 030 — 6 ten (or 60) millions versus 6 thousands
- Pathway 1: **c)** **6**.2**6** million — 6 millions versus 6 ten (or 60) thousands

8 Representing Large Whole Numbers, Diagnostic Tool *Leaps and Bounds* Copyright © 2012 by Nelson Education Ltd.

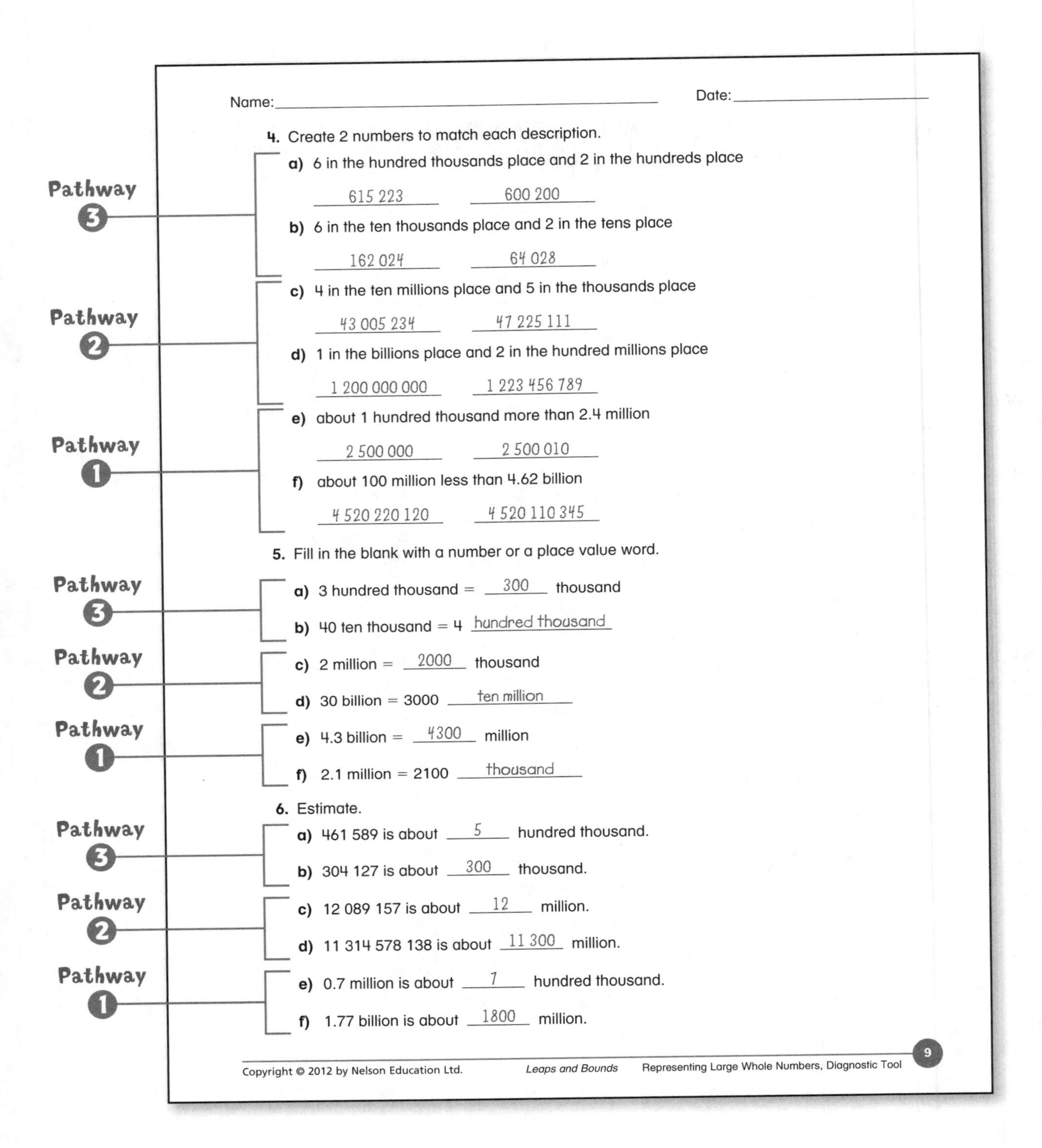

Name:______________________ Date:______________

4. Create 2 numbers to match each description.

Pathway 3

a) 6 in the hundred thousands place and 2 in the hundreds place

615 223   600 200

b) 6 in the ten thousands place and 2 in the tens place

162 024   64 028

Pathway 2

c) 4 in the ten millions place and 5 in the thousands place

43 005 234   47 225 111

d) 1 in the billions place and 2 in the hundred millions place

1 200 000 000   1 223 456 789

Pathway 1

e) about 1 hundred thousand more than 2.4 million

2 500 000   2 500 010

f) about 100 million less than 4.62 billion

4 520 220 120   4 520 110 345

5. Fill in the blank with a number or a place value word.

Pathway 3

a) 3 hundred thousand = 300 thousand

b) 40 ten thousand = 4 hundred thousand

Pathway 2

c) 2 million = 2000 thousand

d) 30 billion = 3000 ten million

Pathway 1

e) 4.3 billion = 4300 million

f) 2.1 million = 2100 thousand

6. Estimate.

Pathway 3

a) 461 589 is about 5 hundred thousand.

b) 304 127 is about 300 thousand.

Pathway 2

c) 12 089 157 is about 12 million.

d) 11 314 578 138 is about 11 300 million.

Pathway 1

e) 0.7 million is about 7 hundred thousand.

f) 1.77 billion is about 1800 million.

Copyright © 2012 by Nelson Education Ltd. *Leaps and Bounds* Representing Large Whole Numbers, Diagnostic Tool 9

 Copyright © 2012 by Nelson Education Ltd.

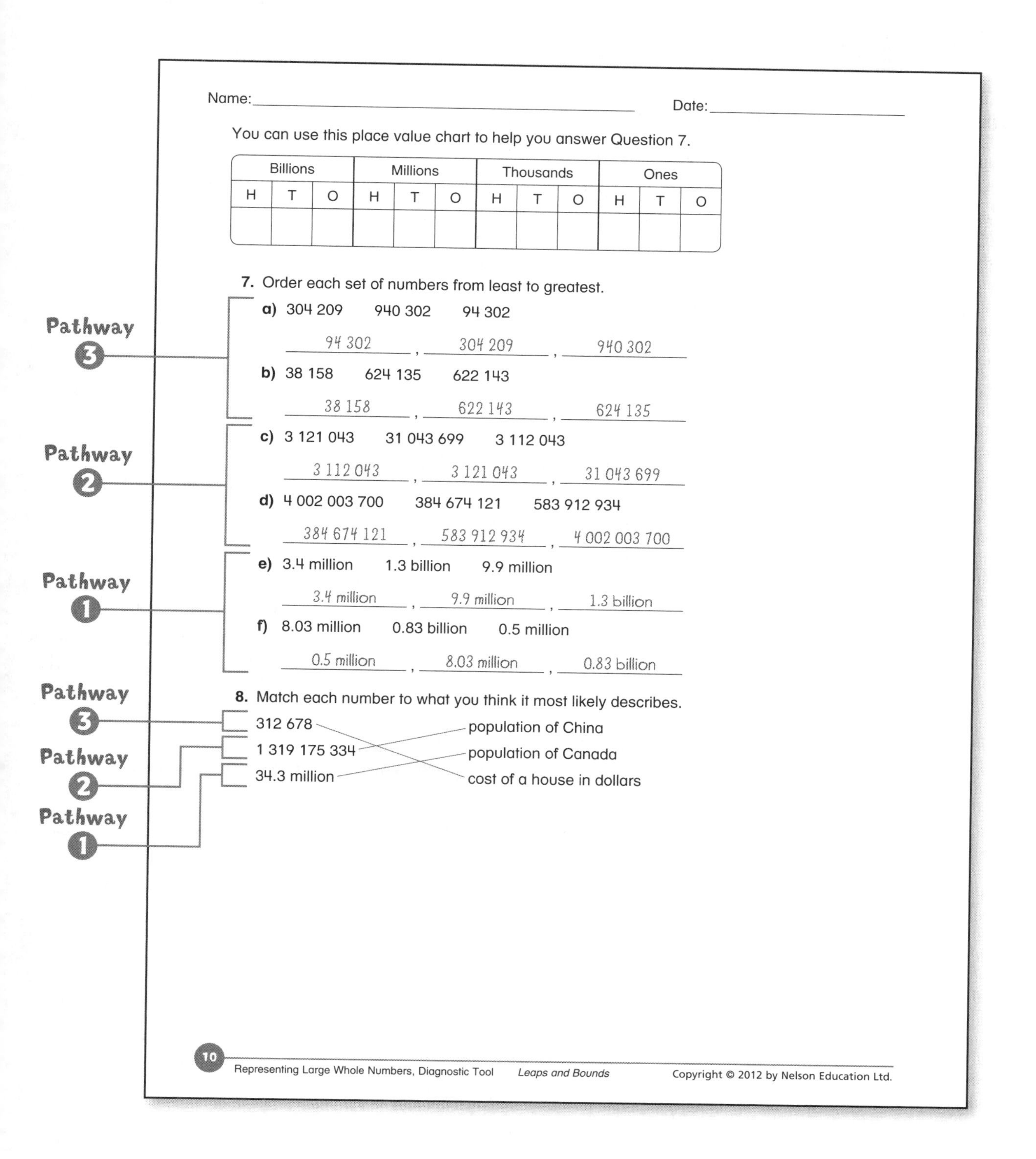

Name:______________________ Date:______________

You can use this place value chart to help you answer Question 7.

Billions			Millions			Thousands			Ones		
H	T	O	H	T	O	H	T	O	H	T	O

Pathway 3

7. Order each set of numbers from least to greatest.

a) 304 209   940 302   94 302

94 302, 304 209, 940 302

b) 38 158   624 135   622 143

38 158, 622 143, 624 135

Pathway 2

c) 3 121 043   31 043 699   3 112 043

3 112 043, 3 121 043, 31 043 699

d) 4 002 003 700   384 674 121   583 912 934

384 674 121, 583 912 934, 4 002 003 700

Pathway 1

e) 3.4 million   1.3 billion   9.9 million

3.4 million, 9.9 million, 1.3 billion

f) 8.03 million   0.83 billion   0.5 million

0.5 million, 8.03 million, 0.83 billion

Pathway 3 / Pathway 2 / Pathway 1

8. Match each number to what you think it most likely describes.

312 678	population of China
1 319 175 334	population of Canada
34.3 million	cost of a house in dollars

10 Representing Large Whole Numbers, Diagnostic Tool *Leaps and Bounds* Copyright © 2012 by Nelson Education Ltd.

Pathway 1 OPEN-ENDED

Using Decimals for Large Whole Numbers

You will need

- base ten blocks (small cube, rod, flat, large cube)
- Place Value Charts (to Billions) (BLM 2)
- Internet access, or Millions and Billions (BLM 1)
- Student Resource pages 1–2

Open-Ended Intervention

Before Using the Open-Ended Intervention

Write the decimal 0.1 and have students read it aloud (*one tenth*). Show a base ten small cube, rod, flat, and large cube and ask:

- Suppose the rod was the whole. What would represent 0.1? Why? (*a small cube; it is one tenth of 10*)
- Suppose the flat was the whole? What would represent 0.1? Why? (*a rod; 10 rods make 1 flat*)
- Why did the decimal value change? (*The value of the whole changed.*)

Draw a place value chart to thousands and have students record 3200 in it. Ask:

- What is the value of the digit 3? the digit 2? (*3 thousands; 2 hundreds*)
- Why might someone write 3200 as 3.2 thousand? (e.g., *If the large cube is the whole, 100 is 0.1 of a thousand, so 200 is 0.2, so the 200 in 3200 is 0.2 and the 3000 is 3.*)

Using the Open-Ended Intervention Student Resource pages 1–2

Read through the tasks on the student pages together. Provide place value charts and allow for Internet access (using search words such as "populations," "popular products," "sales," "space distances," "sports salaries," and "downloads") or provide Millions and Billions (BLM 1). Give students time to work, ideally in pairs.

Observe whether students

- can use standard form to rename a large whole number expressed in decimal form and vice versa
- can order large whole numbers expressed in decimal form
- understand that a number such as 12.3 million (a large whole number expressed in decimal form) is a concise way to write a large whole number
- recognize real-life contexts for large whole numbers expressed in decimal form

Consolidating and Reflecting

Ensure understanding by asking questions such as these based on students' work:

- How did you decide what a decimal portion of a million represented? (e.g., *If it was 1 place to the right of the millions, I knew it was tenths of it, but 2 places to the right meant it was hundredths of it.*)
- Why did the numbers in standard form have a lot of zeros? (e.g., *A number like 2.3 million has non-zero digits only in the millions and hundred thousands places. It's like writing 2.300 000 million.*)
- How did you order the 10 numbers? (e.g., *First, I ordered the millions by putting them in decimal form, then I ordered the billions in decimal form, and I then realized all the million numbers were less than all the billion numbers.*)

 Copyright © 2012 by Nelson Education Ltd.

Using Decimals for Large Whole Numbers

Pathway 1 GUIDED

Guided Intervention

Before Using the Guided Intervention

Write the decimal 0.1 and have students read it aloud (*one tenth*). Show a base ten small cube, rod, flat, and large cube and ask:

- Suppose the rod was the whole. What would represent 0.1? Why? (*a small cube; it is 1 tenth of 10*)
- Suppose the flat was the whole? What would represent 0.1? Why? (*a rod; 10 rods make 1 flat*)
- Why did the decimal value change? (*The value of the whole changed.*)

Draw a place value chart to thousands and have students record 3200 in it. Ask:

- What is the value of the digit 3? the digit 2? (*3 thousands; 2 hundreds*)
- Why might someone write 3200 as 3.2 thousand? (e.g., *If the large cube is the whole, 100 is 0.1 of a thousand, so 200 is 0.2, so the 200 in 3200 is 0.2 and the 3000 is 3.*)

Using the Guided Intervention Student Resource pages 3–5

Provide place value charts and work through the instructional section of the student pages together.

Have students work through the **Try These** questions in pairs or individually. They will need Internet access or Millions and Billions (BLM 1), for Question 7.

Observe whether students

- can change from decimal form to standard form and vice versa (Questions 1, 2, 3)
- can compare and order large whole numbers in decimal form (Question 4)
- recognize the value of a specific digit (Question 5)
- can rename a large whole number in other forms (Question 6)
- recognize contexts for large whole numbers (Question 7)
- recognize the value of writing large whole numbers in decimal form (Question 8)

Consolidating and Reflecting

Ensure understanding by asking questions such as these based on students' work:

- Why did the numbers in Question 1 have a lot of zeros? (e.g., *A number like 4.13 billion has non-zero digits only in the billions, hundred millions, and ten millions places. It's like writing 4.130 000 000 billion.*)
- How did you determine the decimal forms in Question 3? (e.g., *For 12 345 000, I wrote the number of millions, 12, as the whole, and then the other part, 345, as the decimal.*)
- Why do people write large whole numbers as decimal millions or billions? (e.g., *It takes up less space and you're usually interested only in the digits that have big place values anyway, because they tell you how big the number is.*)

You will need

- base ten blocks (small cube, rod, flat, large cube)
- Place Value Charts (to Billions) (BLM 2)
- Internet access, or Millions and Billions (BLM 1)
- Student Resource pages 3–5

Copyright © 2012 by Nelson Education Ltd.

Representing Millions and Billions

You will need

- Place Value Charts (to Billions) (BLM 2)
- Student Resource pages 6–7

Open-Ended Intervention

Before Using the Open-Ended Intervention

Write the numeral 739 907 and read it aloud (*739 thousand 907*). Sketch a place value chart to hundred thousands and label the thousands period. Write 739 907 in the chart and read it aloud. Ask:

- How does the chart help you read the number? (e.g., *The chart shows the number in 2 parts: a number of thousands and a number less than a thousand. Each part has 3 place value columns.*)
- How do you know 740 907 is more than 739 907 but 739 897 is less than 739 907? (e.g., *740 907 has more ten thousands than 739 897, and 739 897 has fewer hundreds than 739 907.*)
- How would you show 999 thousand 999? (*9 in each of the 6 columns*)
- What would happen if you added 1 to 999 999? (e.g., *I'd have 1000 thousand, and I would need a new column on the left.*)

Using the Open-Ended Intervention Student Resource pages 6–7

Read through the tasks on the student pages together. Ensure that students understand the periods shown on the place value chart. Provide place value charts and give students time to work, ideally in pairs.

Observe whether students

- can create numbers that are about 111 million and write them in standard form
- can create numbers very close to 278 billion and write them in standard form
- can order large numbers and explain their thinking
- can read, estimate, and relate large numbers

Consolidating and Reflecting

Ensure understanding by asking questions such as these based on students' work:

- How did you decide that a number was close to 111 million? (e.g., *It was close but greater if it was 111 million and not too many thousands more. It was close but less if it was 110 million and more than 900 thousand more.*)
- How did you decide if a number was close to 278 billion? (e.g., *It was close if it was less than 1 million more than 278 billion or if it was a lot of hundreds of millions more than 277 billion.*)
- How did you arrange the numbers in order? (e.g., *I started with the numbers in millions—110 million is less than 111 million. If the millions parts were the same, then I compared the thousands. Then I used the billions numbers and compared their millions parts, and sometimes their thousands or even their ones parts.*)
- In order, all the million values come first and then all the billion values. Why does that make sense? (e.g., *278 billion is much more than 111 million, so anything close to it has to be more than something close to 111 million.*)

 Copyright © 2012 by Nelson Education Ltd.

Representing Millions and Billions

Pathway 2 GUIDED

Guided Intervention

Before Using the Guided Intervention

Write the numeral 739 907 and read it aloud (*739 thousand 907*). Sketch a place value chart to hundred thousands and label the thousands period. Ask:

- ▸ Write 739 907 in the chart and read it aloud. How does the chart help you read the number?
  (e.g., *The chart shows that the number is in 2 parts: a number of thousands and a number less than a thousand. Each part has 3 place value columns.*)

You will need

- Place Value Charts (to Billions) (BLM 2)
- Millions and Billions (BLM 1)
- Student Resource pages 8–11

Using the Guided Intervention (Student Resource pages 8–11)

Provide place value charts and work through the instructional section of the student pages together. Ensure that students are comfortable using the place value chart and reading and representing numbers to billions. Students will need Internet access or Millions and Billions (BLM 1) for Question 11.

Have students work through the **Try These** questions in pairs or individually.

Observe whether students

- can read large numbers (Questions 1, 6)
- can rename large numbers (Questions 2, 9)
- can compare large numbers and explain their thinking (Questions 3, 8, 10)
- can estimate large numbers (Question 4)
- can relate 2 large numbers (Questions 5, 12)
- recognize the value of a digit in a large number (Question 7)
- recognize real-life contexts for large numbers (Question 11)

Consolidating and Reflecting

Ensure understanding by asking questions such as these based on students' work:

- ▸ How did you know 153 million + 32 thousand was less than 1 billion in Question 3?
  (e.g., *A billion is 1000 million, and 153 million + 32 thousand is not even 154 million.*)
- ▸ How did you decide that 4 128 756 was about 4100 thousand in Question 4?
  (e.g., *4 million = 4000 thousand, so I added that to 100 000 in the next place to the right.*)
- ▸ Which digit changed in Question 5d)? Why?
  (8; e.g., *If you add 2 million to 8 million, you get 10 million, so the million and ten million digits changed.*)
- ▸ Why was the number of hundred thousands less than the number of thousands in Question 9a)?
  (e.g., *Hundred thousands is a bigger place value unit than thousands, and if the unit is bigger, fewer units are needed to represent the same number.*)

Copyright © 2012 by Nelson Education Ltd.

Pathway 3 OPEN-ENDED

Representing Six-Digit Numbers

You will need

- Place Value Charts (to Hundred Thousands) (BLM 3)
- Student Resource page 12

Open-Ended Intervention

Before Using the Open-Ended Intervention

Sketch a place value chart to hundred thousands. Ask:

- ▸ How would you show 20 000 on the chart?
  (*2 in the ten thousands column and then all zeros*)
- ▸ Write 3 numbers that are close to 20 000. Make 1 number less than 20 000.
  (e.g., *20 001, 20 100, 19 999*)
- ▸ Order the 3 numbers and explain how you did it.
  (e.g., *19 999, 20 001, 20 100; 19 999 is 1 less than 20 000, so it's lowest; 20 001 is 1 more and 20 100 is 100 more.*)
- ▸ How would you show 90 000 on the chart?
  (*9 in the ten thousands column and then all zeros*)
- ▸ What would happen if you added 10 000 to 90 000?
  (e.g., *There would be 10 in the ten thousands column, so I would trade it for 1 hundred thousand.*)

Using the Open-Ended Intervention (Student Resource page 12)

Read through the tasks on the student page together.

Provide place value charts and give students time to work, ideally in pairs.

Observe whether students

- can create large numbers of a given size and write them in standard form
- can estimate large numbers
- can order large numbers in standard form and explain their thinking

Consolidating and Reflecting

Ensure understanding by asking questions such as these based on students' work:

- ▸ How did you decide that a number was close to 750 thousand?
  (e.g., *It was close but greater if it was less than 1 thousand more than 750 thousand, or it was close but less if it was a lot of hundreds more than 749 thousand.*)
- ▸ What did you do to make the number close to but less than 750 thousand? Why?
  (e.g., *I used 749 thousand instead of 750 thousand, plus a lot of hundreds and some tens and ones.*)
- ▸ How did you arrange the numbers in order?
  (e.g., *I started with the number of thousands—749 is less than 750. Then within each group, I looked at the rest of the number to see which had a greater group of hundreds, tens, and ones.*)

Copyright © 2012 by Nelson Education Ltd.

Representing Six-Digit Numbers

Pathway 3 GUIDED

Guided Intervention

Before Using the Guided Intervention

Sketch a place value chart to hundred thousands. Ask:

- ▸ How would you show 20 000 on the chart? (*2 in the ten thousands column and then all zeros*)
- ▸ Write 3 numbers that are close to 20 000. Make 1 number less than 20 000. (e.g., *20 001, 20 100, 19 999*)
- ▸ Order the 3 numbers. How did you do it? (e.g., *19 999, 20 001, 20 100; 19 999 is 1 less than 20 000, so it's lowest; 20 001 is 1 more and 20 100 is 100 more.*)
- ▸ How would you show 90 000 on the chart? (*9 in the ten thousands column and then all zeros*)
- ▸ What would happen if you added 10 000 to 90 000? (e.g., *There would be a 10 in the ten thousands column, so I would trade it for 1 hundred thousand.*)

You will need

- Place Value Charts (to Hundred Thousands) (BLM 3)
- Student Resource pages 13–15

Using the Guided Intervention Student Resource pages 13–15

Provide place value charts and work through the instructional section on the student page together.

Have students work through the **Try These** questions in pairs or individually.

Observe whether students

- can read large numbers (Questions 1, 8)
- can write a number in standard form (Question 2)
- can compare and relate numbers (Questions 3, 5, 7)
- can estimate large numbers (Question 4)
- recognize the value of a digit in a large number (Question 6)
- can rename large whole numbers (Question 9)
- recognize real-life contexts for large numbers (Question 10)

Consolidating and Reflecting

Ensure understanding by asking questions such as these based on students' work:

- ▸ How did you know that 80 thousand was less than 20 ten thousand in Question 3b)? (e.g., *80 thousand is 8 ten thousand.*)
- ▸ How did you decide that 493 127 was about 5 hundred thousand in Question 4b)? (e.g., *I knew that 493 thousand was closer to 500 thousand than to 400 thousand and that 500 thousand is 5 hundred thousand.*)
- ▸ Which digits changed in Question 5c)? Why? (*8 and 3*; e.g., *If you add 8 thousand to 3 thousand, you get 11 thousand, or 1 ten thousand and 1 thousand, so 3 thousand changed to 1 thousand and 8 ten thousand plus the 1 ten thousand is 9 ten thousand.*)
- ▸ How did you pick the greater number in Question 7d)? (e.g., *711 234 is a lot of hundred thousands and 79 929 is not even 1 hundred thousand.*)

Copyright © 2012 by Nelson Education Ltd.

Whole Number Operations

Planning For This Topic

Materials for assisting students with whole number operations, particularly multiplication and division, consist of a diagnostic tool and 3 intervention pathways. Pathway 1 involves the use of order of operations with whole numbers. Pathway 2 focuses on division of three-digit and four-digit numbers by one-digit and two-digit numbers. Pathway 3 involves multiplication of two-digit by two-digit numbers.

Each pathway has an open-ended intervention and a guided intervention. Choose the type of intervention more suitable for your students' needs and your particular circumstances.

Curriculum Connections

Grades 5 to 8 curriculum connections for this topic are provided online. See www.nelson.com/leapsandbounds. Both the Ontario and WNCP curricula include multiplying and dividing with whole numbers in Grades 5 and 6. The strategies mentioned include using technology, mental math strategies, and paper and pencil or pictorial/concrete models. The technology is not specifically addressed here, but the conceptual understanding involved in mental math and other strategies is. Both curricula refer to the need for students to understand and apply the order of operations rules with whole numbers. Although calculator use would be expected with complicated numbers, the values used here are simple enough that calculators are not required. There is no specific mention of multiplication and division of whole numbers in Grade 7 or 8; however, students continue to use these operations to solve problems.

Professional Learning Connections

PRIME: Number and Operations, Background and Strategies (Nelson Education Ltd., 2005), pages 82–97

Making Math Meaningful to Canadian Students K–8 (Nelson Education Ltd., 2008), pages 173–189

Big Ideas from Dr. Small Grades 4–8 (Nelson Education Ltd., 2010), pages 25–41

Good Questions (dist. by Nelson Education Ltd., 2009), pages 27, 29

Why might students struggle with operations?

Students might struggle with multiplying and dividing whole numbers and applying the order of operations for any of the following reasons:

- They might not know all of their multiplication facts, making it difficult to perform more complex multiplication or division calculations and making it difficult to estimate or use mental strategies.
- They might multiply 2 two-digit numbers "in columns" like addition (e.g., they might multiply 34×25 by multiplying 3×2 and then appending 4×5 to get 620 instead of 850).
- They might recall procedures learned incorrectly (e.g., multiplying 24×63 by multiplying 4×63 and then putting a 2 in front).
- They might not relate the operations to their various meanings (e.g., division as a sharing or an equal group situation, and multiplication as the area of a rectangle).

 Copyright © 2012 by Nelson Education Ltd.

- They might not recognize division as the inverse operation of multiplication (e.g., that solving 455 ÷ 5 = ■ is the same as solving 5 × ■ = 455).
- They might not recognize how to separate the dividend into convenient parts for dividing (e.g., they may not realize that to divide 120 by 9, it might be convenient to think of 90 ÷ 9 + 30 ÷ 9).
- They might not know what to do with remainders in particular contexts.
- They might not realize that different values can result when a sequence of calculations is performed in a different order and that there are accepted conventions for the correct order.

Diagnostic Tool: Whole Number Operations

Use the diagnostic tool to determine the most suitable intervention pathway for whole number operations. Provide Diagnostic Tool: Whole Number Operations, Teacher's Resource pages 22 and 23, and have students complete it in writing or orally. You might provide base ten materials.

Intervention Pathways

The purpose of the intervention pathways is to help students multiply and divide whole numbers and perform operations correctly in a sequence of calculations. The focus is to prepare them for solving problems and working with other types of numbers, such as integers and fractions.

There are 3 pathways:
- Pathway 1: Order of Operations
- Pathway 2: Dividing Whole Numbers
- Pathway 3: Multiplying Whole Numbers

Use the chart below (or the Keys to Pathways on Teacher's Resource pages 24 and 25) to determine which pathway is most suitable for each student or group of students.

Diagnostic Tool Results	Intervention Pathway
If students struggle with Questions 9 to 11	use Pathway 1: Order of Operations *Teacher Resource pages 26–27* *Student Resource pages 16–19*
If students struggle with Questions 5 to 8	use Pathway 2: Dividing Whole Numbers *Teacher Resource pages 28–29* *Student Resource pages 20–25*
If students struggle with Questions 1 to 4	use Pathway 3: Multiplying Whole Numbers *Teacher Resource pages 30–31* *Student Resource pages 26–31*

Name: ______________________ Date: ______________

Whole Number Operations

Diagnostic Tool

1. Circle the calculation with an answer closest to 400.

   21×24 12×34 50×81 17×25

2. Multiply. Show your work.

   a) $32 \times 21 =$ __________

   c) $17 \times 73 =$ __________

   b) $43 \times 46 =$ __________

   d) $15 \times 15 =$ __________

3. Create a real-life problem that could be solved using 12×38.

   ____

   ____

4. Show or describe 2 ways to calculate 35×17.

5. Circle the calculation with an answer closest to 120.

   $847 \div 7$ $810 \div 6$ $1368 \div 9$ $56 \div 4$

Copyright © 2012 by Nelson Education Ltd.

Name: ______________________________ Date: ______________

6. Calculate. Show your work.

a) $512 \div 6 =$ ________

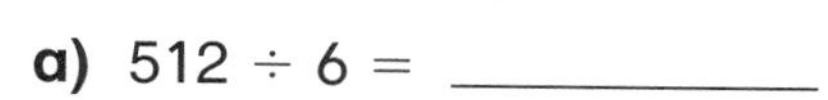

c) $812 \div 9 =$ ________

b) $3009 \div 4 =$ ________

d) $4220 \div 4 =$ ________

7. Create a real-life problem that could be solved using $365 \div 7$.

____

____

8. Show or describe 2 ways to calculate $390 \div 6$.

9. In each expression, circle the sign for the calculation you would do first.

a) $4 + 6 \times 5$

b) $6 \times (4 + 5)$

c) $3 + 12 \div 3$

d) $6 + 3 - 4 \times 2$

10. Is $(3 + 5) \times 2$ equal to $3 + 5 \times 2$? How do you know?

____

____

11. Use $>$, $<$, or $=$ to make each statement true.

a) $4 \times 5 \times 6$ ☐ $4 \times (5 \times 6)$

b) $4 \times 5 + 6$ ☐ $4 \times (5 + 6)$

c) $4 + 5 \times 6$ ☐ $(4 + 5) \times 6$

d) $4 \times (3 + 2) \times 5$ ☐ $(4 \times 3) + (2 \times 5)$

Solutions and Key to Pathways

Name: ______________________ Date: ______________

Whole Number Operations — Diagnostic Tool

Pathway 3

1. Circle the calculation with an answer closest to 400.

   21×24 (12×34) 50×81 17×25

2. Multiply. Show your work.

   a) $32 \times 21 =$ 672

   e.g., 32×21
   $= 32 \times 20 + 32 \times 1$
   $= 640 + 32$
   $= 672$

   c) $17 \times 73 =$ 1241

   e.g., $17 \times 73 = 20 \times 73 - 3 \times 73$
   $= 10 \times 2 \times 73 - 219$
   $= 10 \times 146 - 219$
   $= 1460 - 219$
   $= 1241$

   b) $43 \times 46 =$ 1978

   e.g.,
   ```
     43
   × 46
    240
     18
   1600
    120
   1978
   ```

   d) $15 \times 15 =$ 225

   e.g.,
   10 x 10
   10 x 5
   5 x 10
   5 x 5

3. Create a real-life problem that could be solved using 12×38.

   e.g., You sell 12 magazine subscriptions and each costs $38.
   How much money did you bring in?

4. Show or describe 2 ways to calculate 35×17.

   e.g., Multiply 10×17, then triple it. Then take half of 10×17 and add the 2 numbers together.

   e.g., Make a rectangle that is 35 units long and 17 units wide and fill it with base ten blocks.

Pathway 2

5. Circle the calculation with an answer closest to 120.

   ($847 \div 7$) $810 \div 6$ $1368 \div 9$ $56 \div 4$

22 Whole Number Operations, Diagnostic Tool *Leaps and Bounds* Copyright © 2012 by Nelson Education Ltd.

 Copyright © 2012 by Nelson Education Ltd.

Pathway 2

Name:________________________ Date:______________

6. Calculate. Show your work.

a) 512 ÷ 6 = 85 R2 or 85.33

e.g.,

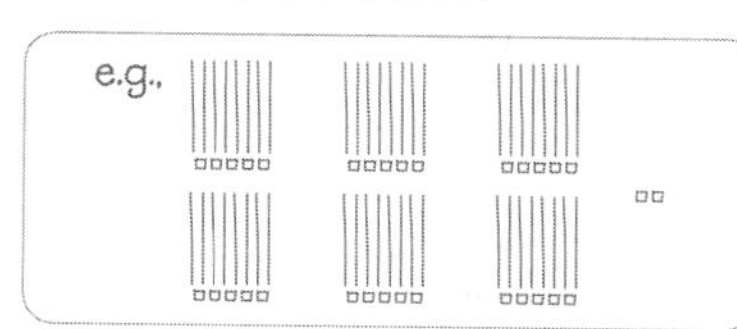

c) 812 ÷ 9 = 90 R2 or 90.22

e.g., 812 ÷ 9
= 810 ÷ 9 + 2 ÷ 9
= 90 R2

b) 3009 ÷ 4 = 752 R1 or 752.25

e.g., 4 × 750 = 3000
4 × 2 = 9
So, 3009 4 = 752 R1

d) 4220 ÷ 4 = 1055

e.g.,

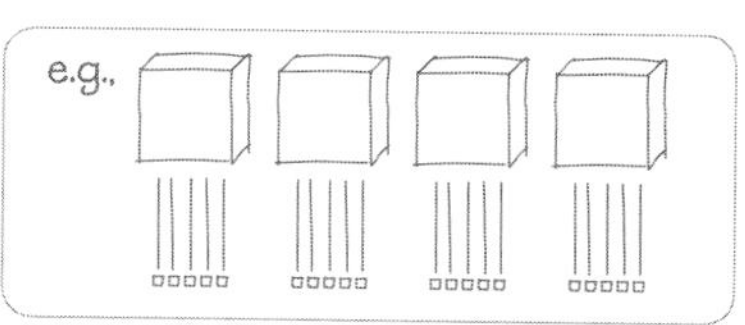

7. Create a real-life problem that could be solved using 365 ÷ 7.

e.g., How many weeks are there in 365 days?

8. Show or describe 2 ways to calculate 390 ÷ 6.

e.g., I divided 390 by 3 in my head. It's 130. Then I divided that in half. 130 is 120 + 10, so half is 60 + 5. The answer is 65.

e.g., Write 390 as 360 + 30 and divide each part by 6 and then add the answers. It would be 60 + 5 = 65.

Pathway 1

9. In each expression, circle the sign for the calculation you would do first.

a) 4 + 6 (×) 5

b) 6 × (4 (+) 5)

c) 3 + 12 (÷) 3

d) 6 + 3 − 4 (×) 2

10. Is (3 + 5) × 2 equal to 3 + 5 × 2? How do you know?

No. e.g., In (3 + 5) × 2, you are doubling both 3 and 5 but in 3 + 5 × 2 you are only doubling 5 and then adding it to 3.

11. Use >, <, or = to make each statement true.

a) 4 × 5 × 6 [=] 4 × (5 × 6)

b) 4 × 5 + 6 [<] 4 × (5 + 6)

c) 4 + 5 × 6 [<] (4 + 5) × 6

d) 4 × (3 + 2) × 5 [>] (4 × 3) + (2 × 5)

Copyright © 2012 by Nelson Education Ltd.

Copyright © 2012 by Nelson Education Ltd.

Order of Operations

You will need

- Student Resource page 16

Open-Ended Intervention

Before Using the Open-Ended Intervention

Record the numbers 9, 2, and 4. Tell students to subtract 2 from 9 and then multiply by 4. Ask:

- ▸ What is the result? (*28*)

Introduce the term *expression* as a series of calculations involving numbers. Ask:

- ▸ Why can you write what you did in an expression as $(9 - 2) \times 4$, or as $4 \times (9 - 2)$?
  (e.g., *The brackets tell you to calculate 9 − 2 first so you'd get the same answer for both, since* $4 \times 7 = 7 \times 4$*.*)
- ▸ Why is $9 - 2 \times 4$ equal to 1 and not 28?
  (e.g., *Since there are no brackets, you multiply 2 × 4 first and then subtract from 9.*)
- ▸ Describe how to calculate $4 + 3 \times (8 - 6)$.
  (e.g., *Subtract 6 from 8 to get 2, then multiply the 2 by 3 to get 6, and then add the 6 to 4 to get 10.*)

Using the Open-Ended Intervention Student Resource page 16

There is no mention of the E (for exponents) in BEDMAS in the Student Resource. However, you might let students know that later they will use special symbols called exponents, involving repeated multiplication.

Read through the tasks on the student page together. Note that simple numbers are used so that students do not need calculators (since some calculators automatically follow the order of operations for the student). Have students who need calculators use them for individual calculations, not the entire expression.

Give students time to work, ideally in pairs.

Observe whether students

- can describe how to perform a sequence of calculations
- can insert brackets into an expression to achieve a given result
- recognize that sometimes the use of brackets is optional
- can communicate clearly how to perform a sequence of operations

Consolidating and Reflecting

Ensure understanding by asking questions such as these based on students' work:

- ▸ Did you write any expressions where the order from left to right would have been the correct order for calculation?
  (*yes;* e.g., *for* $(4 \times 5) \div 2 + 8 - 3$)
- ▸ Why did your strategy for moving brackets to get a greater answer make sense?
  (e.g., *By taking the 36 out of the brackets, it meant that I divided by a smaller number, and I knew that would result in a bigger answer.*)
- ▸ Why are order of operations rules needed?
  (e.g., *Different people might get different answers for an expression involving a lot of operations when that expression was written down and not said.*)

 Copyright © 2012 by Nelson Education Ltd.

Order of Operations

Pathway 1 GUIDED

Guided Intervention

Before Using the Guided Intervention

Record the numbers 9, 2, and 4. Tell students to subtract 2 from 9 and then multiply by 4. Ask:

- ▸ What is the result? (*28*)

Introduce the term *expression* as a series of calculations involving numbers. Ask:

- ▸ Why can you write what you did in an expression as $(9 - 2) \times 4$, or as $4 \times (9 - 2)$? (e.g., *The brackets tell you to calculate* $9 - 2$ *first so you'd get the same answer for both, since* $4 \times 7 = 7 \times 4$.)
- ▸ Why is $9 - 2 \times 4$ equal to 1 and not 28? (e.g., *Since there are no brackets, you multiply* 2×4 *first and then subtract from 9.*)

Using the Guided Intervention (Student Resource pages 17–19)

Read about the skill-testing question and BEDMAS together. Then work through the prompts to complete the calculation together. Have students who need calculators use them for individual calculations, not the entire expression, since some calculators use the order of operations.

Have students work through the **Try These** questions in pairs or individually.

Observe whether students

- can apply the order of operations (Questions 1, 4, 5, 6, 7, 8)
- can communicate how to perform a series of calculations (Questions 2, 7)
- recognize when there are options in evaluating an expression (Question 3)
- can insert brackets into an expression to achieve a given result (Question 8)
- recognize why the order of operations rules are needed (Questions 6, 7, 9, 10)

Consolidating and Reflecting

Ensure understanding by asking questions such as these based on students' work:

- ▸ Why is there a choice about which calculation to perform first in Question 1b)? (e.g., *If there is more than one pair of brackets, it doesn't matter which you do first.*)
- ▸ Why were some answers to Question 5 the same? (e.g., *For parts d) and e), if you divide and then multiply, you get the same result as multiplying and then dividing.*)
- ▸ What strategy did you use in Question 8 to decide where to place the brackets? (e.g., *For part b), I created 2 numbers that, when divided, had a quotient of 5;* $4 \times 30 + 5 = 125$ *and* $17 + 8 = 25$ *and* $125 \div 25 = 5$.)
- ▸ Are there situations where people would likely do the correct order of operations even if they didn't know the rules? (*yes;* e.g., *If all of the operations were the same, you would get the right answer anyway.*)
- ▸ Why are the order of operations rules more important when expressions are written rather than said? (e.g., *When you say an expression like* $8 + 3 \times 2$*, you can say, "Add 8 to the product of* 3×2*" or "Multiply 3 by 2 first and then add 8."*)

You will need

- Student Resource pages 17–19

There is no mention of the E (for exponents) in BEDMAS in the Student Resource. However, you might let students know that later they will use special symbols called exponents, involving repeated multiplication.

Copyright © 2012 by Nelson Education Ltd.

Pathway 2 OPEN-ENDED

Dividing Whole Numbers

You will need

- base ten blocks (optional)
- Student Resource pages 20–21

Open-Ended Intervention

Before Using the Open-Ended Intervention

Tell students that 4 friends shared $84. Ask:

- How would you figure out how much each gets?
  (e.g., *If they each got $20 that would be $80, so they can each get $1 more.*)
- There are 84 books to be put in packages of 4. How would you figure out the number of packages? Why? (e.g., *I'd divide 84 by 4. I'm making a lot of equal packages, or groups, and you divide to find out how many equal groups.*)
- What is 84 ÷ 4? How did you get your answer?
  (*21;* e.g., *There are 4 groups of 20 in 80, and then there is 1 more group of 4.*)
- Why were the answers the same when you shared $84 and packaged 84 books?
  (e.g., *Both times it was about how many 4s are in 84.*)

Using the Open-Ended Intervention Student Resource pages 20–21

Read through the tasks on the student page together. You might provide base ten blocks. If students seem unfamiliar with how to model, you might refer them to the instructional part of the guided intervention. Give students time to work, ideally in pairs.

Observe whether students

- can estimate quotients
- relate a division statement to a problem situation
- can divide three- or four-digit numbers by one-digit numbers
- use different strategies to divide
- relate multiplication to division

Consolidating and Reflecting

Ensure understanding by asking questions such as these based on students' work:

- Why could you solve the problems by multiplying or dividing?
  (e.g., *You can divide when forming equal groups from a total amount or you can multiply to figure out if you have the total you want.*)
- How did you check your answers?
  (e.g., *I multiplied my answer by the team size and added the remainder to see if I got the right number of students.*)
- Which calculations did you find easiest to do? Why?
  (e.g., *dividing by 5, since I divided by 10 and then multiplied by 2*)
- What strategies did you use to solve the team size problems?
  (e.g., *I usually broke up the number into parts that were easy to divide by the team size. Sometimes I used base ten blocks.*)
- What did you decide to do with "leftover" students?
  (e.g., *It depended on how many were left over. If it was a small number, I made a couple of teams a bit bigger. For a bigger number, I made another, smaller team.*)

 Copyright © 2012 by Nelson Education Ltd.

Dividing Whole Numbers

Pathway 2
GUIDED

Guided Intervention

Before Using the Guided Intervention

Tell students that 4 friends shared $84. Ask:

- How would you figure out how much each gets?
  (e.g., *If they each got $20 that would be $80, so they can each get $1 more.*)
- There are 84 books to be put in packages of 4. How would you figure out the number of packages? Why?
  (e.g., *I'd divide 84 by 4. I'm making a lot of equal packages, or groups, and you divide to find out how many equal groups.*)
- What is 84 ÷ 4? How did you get your answer?
  (*21.* e.g., *There are 4 groups of 20 in 80, and then there is 1 more group of 4.*)
- Why were the answers the same when you shared $84 and packaged 84 books?
  (e.g., *Both times it was about how many 4s are in 84*).

Using the Guided Intervention Student Resource pages 22–25

Work through the instructional section together, having students model and complete each strategy for each type of division (sharing and grouping). You might provide base ten blocks.

Have students work through the **Try These** questions in pairs or individually.

Observe whether students

- can estimate quotients (Questions 1, 3, 9)
- can model division calculations (Question 2)
- use different strategies to divide (Questions 2, 5, 8)
- relate division to problem situations (Question 4)
- can calculate quotients (Questions 6, 7)
- relate multiplication to division (Question 10)

Consolidating and Reflecting

Ensure understanding by asking questions such as these based on students' work:

- How did you estimate 3021 ÷ 7 in Question 3?
  (e.g., *I calculated 2800 ÷ 7 = 400, since 28 is easy to divide by 7 and 2800 is close to 3021.*)
- What did the problem situations you created for Question 4 have in common?
  (e.g., *They were both about taking an amount and making equal groups.*)
- What strategies did you use to divide in Question 6?
  (e.g., *I usually broke up the number I was dividing into parts that are easy to divide by the second number. Sometimes I used base ten blocks.*)
- How could you use multiplication to help you answer Question 7?
  (e.g., *For part a), you could multiply 83 by one-digit numbers to get a product greater than 500.*)

You will need

- base ten blocks (optional)
- Student Resource pages 22–25

Pathway 3 OPEN-ENDED

Multiplying Whole Numbers

You will need

- empty egg cartons
- base ten blocks (optional)
- Student Resource pages 26–27

Open-Ended Intervention

Before Using the Open-Ended Intervention

Show students an empty egg carton and establish that it holds 12 eggs. Tell them to imagine that there are 9 full cartons. Ask:

- How would you figure out how many eggs there would be, without counting every one?
  (e.g., *I'd multiply. I know* $9 \times 10 = 90$ *and* $9 \times 2 = 18$*, so I'd add* $90 + 18$*.*)
- How would you figure out the number of eggs in 21 cartons?
  (e.g., *There are 120 eggs in 10 cartons, 120 eggs in the next 10, and 12 more. I'd add* $120 + 120 + 12$*.*)
- What models might help you multiply 21 by 12?
  (e.g., *I'd use base ten blocks and show 21 groups of 12, using 21 ten blocks and 21 pairs of one blocks.*)

Using the Open-Ended Intervention (Student Resource pages 26–27)

Read through the tasks on the student pages together. You might provide base ten blocks. If students seem unfamiliar with how to model multiplication, you might refer them to the instructional part of the guided intervention.

Give students time to work, ideally in pairs.

Observe whether students

- can estimate products
- can calculate products of 2 two-digit numbers
- relate multiplication to a problem situation
- use different strategies to multiply

Consolidating and Reflecting

Ensure understanding by asking questions such as these based on students' work:

- Why did you choose to solve the problems by multiplying?
  (e.g., *You multiply when you are counting the total of a lot of equal groups.*)
- What strategies did you use to multiply?
  (e.g., *I made a base ten block rectangle with one number as the length and one number as the width, and counted the blocks. I also broke the number into smaller parts and multiplied each part and then added them together.*)
- How did you check to see if your answers were reasonable?
  (e.g., *I estimated by using close-by friendly numbers, e.g.,* 38×37 *is almost* $40 \times 40 = 1600$*, so I knew my answer had to be less than but close to 1600.*)

Copyright © 2012 by Nelson Education Ltd.

Multiplying Whole Numbers

Pathway 3 GUIDED

Guided Intervention

Before Using the Guided Intervention

Show students an empty egg carton and establish that it holds 12 eggs. Tell them to imagine that there are 9 full cartons. Ask:

- How would you figure out how many eggs there would be, without counting every one?
  (e.g., *I'd multiply. I know* $9 \times 10 = 90$ *and* $9 \times 2 = 18$*, so I'd add* $90 + 18$*.*)
- How would you figure out the number of eggs in 21 cartons?
  (e.g., *There are 120 eggs in 10 cartons, 120 eggs in the next 10, and 12 more. I'd add* $120 + 120 + 12$*.*)

Using the Guided Intervention (Student Resource pages 28–31)

Read the opening paragraph. Provide base ten blocks and work through the instructional section together, having students model and complete each strategy.

Have students work through the **Try These** questions in pairs or individually.

Observe whether students

- can estimate products (Questions 1, 3, 9)
- can model multiplication (Question 2)
- relate multiplication to problem situations (Question 4)
- use different strategies to multiply (Questions 5, 7)
- can calculate products (Questions 5, 6, 7, 8, 9, 10)
- recognize computational errors (Question 7)
- relate 2 products (Question 8)

Consolidating and Reflecting

Ensure understanding by asking questions such as these based on students' work:

- How did you estimate 35×49 in Question 3?
  (e.g., *I figured it was about halfway between* $30 \times 50 = 1500$ *and* $40 \times 50 = 2000$*.*)
- What sort of problem situations did you create for Question 4?
  (e.g., *I created problems where there were lots of items in equal groups and I needed to know the total.*)
- What model could you create to show what was wrong in Question 7?
  (e.g., *I'd create a base ten block rectangle, and she would see that the 4 parts of the area have to be added.*)
- What strategies did you use to multiply in Question 8?
  (e.g., *I broke the numbers into parts, multiplied the parts, and added them together. I also made base ten block rectangles.*)

You will need

- empty egg cartons
- base ten blocks
- Student Resource pages 28–31

Copyright © 2012 by Nelson Education Ltd.

Strand: Number

Representing and Comparing Decimals

Planning For This Topic

Materials for assisting students with representing decimals consist of a diagnostic tool and 4 intervention pathways. Pathway 1 focuses on working with decimals with more than 3 decimal places (with a particular emphasis on very small numbers less than 1). Pathway 2 concentrates on comparing decimals involving thousandths. Pathway 3 focuses on representing decimal thousandths. Pathway 4 involves multiplying decimals by powers of 10.

Each pathway has an open-ended intervention and a guided intervention. Choose the type of intervention more suitable for your students' needs and your particular circumstance.

Curriculum Connections

Grades 5 to 8 curriculum connections for this topic are provided online. See www.nelson.com/leapsandbounds. Pathway 1 is appropriate for those following the WNCP curriculum. Pathway 4 is primarily for those following the Ontario curriculum. Pathways 2 and 3 are appropriate for both curriculums.

Why might students struggle with decimals?

Students might struggle with representing and comparing decimals for any of the following reasons:

- They might have difficulty switching from modelling whole numbers using base ten blocks to modelling decimals using base ten blocks (particularly when the large cube, instead of the flat, represents 1).
- They might not recognize that a decimal such as 0.060 can be read or renamed as thousandths; that is, 0.060 is 60 thousandths.
- They might think that a number with more digits is always greater than a number with fewer digits, as with whole numbers (e.g., they may think, mistakenly, that $0.8 < 0.423$).
- They neglect to consider the value of the zeros immediately to the right of the decimal point (e.g., they mistakenly think that $0.04 > 0.3$).
- They might not consider renaming decimals as a strategy for comparing them (e.g., to compare 4.1 with 4.123, rename 4.1 as 4.100).
- They might not understand how ten thousandths, hundred thousandths, and millionths relate to tenths, hundredths, and thousandths (e.g., 1000 millionths is the same as 1 thousandth).
- Their number sense, when it comes to very small numbers, may be weak, so they rely solely on rote procedures to compare and order these decimals.
- They might multiply a decimal by a power of 10 by appending 0s to the right of the decimal point instead of moving digits (e.g., writing 10×3.14 as 3.140).
- They might mix up the effects of multiplying and dividing by powers of 10 (e.g., they might multiply 0.234 by 100 to get 0.00234 instead of 23.4).

Professional Learning Connections

PRIME: Number and Operations, Background and Strategies (Nelson Education Ltd., 2005), pages 117–122, 127–130

Making Math Meaningful to Canadian Students K–8 (Nelson Education Ltd., 2008), pages 227–234, 241–244

Big Ideas from Dr. Small Grades 4–8 (Nelson Education Ltd., 2009), pages 61–67

Good Questions (dist. by Nelson Education Ltd., 2009), pages 35, 52

More Good Questions (dist. by Nelson Education Ltd., 2010), pages 66, 81

Copyright © 2012 by Nelson Education Ltd.

Diagnostic Tool: Representing and Comparing Decimals

Use the diagnostic tool to determine the most suitable intervention pathway for representing and comparing decimals. Provide Diagnostic Tool: Representing and Comparing Decimals, Teacher's Resource pages 34 to 36, and have students complete it in writing or orally. Make Thousandths Grids (BLM 7) available.

See solutions on Teacher's Resource pages 37 to 39.

Intervention Pathways

The purpose of the intervention pathways is to help students work comfortably with decimals with many decimal places, which they are less accustomed to in their everyday lives, and relate multiplying by powers of 10 to place value.

There are 4 pathways:

- Pathway 1: Decimals with Many Places
- Pathway 2: Comparing Decimals
- Pathway 3: Representing Decimal Thousandths
- Pathway 4: Multiplying and Dividing by 10s

Use the chart below (or the Key to Pathways on Teacher's Resource pages 37 to 39) to determine which pathway is most suitable for each student or group of students.

Diagnostic Tool Results	Intervention Pathway
If students struggle with Questions 11 to 13	use Pathway 1: Decimals with Many Places *Teacher's Resource pages 40–41* *Student Resource pages 32–36*
If students struggle with Questions 8 to 10	use Pathway 2: Comparing Decimals *Teacher's Resource pages 42–43* *Student Resource pages 37–42*
If students struggle with Questions 4 to 7	use Pathway 3: Representing Decimal Thousandths *Teacher's Resource pages 44–45* *Student Resource pages 43–48*
If students struggle with Questions 1 to 3	use Pathway 4: Multiplying and Dividing by 10s *Teacher's Resource pages 46–47* *Student Resource pages 49–55*

If students successfully complete Pathway 3 (or 2), they may or may not need the additional intervention provided by Pathway 2 (or 1). Either re-administer Pathway 2 (or 1) questions from the diagnostic tool or encourage students to do a portion of the open-ended intervention for Pathway 2 (or 1) to decide if more work in that pathway would be beneficial. Pathway 4 is somewhat independent of the other pathways.

Copyright © 2012 by Nelson Education Ltd.

Name: ______________________ Date: ______________

Representing and Comparing Decimals

Diagnostic Tool

You will need

- Thousandths Grids (BLM 7)

1. Calculate, without using a calculator.

a) $10 \times 2.345 =$ ________

b) $100 \times 0.819 =$ ________

c) $1000 \times 0.006 =$ ________

d) $100 \times 0.235 =$ ________

e) $512.4 \div 10 =$ ________

f) $123.8 \div 100 =$ ________

2. Draw a picture to show why $100 \times 2.13 = 213$.
Explain your picture.

3. Why does it make sense that $48.7 \div 10 = 4.87$?

____

____

____

Copyright © 2012 by Nelson Education Ltd.

Name: ______________________ Date: ______________

4. Represent each decimal on the thousandths grid.
Label each part with the decimal.

a) 0.004

b) 0.125

c) 0.2

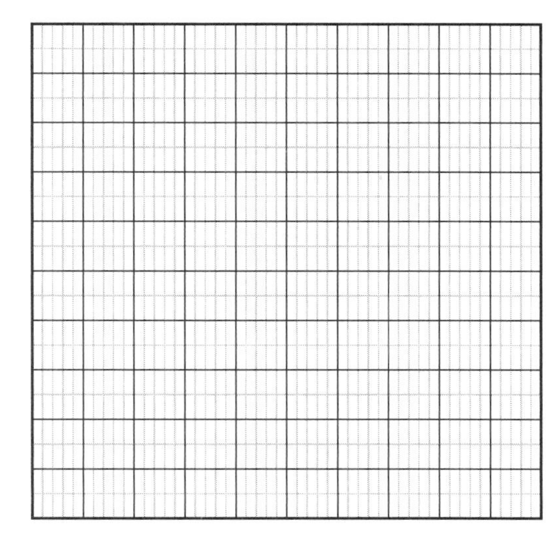

5. Write each number in decimal form.

a) 14 out of 1000 = ______________

b) $\frac{21}{1000}$ = ______________

c) 2 and 4 thousandths = ______________

d) 182 thousandths = ______________

e) 104 and 95 thousandths = ______________

f) 64 thousandths = ______________

6. Write how you would read each number. For example, 0.45 is *45 hundredths.*

a) 0.004 ______________________________

b) 0.125 ______________________________

c) 0.2 ______________________________

7. Circle the correct decimal for each description.

a) the decimal closest to 0.45	0.4	0.449	0.5
b) the decimal equivalent to 0.83	0.84	0.838	0.830
c) the decimal closest to 2	2.101	2.002	2.121

Copyright © 2012 by Nelson Education Ltd.

Name: ______________________ Date: ______________

8. Circle the greater number in each pair.

a) 3.8 or 3.080

c) 0.289 or 0.298

b) 4.16 or 4.147

d) 1.42 or 1.402

9. Use $>$ or $<$ to complete each comparison.

a) 62 hundredths ☐ 143 thousandths

b) 150 thousandths ☐ 34 hundredths

c) 150 thousandths ☐ 12 hundredths

d) 218 thousandths ☐ 22 hundredths

10. Explain how you know that $0.2 > 0.148$.

____

____

11. Why is 0.000 002 the same as 2 out of a million?

____

____

____

____

12. Write how you would read each number.

a) 0.000 031 ______________________________

b) 0.0042 ______________________________

c) 0.001 325 ______________________________

d) 2.124 001 ______________________________

13. Circle the statement that is true. Explain your thinking.

$0.213 = 0.213\,000$ $0.213 = 0.0213$ $0.213 = 0.20013$

____

____

 Copyright © 2012 by Nelson Education Ltd.

Solutions and Key to Pathways

Name:______________________ Date:______________

Representing and Comparing Decimals

Diagnostic Tool

You will need

- Thousandths Grids (BLM 7)

1. Calculate, without using a calculator.

   a) $10 \times 2.345 =$ 23.45

   b) $100 \times 0.819 =$ 81.9

   c) $1000 \times 0.006 =$ 6

   d) $100 \times 0.235 =$ 23.5

   e) $512.4 \div 10 =$ 51.24

   f) $123.8 \div 100 =$ 1.238

2. Draw a picture to show why $100 \times 2.13 = 213$. Explain your picture.

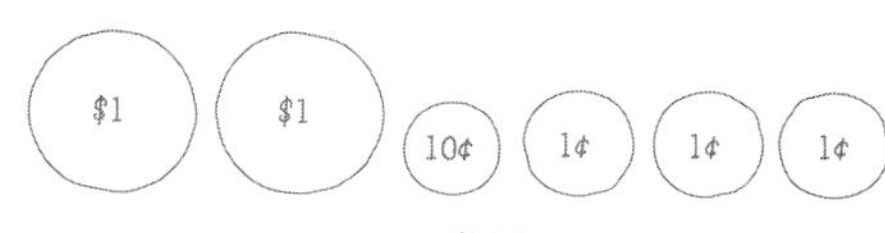

$2.13

e.g., If you have 100 of each of these coins, you have $100, $100, $10, $1, $1, and $1.
That is $213.
So $100 \times 2.13 = 213$.

3. Why does it make sense that $48.7 \div 10 = 4.87$?

   e.g., If you have almost $50 and 10 people share it, they each get almost $5 and that's what $4.87 is.

Pathway 4

34 Representing and Comparing Decimals, Diagnostic Tool *Leaps and Bounds*

Copyright © 2012 by Nelson Education Ltd.

Name:____________________ Date:____________

4. Represent each decimal on the thousandths grid. Label each part with the decimal.

a) 0.004

b) 0.125

c) 0.2

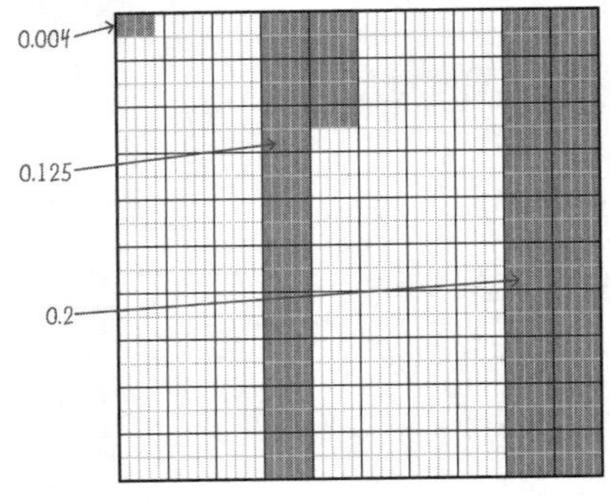

5. Write each number in decimal form.

a) 14 out of 1000 = 0.014

b) $\frac{21}{1000}$ = 0.021

c) 2 and 4 thousandths = 2.004

d) 182 thousandths = 0.182

e) 104 and 95 thousandths = 104.095

f) 64 thousandths = 0.064

Pathway 3

6. Write how you would read each number. For example, 0.45 is *45 hundredths*.

a) 0.004 4 thousandths

b) 0.125 125 thousandths

c) 0.2 2 tenths

7. Circle the correct decimal for each description.

a) the decimal closest to 0.45	0.4	(0.449)	0.5
b) the decimal equivalent to 0.83	0.84	0.838	(0.830)
c) the decimal closest to 2	2.101	(2.002)	2.121

Copyright © 2012 by Nelson Education Ltd.

Copyright © 2012 by Nelson Education Ltd.

Name:______________________ Date:______________

Pathway 2

8. Circle the greater number in each pair.

a) (3.8) or 3.080

b) (4.16) or 4.147

c) 0.289 or (0.298)

d) (1.42) or 1.402

9. Use > or < to complete each comparison.

a) 62 hundredths [>] 143 thousandths

b) 150 thousandths [<] 34 hundredths

c) 150 thousandths [>] 12 hundredths

d) 218 thousandths [<] 22 hundredths

10. Explain how you know that 0.2 > 0.148.

e.g., On a thousandths grid, 0.2 fills 2 full columns, but 0.148 fills only one column and part of another column.

Pathway 1

11. Why is 0.000 002 the same as 2 out of a million?

e.g., Each place value is 1 tenth as much as the one to its left. So the fourth place after the decimal point is 1 out of ten thousand, the fifth place is 1 out of a hundred thousand, and the last place is 1 out of a thousand thousand, which is a million.

12. Write how you would read each number.

a) 0.000 031 31 millionths

b) 0.0042 42 ten thousandths

c) 0.001 325 1325 millionths

d) 2.124 001 2 and 124 001 millionths

13. Circle the statement that is true. Explain your thinking.

(0.213 = 0.213 000) 0.213 = 0.0213 0.213 = 0.20013

e.g., If you add 0s to the end of a decimal number, the value of the number doesn't change.

 Copyright © 2012 by Nelson Education Ltd.

Copyright © 2012 by Nelson Education Ltd.

Pathway 1 OPEN-ENDED

Decimals with Many Places

You will need

- Place Value Charts (to Millionths) (BLM 4)
- Student Resource pages 32–33

Open-Ended Intervention

Before Using the Open-Ended Intervention

Show a place value chart (BLM 4) and ask:

- ▸ Why are the tenths to the right of the ones in the chart?
  (e.g., *It takes 10 tenths to make a 1, and you trade 10 for 1 as you move left.*)
- ▸ What place would you expect to be to the right of the thousandths place? Why?
  (*The ten thousandths place;* e.g., *ten thousands are to the left of thousands on the other side of the decimal point.*)
- ▸ Where do you think the millionths place would be? Why do you think that?
  (e.g., *3 places to the right of the thousandths, to match the millions on the left*)
- ▸ Write 0.0003 ("3 ten thousandths") in the chart. How does the chart help you see why you could write 0.0003 as 0.000 300?
  (e.g., *0.000 300 means there are 3 ten thousandths, no hundred thousandths, and no millionths, and so does 0.0003.*)

Using the Open-Ended Intervention (Student Resource pages 32–33)

Read through the introduction and the tasks on the student pages together. Provide place value charts. Give students time to work, ideally in pairs.

Observe whether students

- can read and make sense of numbers with many decimal places
- can create reasonable descriptions of decimals and provide examples
- recognize that there are many possible numbers to match each description

Consolidating and Reflecting

Ensure understanding by asking questions such as these based on students' work:

- ▸ How can you write 23 millionths as a decimal? (e.g., *0.000 023*)
- ▸ Why does that make sense?
  (e.g., *Millionths are really small, so you need a lot of decimal places.*)
- ▸ Why does it make sense that the place to the right of the hundred thousandths is millionths?
  (e.g., *Since millions are just left of hundred thousands before the decimal point, then millionths should be just right of hundred thousandths after the decimal point.*)
- ▸ How many millionths make 1 ten thousandth? How do you know?
  (*100,* e.g., *ten thousandths is 2 columns to the left of millionths*)
- ▸ Why would a number less than 1 hundred thousandth but more than 0 need 6 or more decimal places?
  (e.g., *1 hundred thousandth is 0.000 01, and there is nothing less than 1 except 0, so you have to include the 0 in front of the 1.*)
- ▸ Explain why 540.000 123 has many decimal places even though it's not a small number. (e.g., *Many decimal places are also used to show precision (really exact amounts) with big and small numbers.*)

Copyright © 2012 by Nelson Education Ltd.

Decimals with Many Places

Pathway 1 GUIDED

Guided Intervention

Before Using the Guided Intervention

Show a decimal place value chart with the ones, tenths, hundredths, and thousandths places labelled and with 3 more empty columns to the right. Ask:

- Why are the tenths to the right of the ones in the chart?
  (e.g., *It takes 10 tenths to make a 1, and you trade 10 for 1 as you move left.*)
- What place would you expect to be right of the thousandths place? Why?
  (*The ten thousandths place;* e.g., *ten thousands are to the left of thousands on the other side of the decimal point.*)
- Where do you think the millionths place would be? Why do you think that?
  (e.g., *3 places to the right of the thousandths, to match the millions on the left*)

Using the Guided Intervention **Student Resource pages 34–36**

Work through the instructional section of the student pages together. Have students write the numbers in place value charts (BLM 4) and read them as described.

Have students work through the **Try These** questions individually or in pairs.

Observe whether students

- can read and record numbers with 4, 5, and 6 decimal places (Questions 1, 2, 3, 4, 5)
- can create numbers to fit descriptions of decimals (Question 4)
- recognize why extremely small items are described using numbers with many decimal places (Question 6)

Consolidating and Reflecting

Ensure understanding by asking questions such as these based on students' work:

- When did the numbers in Question 2 need 6 decimal places to the right of the decimal point?
  (e.g., *when the numbers were written as millionths in parts a), b), and f)*)
- Why did it make sense that the first 3 digits to the right of the decimal point were 0s for Question 5c)?
  (e.g., *150 cm^3 is 150 millionths cubic metres, which is 0.000 150 m^3.*)
- Question 6 says that you need a lot of decimal places to write a very small number. Explain why the number 540.000 123 has many decimal places even though it's not a small number.
  (e.g., *Many decimal places are also used to show precision (really exact amounts) with big and small numbers.*)
- Why does it make sense that the place to the right of the hundred thousandths is millionths?
  (e.g., *Millions are just left of hundred thousands before the decimal point, so millionths should be just right of hundred thousandths after the decimal point.*)

You will need

- Place Value Charts (to Millionths) (BLM 4)
- Student Resource pages 34–36

Comparing Decimals

You will need

- an apple
- Thousandths Grids (BLM 7)
- Place Value Charts (to Thousandths) (BLM 6)
- Internet access, or kilogram and gram masses and a balance scale
- Student Resource pages 37–38

Sample items
package of meat: 0.492 kg
tennis racquet: 0.487 kg
apple: 0.15 kg
textbook: 1.1 kg
newborn baby: 3.1 kg
big bag of flour: 5.0 kg
cat: 4.52 kg
backpack: 5.334 kg

Open-Ended Intervention

Before Using the Open-Ended Intervention

Show students an apple and tell them the mass is about 150 g. Ask:

- Why can you write 150 g as 0.150 kg?
  (e.g., *There are 1000 g in 1 kg, so 150 g would be $\frac{150}{1000}$ of a kilogram.*)
- If I took a very, very small bite out of it, what might the mass of the apple be then? Express the mass as a decimal amount of kilograms. (e.g., *0.142 kg*)
- Why might you describe the mass of the remaining apple as about 0.1 kg ("1 tenth of a kilogram")?
  (e.g., *It's 0.1 kg plus a bit more, $\frac{42}{1000}$ of a kilogram, which is not very much.*)
- How might you model 0.150 using a thousandths grid?
  (e.g., *I'd shade 1 big column and 5 squares; that's 150 small sections of the grid.*)

Using the Open-Ended Intervention (Student Resource pages 37–38)

Read through the tasks on the student pages together. Provide the necessary materials or Internet access for students to find the mass of the 8 items, or provide the list in the margin. Provide Thousandths Grids (BLM 7) and Place Value Charts (to Thousandths) (BLM 6). Give students time to work, ideally in pairs.

Observe whether students

- can use models to compare decimals
- can estimate how far apart 2 decimals are
- can compare and order decimals and explain their strategies

Consolidating and Reflecting

Ensure understanding by asking questions such as these based on students' work:

- How did you decide that 0.487 was less than 0.492?
  (e.g., *I used what I know about place value. 487 thousandths is less than 492 thousandths.*)
- How do you know that 0.492 and 0.487 are less than 0.1 apart?
  (e.g., *If you model decimals that are 0.1 apart, you use an extra column of the thousandths grid for the greater decimal. This doesn't happen with 0.492 and 0.487, since both have more than 4 but less than 5 columns shaded.*)
- Some people might think that 1.13 is greater than 1.2 because 1.13 has more digits. What would you tell them?
  (e.g., *You can't compare decimals by counting digits as you do with whole numbers. 1.13 is less than 1.15, but 1.2, which is 1.20, is more than 1.15.*)
- How did you decide which mass was greatest?
  (e.g., *I knew that the mass with the greatest whole number part had to be greatest, but since the whole number parts were the same for several of them, I compared the tenths, hundredths, or thousandths.*)

 Copyright © 2012 by Nelson Education Ltd.

Comparing Decimals

Pathway 2
GUIDED

Guided Intervention

Before Using the Guided Intervention

Find an item (or distance) that is between 2 m and 2.5 m long. Ask:

- ▸ What do you think is the length of this item (or distance)? (e.g., *about 2 m*)
- ▸ What is the actual length of the item? (e.g., *2.2 m*)
- ▸ What is a length between your estimate and the actual length? (e.g., *2.1 m*)
- ▸ What is a decimal thousandth length that is between the 2 measurements? (e.g., *2.104 m*)
- ▸ What length might be a little shorter than 2.104 m but a decimal hundredth? (e.g., *2.10 m*)

You will need

- a tape measure or metre stick
- Thousandths Grids (BLM 7)
- Place Value Charts (to Thousandths) (BLM 6)
- Student Resource pages 39–42

Using the Guided Intervention (Student Resource pages 39–42)

Work through the instructional section of the student pages together and have students use the strategies shown to compare decimals using Thousandths Grids (BLM 7) and Place Value Charts (to Thousandths) (BLM 6).

Have students work through the **Try These** questions individually or in pairs.

Observe whether students

- can use models to compare decimals (Questions 1, 2)
- can compare and order decimals and explain their strategies (Questions 3, 4, 5, 7, 8, 9, 10)
- can create decimals between given decimals (Question 6)

Consolidating and Reflecting

Ensure understanding by asking questions such as these based on students' work:

- ▸ How did you decide that 0.289 was less than 0.3 in Question 2b)?
(e.g., *I know that 0.289 is 28 hundredths and a bit more, so it's 28 squares and a bit, which is not as much as 30 hundredths, which is 30 squares.*)
- ▸ How did you know that 32 hundredths was more than 123 thousandths in Question 3b)?
(e.g., *32 hundredths is 320 thousandths, and 320 is more than 123.*)
- ▸ How do you compare decimals like 4.1 and 4.123 (Question 5) when they have a different number of digits?
(e.g., *I think of 4.1 as 4.100 and then compare 100 thousandths and 123 thousandths.*)
- ▸ How did you decide which number was greatest in Question 7?
(e.g., *It was 5.1, since the whole number part was the greatest.*)
- ▸ How did you decide on the next greatest number in Question 7?
(e.g., *I knew it was either 4.97 or 4.909. After I thought of 4.97 as 4 and 970 thousandths, and 4.909 as 4 and 909 thousandths, I knew 4.97 was greater than 4.090.*)

Copyright © 2012 by Nelson Education Ltd.

Pathway 3 OPEN-ENDED

Representing Decimal Thousandths

You will need

- Hundredths Grids (BLM 8)
- Thousandths Grids (BLM 7) or base ten blocks
- coloured pencils
- Student Resource pages 43–44

Open-Ended Intervention

Before Using the Open-Ended Intervention

Show students a hundredths grid. Ask:

- ▸ How do you know that each square in this grid is worth 1 hundredth? (e.g., *There are 100 sections, and 1 out of 100 is 1 hundredth.*)
- ▸ How would you write that as a decimal? (*0.01*)
- ▸ What part of the grid is worth $\frac{1}{10}$ (or 0.1)? (e.g., *1 column*)

Show students a thousandths grid. Ask:

- ▸ How do you know this grid has 1000 equal sections? (e.g., *There are 100 squares and 10 small rectangles in each square, and* $10 \times 100 = 1000$.)
- ▸ Why does it make sense to call each small rectangle a thousandth? (e.g., *A thousandth means 1 in 1000, and each small rectangle is 1 out of 1000 equal sections in the grid.*)

Using the Open-Ended Intervention

Student Resource pages 43–44

Read through the tasks on the student pages together. Provide Thousandths Grids (BLM 7) and/or base ten blocks, and coloured pencils. Give students time to work, ideally in pairs.

Observe whether students

- relate a decimal thousandth to a model
- relate a decimal to a fraction
- relate symbolic and verbal descriptions for a decimal
- recognize the relationships among decimal thousandths, hundredths, and tenths

Consolidating and Reflecting

Ensure understanding by asking questions such as these based on students' work:

- ▸ Why can you describe a column of a thousandth grid using tenths, hundredths, or thousandths? (e.g., *Each column is 1 tenth of the grid, but it is also 10 hundredths [10 squares] and 100 thousandths [100 small rectangles].*)
- ▸ How did you know that 0.399 is almost 0.4? (e.g., *399 little rectangles is almost 400 of them, and 0.4 is* $\frac{400}{1000}$.)
- ▸ Why is it useful to use a thousandths grid instead of a hundredths grid to represent a decimal between 0.29 and 0.3? (e.g., *There are no hundredths between 29 hundredths and 30 hundredths, so you need smaller pieces.*)
- ▸ Which of your own descriptions was easiest to represent? Why? (e.g., *almost 0.23, since I just coloured in 22 squares and most of a 23rd square*)

Copyright © 2012 by Nelson Education Ltd.

Representing Decimal Thousandths

Pathway 3 GUIDED

Guided Intervention

Before Using the Guided Intervention

Show students a hundredths grid. Ask:

- How do you know that each square in this grid is worth 1 hundredth? (e.g., *There are 100 sections, and 1 out of 100 is 1 hundredth.*)
- How would you write that as a decimal? (*0.01*)
- What part of the grid is worth $\frac{1}{10}$ (or 0.1)? (e.g., *1 column*)

Show students a thousandths grid. Ask:

- How do you know this grid has 1000 equal sections? (e.g., *There are 100 squares and 10 small rectangles in each square, and* $10 \times 100 = 1000$.)
- Why does it make sense to call each small rectangle a thousandth? (e.g., *A thousandth means 1 in 1000, and each small rectangle is 1 out of 1000 equal sections in the grid.*)

You will need

- Hundredths Grids (BLM 8)
- Thousandths Grids (BLM 7)
- base ten blocks (optional)
- Place Value Charts (to Thousandths) (BLM 6)
- coloured pencils (optional)
- Student Resource pages 45–48

Using the Guided Intervention **Student Resource pages 45–48**

Work through the instructional section of the student pages together using Thousandths Grids (BLM 7) and Place Value Charts (to Thousandths) (BLM 6).

Have students work through the **Try These** questions individually or in pairs. Provide base ten blocks if available.

Observe whether students

- relate a decimal thousandth to a model (Questions 1, 2, 3, 4, 10)
- relate a decimal to a fraction (Question 5)
- relate symbolic and verbal descriptions for a decimal (Questions 6, 8, 9)
- recognize the relationships among decimal thousandths, hundredths, and tenths (Question 7)

Consolidating and Reflecting

Ensure understanding by asking questions such as these based on students' work:

- In Question 3, how can you tell that 0.148 is between 0.1 and 0.2? (e.g., *It's more than* $\frac{1}{10}$, *since* $\frac{148}{1000} > \frac{100}{1000}$, *and it's less than* $\frac{2}{10}$, *since* $\frac{148}{1000} < \frac{200}{1000}$.)
- For Question 4, how do you know there were many decimal thousandths between 0.4 and 0.5? (e.g., *0.4 is* $\frac{400}{1000}$ *and 0.5 is* $\frac{500}{1000}$, *and there are lots of numbers between 400 and 500.*)
- Why can you describe a column of a thousandth grid using tenths, hundredths, and thousandths? (e.g., *Each column is 1 tenth of the grid, but it is also 10 hundredths [10 squares] and 100 thousandths [100 small rectangles].*)

Multiplying and Dividing by 10s

You will need

- Place Value Charts (to Ten Thousandths) (BLM 5)
- rulers, or Internet access
- Student Resource pages 49–50

Small items
ant: 1.74 cm
pencil: 0.17 m
thickness of paper: 0.01 cm
cellphone: 0.09 m
width of tape: 0.02 m
width of rubber band: 0.003 m

Large items
desk: 152.45 cm
house: 10.2 m
car race track: 5.451 km
oak tree: 18.462 m
refrigerator height: 1.659 m

Open-Ended Intervention

Before Using the Open-Ended Intervention

- About how long is 3 cm? (e.g., *about as wide as 3 fingers*)
- What if something were 10 times as long? How long would it be? (*30 cm*)
- A length is 10 times as long as 3.1 cm. Is it closer to 30 cm or 40 cm? How do you know? (*30 cm.* e.g., *3.1 is a lot closer to 3 than to 4, so 10 times as much is closer to 30 than 40.*)
- How do you know that 10 times 3.1 cm is 31 cm? (e.g., *3 × 10 is 30 cm, and since 0.1 cm is $\frac{1}{10}$ of a centimetre, 10 of them is another whole centimetre.*)
- How many centimetres is 100 times as long as 3.1 cm? How do you know? (*310 cm.* e.g., *It's 10 times as long as 31, and 10 × 31 is 310.*)

Using the Open-Ended Intervention Student Resource pages 49–50

Read through the tasks on the student pages together. Note that students do not draw the items; they find lengths by measuring or from the Internet, and calculate the lengths of various scale drawings. Provide rulers or Internet access, and Place Value Charts (BLM 5). If students need ideas for lengths of items, provide the list in the margin. Ensure students know that they can adjust the measurements to fit the requirements. Give students time to work, ideally in pairs.

Observe whether students

- can model products and quotients of decimals multiplied and divided by powers of 10
- can multiply and divide decimals by powers of 10
- relate problem situations to multiplying/dividing decimals by powers of 10
- can explain why the products/quotients of decimals and powers of 10 are what they are

Consolidating and Reflecting

Ensure understanding by asking questions such as these based on students' work:

- How is a place value chart useful for multiplying and dividing by 10, 100, and 1000?
  (e.g., *You can see how many places to move the digits and in what direction.*)
- What do you know about the product of a decimal thousandth and 100?
  (e.g., *It could have non-zero digits in the tenths, but not in the hundredths or thousandths.*)
- What do you notice about the digits in a number when you divide by 10?
  (e.g., *The digits move 1 column to the right, since you're trading 1 for 10 and then taking one of those 10.*)
- How is multiplying by 10 like multiplying by 1000? How is it different?
  (e.g., *It's the same since the digits move to the left. It's different since they move a different number of columns.*)

 Copyright © 2012 by Nelson Education Ltd.

Multiplying and Dividing by 10s

Guided Intervention

Before Using the Guided Intervention

Ask:

- What does 10 × 4.12 mean? (e.g., *10 groups of 4.12*)

Write "10 groups of 4 + 10 groups of 0.1 + 10 groups of 0.02" on the board. Ask:

- How do you know this is the same as 10 × 4.12? (e.g., *If you have 10 groups of something, you can split up the something into parts and have 10 of each part.*)
- Why does it make sense that 10 × 4.12 is a bit more than 40? (e.g., *4.12 is just a bit more than 4, and 10 × 4 is 40.*)
- What does 52.1 ÷ 10 mean? (e.g., *How many groups of 10 are in 52.1?*)
- Why does it make sense that the quotient of 52.1 ÷ 10 is about 5? (e.g., *50 ÷ 10 = 5, and 52.1 is close to 50.*)

You will need

- play coins (pennies, dimes, loonies, and $10 bills)
- base ten blocks (optional)
- Hundredths Grids (BLM 8)
- Place Value Charts (to Ten Thousandths) (BLM 5)
- Student Resource pages 51–55

Using the Guided Intervention **Student Resource pages 51–55**

Work through the instructional section of the student pages together. Provide place value charts and coins and have students model the multiplying and dividing shown.

Have students work through the **Try These** questions individually or in pairs. Provide hundredths grids, and base ten blocks if available.

Observe whether students

- can model products/quotients of decimals multiplied/divided by powers of 10 (Questions 1, 3, 4, 5)
- can explain why the products/quotients of decimals and powers of 10 are what they are (Questions 1, 3, 4, 5, 8, 10, 11)
- can multiply/divide decimals by powers of 10 (Questions 2, 7)
- relate measurement conversions to multiplying by 1000 (Question 6)
- relate problem situations to multiplying/dividing by powers of 10 (Question 9)

Consolidating and Reflecting

Ensure understanding by asking questions such as these based on students' work:

- In Question 8, you described the product of a decimal thousandth and 100. What do you know about the product of a decimal thousandth and 1000? (e.g., *It's a whole number.*)
- How is multiplying by 10 like multiplying by 1000? How is it different? (e.g., *It's the same since the digits move to the left. It's different since they move a different number of columns.*)
- How did the place value chart help you multiply and divide by 10, 100, and 1000? (e.g., *You can see how many columns to move the digits over and in what direction.*)

Copyright © 2012 by Nelson Education Ltd.

Strand: Number

Decimal Operations

Planning For This Topic

Materials for assisting students with decimal operations consist of a diagnostic tool and 4 intervention pathways. Pathway 1 focuses on dividing whole numbers by decimals to hundredths. Pathway 2 deals with dividing decimals by whole numbers. Pathway 3 involves multiplying decimals by whole numbers. Pathway 4 addresses addition and subtraction of decimals to thousandths.

As adults, we tend to perform calculations involving decimals with calculators, or we estimate. However, the curriculum outcomes/expectations specify using a variety of tools—including algorithms, concrete materials, mental strategies, and estimation—with single-digit divisors and double-digit factors. The pathways focus on these non-calculator strategies.

Each pathway has an open-ended intervention and a guided intervention. Choose the type of intervention more suitable for your students' needs and your particular circumstances.

Curriculum Connections

Grades 5 to 8 curriculum connections for this topic are provided online. See www.nelson.com/leapsandbounds. Pathway 1 covers content from the Ontario curriculum that is not specifically mentioned in the WNCP curriculum. Pathways 2 to 4 are relevant to both Ontario and WNCP curriculums. For students who are struggling significantly with adding and subtracting decimals, it may be necessary to look at *Leaps and Bounds Toward Math Understanding 5/6*, which separates addition and subtraction of tenths, hundredths, and thousandths.

Professional Learning Connections

PRIME: Number and Operations, Background and Strategies (Nelson Education Ltd., 2005), pages 122–126, 128–130

Making Math Meaningful to Canadian Students K–8 (Nelson Education Ltd., 2008), pages 236–240, 242–244

Big Ideas from Dr. Small Grades 4–8 (Nelson Education Ltd., 2009), pages 68–73

Good Questions (dist. by Nelson Education Ltd., 2009), pages 36, 56

More Good Questions (dist. by Nelson Education Ltd., 2010), page 72

Why might students struggle with decimal operations?

Students might struggle with decimal operations for any of the following reasons:

- They might have difficulty with whole number calculations, which then affects their performance with decimals.
- They might not have a strong enough understanding of thousandths and, therefore, might have difficulty computing with them.
- They might struggle with adding or subtracting decimals with different numbers of digits (e.g., they might line up $1.34 + 12.3$ by the 3 digits instead of by the decimal points and digits of equivalent place value).
- They might struggle when the number of tenths, hundredths, or thousandths in the minuend is fewer than the number of tenths, hundredths, or thousandths in the subtrahend (e.g., they calculate $0.327 - 0.269$ incorrectly as 0.142).
- They might misplace the decimal point in the answer because they do not estimate and/or they might have limited number sense.
- They might struggle with remainders (e.g., when dividing 4.1 by 3, they get 1.3 and do not know what to do with the 2 tenths remainder).

 Copyright © 2012 by Nelson Education Ltd.

- They might not recognize that division can mean counting the number of groups in an amount and not just sharing. This would make it difficult to understand dividing a whole number by a decimal.

Diagnostic Tool: Decimal Operations

Use the diagnostic tool to determine the most suitable intervention pathway for calculating with decimals. Provide Diagnostic Tool: Decimal Operations, Teacher's Resource pages 50 to 52, and have students complete it in writing or orally. Provide materials for modelling decimal operations (e.g., Hundredths Grids (BLM 8), Thousandths Grids (BLM 7), base ten blocks, or play coins).

See solutions on Teacher's Resource pages 53 to 55.

Intervention Pathways

The purpose of the intervention pathways is to help students work with a variety of ways to represent operations with decimals. The focus is to prepare them for multiplication and division involving decimals as well as to solidify their understanding of adding and subtracting. Therefore, calculators should not be used.

There are 4 pathways:
- Pathway 1: Dividing Whole Numbers by Decimals
- Pathway 2: Dividing Decimals by Whole Numbers
- Pathway 3: Multiplying with Decimals
- Pathway 4: Adding and Subtracting Decimals

Use the chart below (or the Key to Pathways on Teacher's Resource pages 53 to 55) to determine which pathway is most suitable for each student or group of students.

Diagnostic Tool Results	Intervention Pathway
If students struggle with Questions 10 to 12	use Pathway 1: Dividing Whole Numbers by Decimals *Teacher's Resource pages 56–57* *Student Resource pages 56–61*
If students struggle with Questions 7 to 9	use Pathway 2: Dividing Decimals by Whole Numbers *Teacher's Resource pages 58–59* *Student Resource pages 62–67*
If students struggle with Questions 4 to 6	use Pathway 3: Multiplying with Decimals *Teacher's Resource pages 60–61* *Student Resource pages 68–73*
If students struggle with Questions 1 to 3	use Pathway 4: Adding and Subtracting Decimals *Teacher's Resource pages 62–63* *Student Resource pages 74–79*

Name: ______________________ Date: ______________

Decimals Operations

Diagnostic Tool

You will need

- materials for modelling decimal operations (e.g., Hundredths Grids (BLM 8), Thousandths Grids (BLM 7), base ten blocks, or play coins)

1. Estimate each sum or difference as a whole number.

a) $3.45 + 2.8$ is about ________.

b) $12.97 + 8.831$ is about ________.

c) $15.2 - 8.941$ is about ________.

d) $25.93 - 11.682$ is about ________.

e) $103.4 - 28.99$ is about ________.

2. You have 10 m of ribbon.
How much is left after each amount is cut off?

a) 2.5 m

b) 7.819 m

3. a) How much more is $18.341 + 50$ than $17.34 + 49.1$? ________

b) How much more is $49.937 + 77.81$ than $39.927 + 66.51$? ________

4. Model each calculation using a grid, money, or base ten blocks.
Sketch your model.

a) 3×1.6

b) 6×4.23

Copyright © 2012 by Nelson Education Ltd.

Name:____________________ Date:______________

5. Solve each problem.

a) What is the total length of 2 rugs, if each rug is 5.2 m long?

b) What is the total distance of 5 trips, if each trip is 72.7 km long?

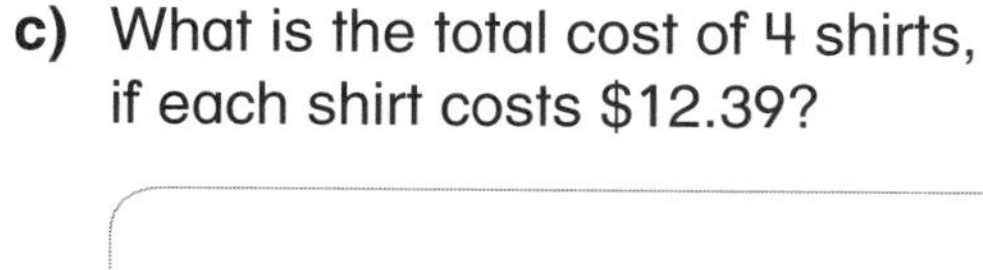

c) What is the total cost of 4 shirts, if each shirt costs $12.39?

d) What is the total length of 5 pieces of string, each 4.238 m long, laid end to end?

6. Fill in the blanks with a decimal thousandth and a whole number.

a) □.□□□ × □ is about 35.

b) □.□□□ × □ is about 24.8.

7. What is each person's share, if they share equally?

a) 5 people sharing 3.45 L of juice

b) 4 people sharing $123.24

8. Fill in the blanks with a decimal hundredth and a whole number.

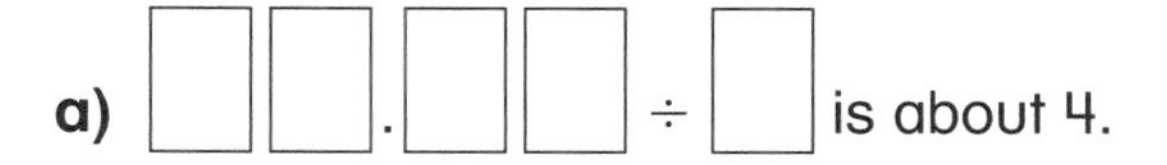

a) □□.□□ ÷ □ is about 4.

b) □□.□□ ÷ □ is about 6.

Name:______________________________ Date:______________

9. Model each calculation using a grid, money, or base ten blocks.
Sketch your model.

a) $9.23 \div 5$

b) $1.456 \div 6$

10. Use the hundredths grids to show each calculation.
Write each quotient.

a) $1 \div 0.05 =$ ________

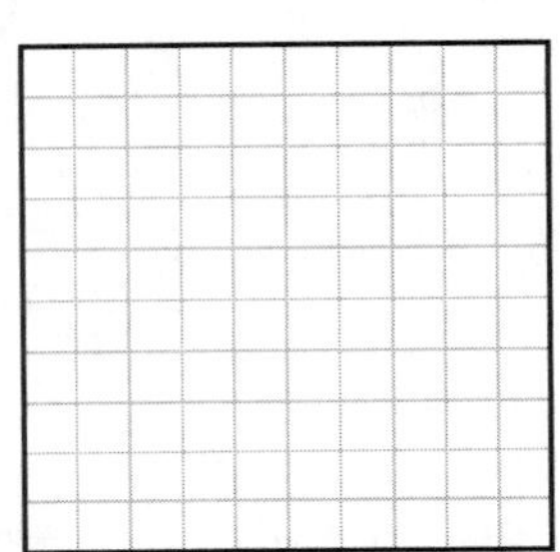

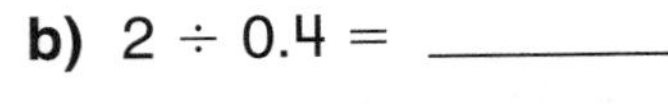

b) $2 \div 0.4 =$ ________

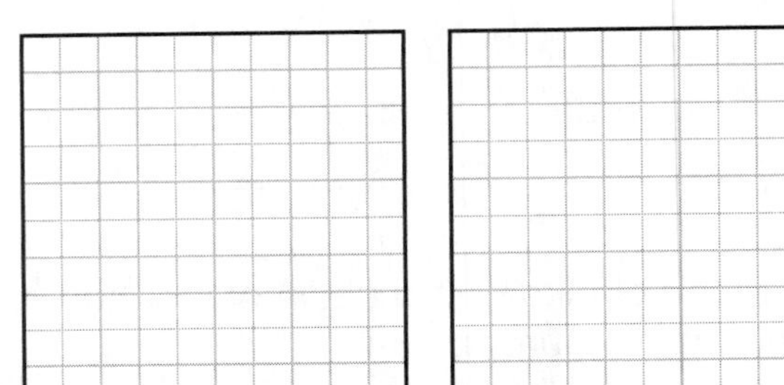

11. Circle the greater quotient:

$5 \div 0.3$ or $5 \div 0.4$

Show or explain how you know it is greater.

12. How can you use $4 \div 0.4 = 10$ to help you calculate $4 \div 0.8$?

____

____

 Copyright © 2012 by Nelson Education Ltd.

Solutions and Key to Pathways

Name:______________________________ Date:__________________

Decimals Operations

Diagnostic Tool

Pathway 4

1. Estimate each sum or difference as a whole number.

   a) 3.45 + 2.8 is about 6.

   b) 12.97 + 8.831 is about 20.

   c) 15.2 − 8.941 is about 6.

   d) 25.93 − 11.682 is about 14.

   e) 103.4 − 28.99 is about 70.

You will need

- materials for modelling decimal operations (e.g., Hundredths Grids (BLM 8), Thousandths Grids (BLM 7), base ten blocks, or play coins)

2. You have 10 m of ribbon.
   How much is left after each amount is cut off?

   a) 2.5 m

   7.5 m

   e.g.,

   2.5 m

   0 1 2 3 4 5 6 7 8 9 10

   b) 7.819 m

   2.181, e.g., I regrouped 10 base ten blocks as 9 ones, 9 tenths, 9 hundredths, and 10 thousandths, and I subtracted 7.819. That left 2 ones, 1 tenth, 8 hundredths, and 1 thousandth.

3. a) How much more is 18.341 + 50 than 17.34 + 49.1? 1.901

   b) How much more is 49.937 + 77.81 than 39.927 + 66.51? 21.31

Pathway 3

4. Model each calculation using a grid, money, or base ten blocks.
   Sketch your model.

   a) 3 × 1.6

   e.g.,

   b) 6 × 4.23

   e.g.,

50 Decimals Operations, Diagnostic Tool *Leaps and Bounds* Copyright © 2012 by Nelson Education Ltd.

Name: ______________________ Date: ______________

Pathway 3

5. Solve each problem.

a) What is the total length of 2 rugs, if each rug is 5.2 m long?

10.4 m
e.g., I can double 5.2 m. 5 m doubled is 10 m and 0.2 m doubled is 0.4 m, so the total length is 10.4 m.

b) What is the total distance of 5 trips, if each trip is 72.7 km long?

363.5 km, e.g., I used base ten blocks and multiplied in parts.
5 groups of 70 is 350.
5 groups of 2 is 10.
5 groups of 0.7 is 3.5.
The total is 363.5.

c) What is the total cost of 4 shirts, if each shirt costs $12.39?

$49.56, e.g., I made 4 groups of $12.39, and then I regrouped to get $49.56

d) What is the total length of 5 pieces of string, each 4.238 m long, laid end to end?

21.19 m, e.g., I thought of 4.238 m as 4238 mm and multiplied.

$$\begin{array}{r} 4238 \\ \times \quad 5 \\ \hline 21\,190 \end{array}$$

21 190 mm = 21.19 m

6. Fill in the blanks with a decimal thousandth and a whole number.

a) 4.999 × 7 is about 35.

b) 3.101 × 8 is about 24.8.

Pathway 2

7. What is each person's share, if they share equally?

a) 5 people sharing 3.45 L of juice

e.g., If you share 3 flats, 4 rods, and 5 small cubes in 5 groups and trade blocks when necessary, each person gets 6 rods and 9 small cubes or 0.69 L.

b) 4 people sharing $123.24

$30.81
e.g., $123.24 = 120 dollars + 32 dimes + 4 pennies. Shared equally 4 ways, that would be 30 dollars, 8 dimes, and 1 penny each.
That's $30.81.

8. Fill in the blanks with a decimal hundredth and a whole number.

a) 24.20 ÷ 6 is about 4.

b) 18.48 ÷ 3 is about 6.

Name: ______________________ Date: ______________

Pathway 2

9. Model each calculation using a grid, money, or base ten blocks. Sketch your model.

a) 9.23 ÷ 5

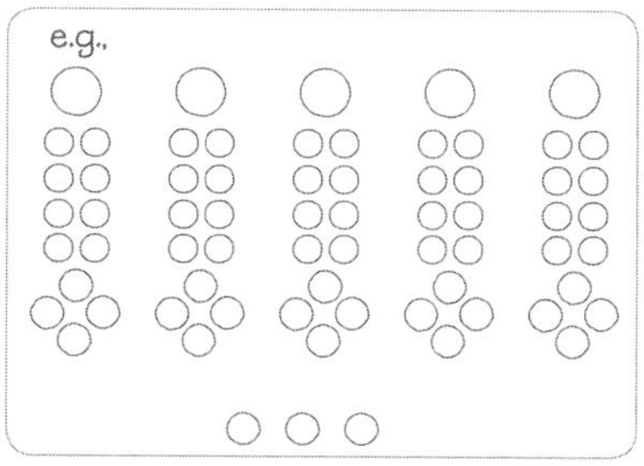

e.g.,

b) 1.456 ÷ 6

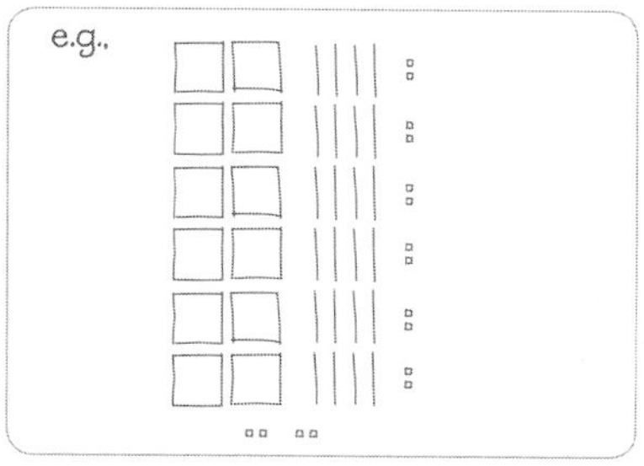

e.g.,

10. Use the hundredths grids to show each calculation. Write each quotient.

a) 1 ÷ 0.05 = 20

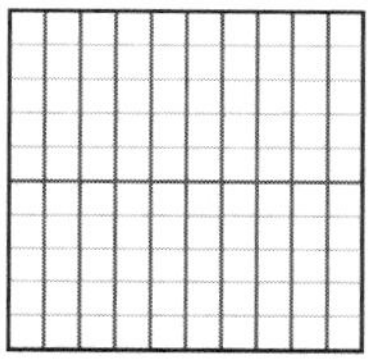

b) 2 ÷ 0.4 = 5

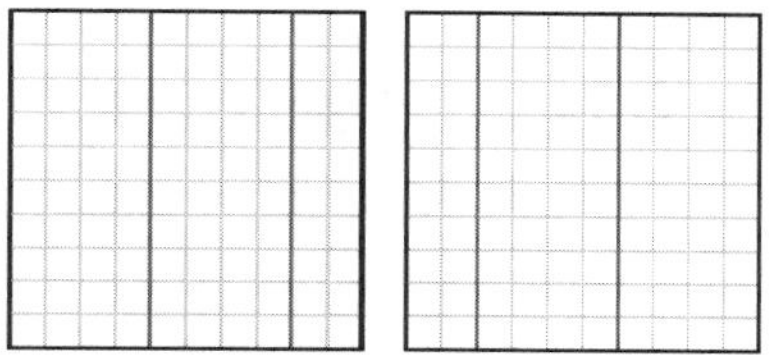

Pathway 1

11. Circle the greater quotient:

(5 ÷ 0.3) or 5 ÷ 0.4

Show or explain how you know it is greater.

e.g., You start with the same amount for both, 5, but for 5 ÷ 0.3, the size of each group is smaller (0.3 is less than 0.4), which means there will be more groups.

12. How can you use 4 ÷ 0.4 = 10 to help you calculate 4 ÷ 0.8?

e.g., The answer would be half as much, 5, since the group size is twice as much (0.8 is twice 0.4).

 Copyright © 2012 by Nelson Education Ltd.

Copyright © 2012 by Nelson Education Ltd.

Pathway 1 OPEN-ENDED

Dividing Whole Numbers by Decimals

You will need

- counters
- materials for modelling decimals (e.g., Hundredths Grids (BLM 8) or base ten blocks)
- containers, spoons, and millilitre and litre measures (optional)
- Student Resource pages 56–57

Open Ended Intervention

Before Using the Open-Ended Intervention

Record 12 ÷ 2 and put out 12 counters. Then ask:

- What does 12 ÷ 2 mean?
  (e.g., *How many 2s are in 12?*)
- What do you think 12 ÷ 0.5 means?
  (e.g., *How many halves are in 12?*)
- How do you know the answer to 12 ÷ 0.5 is more than 12?
  (e.g., *There are 2 halves in each of the 12 wholes.*)
- What do you think 12 L ÷ 0.25 L means?
  (e.g., *How many 0.25 L cans does it take to fill a 12 L container?*)

Show a number line from 0 to 3, with 12 jumps of 0.25 from 0 to 3. Ask:

- How does this show 3 ÷ 0.25? (e.g., *It shows that 0.25 fits 12 times into 3.*)

Using the Open-Ended Intervention (Student Resource pages 56–57)

Read through the tasks on the student pages together. Provide materials for modelling decimal division. Students might like to model some of the division concretely, using containers, spoons, and millilitre and litre measures. Give students time to work, ideally in pairs.

Observe whether students

- relate real-life situations to the division of a whole number by a decimal
- can estimate quotients
- can model the division of a whole number by a decimal
- can explain their strategies for dividing whole numbers by decimals

Consolidating and Reflecting

Ensure understanding by asking questions such as these based on students' work:

- How would you explain to someone what 4 ÷ 0.33 means?
  (e.g., *How many groups of 0.33 are in 4?*)
- What does that mean in the context of the cans and containers?
  (e.g., *How many cans holding 0.33 L are needed to fill a 4 L container?*)
- How did you estimate 3.0 ÷ 0.2?
  (e.g., *There are 4 groups of 0.25 in 1, so there would be 12 groups of 0.25 in 3. 3.0 ÷ 0.2 would be a bit more because 0.2 is smaller than 0.25, so it's about 14.*)
- Why were your answers all more than 1?
  (e.g., *I was always figuring out how many times something less than 1 fit into 1 or more.*)
- What model did you think was most useful? Why?
  (e.g., *a number line; It's easy to see how you fill a container with each possible number of cans.*)

Copyright © 2012 by Nelson Education Ltd.

Dividing Whole Numbers by Decimals

Pathway 1 GUIDED

Guided Intervention

You will need

- counters
- Fraction Strips (BLM 10)
- Hundredths Grids (BLM 8)
- base ten blocks
- play coins
- Student Resource pages 58–61

Before Using the Guided Intervention

Record 12 ÷ 2 and put out 12 counters. Then ask:

- What does 12 ÷ 2 mean?
  (e.g., *How many 2s are in 12?*)
- What do you think 12 ÷ 0.5 means?
  (e.g., *How many halves are in 12?*)
- How do you know the answer to 12 ÷ 0.5 is more than 12?
  (e.g., *There are 2 halves in each of the 12 wholes.*)
- What do you think 12.0 ÷ 0.25 means? (e.g., *How many quarters are in $12?*)

Show a number line from 0 to 3, with 12 jumps of 0.25 from 0 to 3. Ask:

- How does this show 3 ÷ 0.25? (e.g., *It shows that 0.25 fits 12 times into 3.*)

Using the Guided Intervention Student Resource pages 58–61

Provide Fractions Strips (BLM 10) and Hundredths Grids (BLM 8) and work through the instructional section with students as they model each strategy.

Have students work through the **Try These** questions individually or in pairs. Provide play coins and base ten blocks.

Observe whether students

- can explain their strategies for dividing whole numbers by decimals (Questions 1, 2, 7, 8, 9)
- can model the division of a whole number by a decimal (Questions 2, 3)
- can estimate or calculate quotients (Questions 3, 4, 5, 6)
- relate real-life situations to the division of a whole number by a decimal (Question 5)
- relate calculations involving quotients to other calculations (Question 8)

Consolidating and Reflecting

Ensure understanding by asking questions such as these based on students' work:

- In Question 2a), how did you model 2 ÷ 0.25?
  (e.g., *I figured out how many quarters are in 2 dollars.*)
- How could you have used a number line to model 3 ÷ 0.6 in Question 2b)?
  (e.g., *I would have made jumps of 0.6 from 0 to 3.*)
- How would you explain to someone what 4 ÷ 0.7 in Question 2d) means?
  (e.g., *How many groups of 0.7 are in 4?*)
- Describe a problem that you could solve with 12 ÷ 0.4 in Question 8a).
  (e.g., *A bowl holds 0.4 L. How many bowls can you fill with 12 L of soup?*)
- Why were your quotients all more than 1?
  (e.g., *I was always figuring out how many times something less than 1 fit into 1 or more.*)

Copyright © 2012 by Nelson Education Ltd.

Pathway 2
OPEN-ENDED

Dividing Decimals by Whole Numbers

You will need

- materials for modelling decimals (e.g., Tenths Grids (BLM 9), Hundredths Grids (BLM 8), Thousandths Grids (BLM 7), base ten blocks, or play money)
- Student Resource pages 62–63

Open Ended Intervention

Before Using the Open-Ended Intervention

Record 6.00 ÷ 4 and ask:

- What could 6.00 ÷ 4 represent? (e.g., *4 people sharing $6*)
- How much would each get? How do you know?
  (e.g., *$1.50. 2 people would each get $3, so 4 people would get half as much.*)
- How can you check your answer?
  (e.g., *Add $1.50 four times and see if I get $6.*)
- How could you use hundredths grids to model 6.00 ÷ 4?
  (e.g., *I'd share 6 hundredths grids 4 ways. I'd give 1 whole grid to each person and then share the 20 columns in the last 2 grids by giving each person 5 columns.*)
- How could you show it?
  (e.g., *I'd shade each person's share a different colour.*)

Using the Open-Ended Intervention Student Resource pages 62–63

Read through the tasks on the student pages together. Provide materials for modelling decimal division. Give students time to work, ideally in pairs.

Observe whether students

- relate real-life situations to the division of a decimal by a whole number
- can estimate and determine quotients of decimals divided by whole numbers
- use reasonable strategies to divide, including models, and can explain their strategies

Consolidating and Reflecting

Ensure understanding by asking questions such as these based on students' work:

- Why did it make sense to use division for the garbage-zone problem?
  (e.g., *You're dividing up a whole into equal parts.*)
- What strategies did you use to estimate?
  (e.g., *I used only the whole number part of the decimals.*)
- When you divided 78.2 ÷ 6, you had 2 tenths left over. What does that mean, and how did you deal with it?
  (e.g., *It meant each of the 6 zones had 13 km^2, with 0.2 km^2 left to divide up into the 6 zones. I thought of 2 tenths as 2 dimes, or 20 cents, and divided by 6.*)
- Why did you use a hundredths grid for 46.24 ÷ 5?
  (e.g., *I knew 46.24 = 45 + 1.24, so I thought about 1.24 ÷ 5 as 124 hundredths ÷ 5, which would be 124 squares shared 5 ways.*)
- Why might someone say that dividing decimals is no different than dividing whole numbers?
  (e.g., *For a decimal like 2.3, you can think about dividing 23 tenths, 23 dimes, or 230 cents instead.*)

 Copyright © 2012 by Nelson Education Ltd.

Dividing Decimals by Whole Numbers

Pathway 2 GUIDED

Guided Intervention

Before Using the Guided Intervention

- What could 6.00 ÷ 4 represent? (e.g., *4 people sharing $6*)
- How much would each get? How do you know?
  (*$1.50;* e.g., *2 people would each get $3, so 4 people would get half as much.*)
- How can you check your answer?
  (e.g., *Add $1.50 four times and see if I get $6.*)
- How could you use hundredths grids to model 6.00 ÷ 4?
  (e.g., *I'd share 6 hundredths grids 4 ways. I'd give 1 whole grid to each person and then share the 20 columns in the last 2 grids by giving each person 5 columns.*)
- How could you show it? (e.g., *I'd shade each person's share a different colour.*)

Using the Guided Intervention Student Resource pages 64–67

Work through the division strategies in the instructional section of the student pages together. Provide materials for modelling the division of decimals.

Have students work through the **Try These** questions individually or in pairs.

Observe whether students

- relate a model to the division of a decimal by a whole number (Questions 1, 2)
- can explain their division strategies (Questions 2, 4, 6, 8)
- can estimate and calculate quotients (Questions 3, 5, 7, 11)
- relate division calculations to other calculations (Questions 6, 9)
- relate real-life situations to the division of a decimal by a whole number (Questions 7, 9, 10)

Consolidating and Reflecting

Ensure understanding by asking questions such as these based on students' work:

- How did you model 5.2 ÷ 4 in Question 3a)?
  (e.g., *I divided $5 and 2 dimes into 4 equal shares.*)
- How else could you have done it?
  (e.g., *I know 4 ÷ 4 is 1, so I could just model 1.2 as 1 hundredths grid and 2 columns in a second one. Then I could share the 12 columns among 4.*)
- How did you estimate in Question 4?
  (e.g., *I usually only thought about the whole number parts, but sometimes the tenths if the decimal part was near a half.*)
- In what situations might you divide a decimal by a whole number for Question 10?
  (e.g., *when you share a decimal amount like money or a measurement*)
- Why might someone say that dividing decimals is no different than dividing whole numbers?
  (e.g., *For a decimal like 2.3, you really have a whole number of tenths, 23 tenths, which is like 23 dimes.*)

You will need

- materials for modelling decimals (e.g., Tenths Grids (BLM 9), Hundredths Grids (BLM 8), Thousandths Grids (BLM 7), base ten blocks, or play money)
- Student Resource pages 64–67

Copyright © 2012 by Nelson Education Ltd.

Pathway 3 OPEN-ENDED

Multiplying with Decimals

You will need

- materials for modelling decimals (e.g., Tenths Grids (BLM 9), Hundredths Grids (BLM 8), Thousandths Grids (BLM 7), base ten blocks, or play money)
- Student Resource pages 68–69

Open Ended Intervention

Before Using the Open-Ended Intervention

Record 3 × 2.2 and ask:

- What might 2.2 represent? (e.g., *the length of a string in centimetres*)
- What do you think 3 × 2.2 could mean, and what is the product? (e.g., *the total length of three 2.2 cm strings, 6.6 cm*)
- How do you know? (e.g., *3 × 2 = 6 cm and 0.2 + 0.2 + 0.2 = 0.6, so 6.6 altogether*)
- How can you check? (e.g., *I'd add 2.2 + 2.2 + 2.2.*)
- How could you model 3 × 2.2 using tenths or hundredths grids? (e.g., *Make 3 sets of models of 2.2. Each would use 2 shaded grids and 2 columns of a third grid. Count the total amount shaded: 6 whole grids and 6 tenths.*)

Using the Open-Ended Intervention (Student Resource pages 68–69)

Read through the tasks on the student pages together. Provide materials for modelling decimal division. Give students time to work, ideally in pairs.

Observe whether students

- relate real-life situations to the multiplication of a whole number and a decimal
- can estimate and determine products of a whole number by a decimal tenth, hundredth, or thousandth
- use reasonable strategies to multiply, including models, and can explain their strategies

Consolidating and Reflecting

Ensure understanding by asking questions such as these based on students' work:

- Why did it make sense to solve the problem by multiplying? (e.g., *There are many groups of the same size.*)
- Why did you use a different strategy for multiplying by 9 than for 7? (e.g., *It's easy to multiply by 10 in my head and then subtract 1. For 7, I'd have to multiply by 3 to figure out what to subtract.*)
- How can you use tenths or hundredths grids to model 9 × 4.3? (e.g., *Shade 3 columns 9 times to multiply 9 × 0.3. That's 27 columns, or 2.7. I'd add that to 9 × 4 = 36 to get 38.7.*)
- Why might someone say that multiplying decimals is really no different than multiplying whole numbers? (e.g., *A decimal like 1.4 is 14 tenths, which is a whole number of tenths.*)
- Can you ignore the value after the decimal point when you estimate? (e.g., *Yes, if the decimal part is low. If the decimal part is high, you might increase the whole number by 1 to estimate.*)

 Copyright © 2012 by Nelson Education Ltd.

Multiplying with Decimals

Pathway 3 GUIDED

Guided Intervention

Before Using the Guided Intervention

Record 3 × 2.2 and ask:

- What might 2.2 represent? (e.g., *the length of a string in centimetres*)
- What do you think 3 × 2.2 could mean, and what is the product? (e.g., *the total length of three 2.2 cm strings, 6.6 cm*)
- How do you know? (e.g., *3 × 2 = 6 cm and 0.2 + 0.2 + 0.2 = 0.6, so 6.6 altogether*)
- How can you check? (e.g., *I'd add 2.2 + 2.2 + 2.2.*)
- How could you model 3 × 2.2 using tenths or hundredths grids? (e.g., *Make 3 sets of models of 2.2. Each would use 2 shaded grids and 2 columns of a third grid. Count the total amount shaded: 6 whole grids and 6 tenths.*)

Using the Guided Intervention (Student Resource pages 70–73)

Work through the multiplication strategies in the instructional section of the student pages together. Provide materials for modelling decimal division.

Have students work through the **Try These** questions individually or in pairs.

Observe whether students

- relate a model to the multiplication of a whole number and a decimal (Questions 1, 2)
- can estimate and calculate products (Questions 3, 6, 7, 8, 11)
- can explain their multiplication strategies (Question 4)
- relate calculations to other calculations (Questions 5, 10)
- relate real-life situations to the multiplication of a whole number and a decimal (Questions 6, 7, 8, 9, 12)

Consolidating and Reflecting

Ensure understanding by asking questions such as these based on students' work:

- How did you model 6 × 2.103 in Question 2b)? (e.g., *I drew 6 groups with 2 large cubes, 1 flat, and 3 small cubes in each.*)
- When you estimated in Question 3, which digits did you consider? (e.g., *usually only the whole numbers unless the decimal part was close to half, as with 9 × 2.412*)
- In what situations might you multiply a decimal by a whole number for Question 9? (e.g., *when a lot of people have the same amount of money and you want a total, or with measurements*)
- Why might someone say that multiplying decimals is really no different than multiplying whole numbers? (e.g., *A decimal like 1.4 is 14 tenths, which is a whole number of tenths.*)

You will need

- materials for modelling decimals (e.g., Tenths Grids (BLM 9), Hundredths Grids (BLM 8), Thousandths Grids (BLM 7), or base ten blocks)
- Student Resource pages 70–73

Adding and Subtracting Decimals

You will need

- materials for modelling decimals (e.g., Place Value Charts (to Thousandths) (BLM 6), Thousandths Grids (BLM 7), or base ten blocks)
- Student Resource pages 74–75

Open Ended Intervention

Before Using the Open-Ended Intervention

Record 3.204 + 2.224 and 3204 + 2224. Ask:

- ▸ How is adding 3.204 and 2.224 like adding 3204 and 2224?
  (e.g., *Both times you are putting together 3 + 2 large cubes, 2 + 2 flats, 0 + 2 rods, and 4 + 4 small cubes. The difference is that for 3.204 + 2.224, the big cubes represent 1, and for 3204 and 2224, they represent 1000.*)
- ▸ What whole number calculation is 3.204 + 2.224 like? Explain.
  (e.g., *3204 + 2224; 3.204 + 2.224 is 3204 thousandths + 2224 thousandths*)
- ▸ What other strategy could you use to add 3.204 and 2.224?
  (e.g., *Add the thousandths, then the hundredths, and then the tenths.*)
- ▸ How much more is 3.240 than 2.224? How do you know?
  (*1.016*; e.g., *3240 thousandths has 1016 more thousandths than 2224 thousandths does.*)

Using the Open-Ended Intervention **Student Resource pages 74–75**

Read through the introduction and the tasks with students. Provide materials for modelling decimals. Give students time to work, ideally in pairs.

Observe whether students can

- solve real-world measurement problems involving the addition and subtraction of decimal thousandths and hundredths
- add and subtract decimals, with and without models
- explain their strategies for adding and subtracting decimals

Consolidating and Reflecting

Ensure understanding by asking questions such as these based on students' work:

- ▸ How could you calculate the perimeter in different ways?
  (e.g., *I could double the length and double the width and then add the 2 sums together, or I could add the length and width and then double that sum.*)
- ▸ What did you notice about the thousandths digit in the sum when you added?
  (e.g., *It was double, or 10 less than double, the thousandths digit in the width.*)
  Why? (e.g., *There were no thousandths in Javier's length.*)
- ▸ Tell about a model you used. How did it help?
  (e.g., *It was useful to have thousandths grids when I was adding the decimal parts, so I could see how many columns, squares, and small rectangles I had.*)
- ▸ Why might someone say that adding or subtracting decimals is really no different than subtracting whole numbers?
  (e.g., *If you are subtracting, say, 0.003 from 0.08, you can think of subtracting 3 thousandths from 80 thousandths. Then, 3 and 80 are whole numbers.*)

Copyright © 2012 by Nelson Education Ltd.

Adding and Subtracting Decimals

Guided Intervention

Before Using the Guided Intervention

Record 3.204 + 2.224 and 3204 + 2224. Ask:

- How is adding 3.204 and 2.224 like adding 3204 and 2224?
  (e.g., *Both times you are putting together 3 + 2 large cubes, 2 + 2 flats, 0 + 2 rods, and 4 + 4 small cubes. The difference is that for 3.204 + 2.224, the big cubes represent 1, and for 3204 and 2224, they represent 1000.*)
- What whole number calculation is 3.204 + 2.224 like? Explain.
  (e.g., *3204 + 2224; 3.204 + 2.224 is 3204 thousandths + 2224 thousandths*)
- What other strategy could you use to add 3.204 and 2.224?
  (e.g., *Add the thousandths, then the hundredths, and then the tenths.*)
- How much more is 3.240 than 2.224? How do you know?
  (*1.016*; e.g., *3240 thousandths has 1016 more thousandths than 2224 thousandths does.*)

You will need

- materials for modelling decimals (e.g., Place Value Charts (to Thousandths) (BLM 6), Thousandths Grids (BLM 7), or base ten blocks)
- Student Resource pages 76–79

Using the Guided Intervention Student Resource pages 76–79

Work through the instructional section of the student pages together. Have students describe and model each strategy. Provide materials for modelling decimals.

Have students work through the **Try These** questions individually or in pairs.

Observe whether students

- can estimate sums or differences of decimals (Questions 1, 2, 3, 9)
- can model a decimal sum or difference (Question 4)
- recognize which parts of 2 numbers are added or subtracted to calculate sums or differences (Questions 5, 7)
- solve real-world problems involving the addition and subtraction of decimals (Question 6)
- can add and subtract decimals (Questions 8, 9)

Consolidating and Reflecting

Ensure understanding by asking questions such as these based on students' work:

- When you estimated in Question 2, which digits did you consider?
  (e.g., *usually only the whole numbers, unless the decimal part was close to a half or close to a whole number*)
- Which model did you find most useful in Question 4?
  (e.g., *The base ten blocks were easier, since I didn't have to do any shading.*)
- Why might someone say that adding or subtracting decimals is really no different than adding or subtracting whole numbers?
  (e.g., *If you are subtracting, say, 3.872 from 15.1 in Question 4b), you can think of subtracting 3872 thousandths from 15 100 thousandths instead.*)

Strand: Number

Relating Situations to Operations

Planning For This Topic

Materials to help students relate situations to operations consist of a diagnostic tool and 3 intervention pathways. Pathway 1 focuses on division situations involving decimals. Pathway 2 focuses on multiplication and Pathway 3 focuses on subtraction. There is no separate pathway for addition, since students typically have no difficulty recognizing addition situations. Addition problems are, however, integrated into all pathways.

Each pathway has an open-ended intervention and a guided intervention. Choose the type of intervention more suitable for your students' needs and your particular circumstances.

Curriculum Connections

Grades 5 to 8 curriculum connections for this topic are provided online (see www.nelson.com/leapsandbounds). All 3 pathways are relevant to both curricula.

Professional Learning Connections

PRIME: Number and Operations, Background and Strategies (Nelson Education Ltd., 2005), pages 41–44, 49–55, 122

Making Math Meaningful to Canadian Students K–8 (Nelson Education Ltd., 2008), pages 104–107, 118–123, 236–240, 242–244

Big Ideas from Dr. Small Grades 4–8 (Nelson Education Ltd., 2009), pages 25–26, 68

Good Questions (dist. by Nelson Education Ltd., 2009), page 36

Why might students struggle with decimal situations?

Students might struggle with relating situations to operations with decimals for any of the following reasons:

- They might not recognize that operations with decimals hold the same meanings as they do with whole numbers.
- They might know only one meaning for each operation and not recognize other meanings (e.g., they know that sharing a total amount equally is division, but they do not know that making equal groups from a total amount is also division).
- They might not realize that
  - any division situation implicitly involves multiplication (and vice versa)
  - any multiplication situation implicitly involves addition
  - any subtraction situation implicitly involves addition (and vice versa)
- They might not realize that a problem situation that appears to be a certain operation can be solved using a different operation by reframing the problem.
- They might not realize that 2 very different problem situations can be related to the same operation (e.g., 4×2.4 can represent the combined height of 4 towers that are each 2.4 m high or the area of a 4 m by 2.4 m rectangle).

 Copyright © 2012 by Nelson Education Ltd.

Diagnostic Tool: Relating Situations to Operations

Use the diagnostic tool to determine the most suitable intervention pathway for relating situations to operations. Provide Diagnostic Tool: Relating Situations to Operations, Teacher's Resource pages 66 and 67, and have students complete it in writing or orally.

See solutions on Teacher's Resource pages 68 and 69.

Intervention Pathways

The purpose of the intervention pathways is to help students recognize that the meanings of operations involving decimal numbers are no different than those involving whole numbers, and that there are multiple types of situations for each operation.

There are 3 pathways:

- Pathway 1: Recognizing Division Situations
- Pathway 2: Recognizing Multiplication Situations
- Pathway 3: Recognizing Subtraction Situations

Use the chart below (or the Key to Pathways on Teacher's Resource pages 68 and 69) to determine which pathway is most suitable for each student or group of students.

Diagnostic Tool Results	Intervention Pathway
If students struggle with Questions 5, 6, 11, 13	use Pathway 1: Recognizing Division Situations *Teacher's Resource pages 70–71* *Student Resource pages 80–84*
If students struggle with Questions 4, 9, 10, 12	use Pathway 2: Recognizing Multiplication Situations *Teacher's Resource pages 72–73* *Student Resource pages 85–89*
If students struggle with Questions 1, 2, 3, 7, 8	use Pathway 3: Recognizing Subtraction Situations *Teacher's Resource pages 74–75* *Student Resource pages 90–95*

If students successfully complete Pathway 2, they may or may not need the additional intervention provided by Pathway 1. Either re-administer Pathway 1 questions from the diagnostic tool or encourage students to do a portion of the open-ended intervention for Pathway 1 to decide if more work in that pathway would be beneficial.

Name: ____________________ Date: ____________________

Relating Situations to Operations

Diagnostic Tool

For each problem, write the addition, subtraction, multiplication, or division expression you would use to solve it.

- You can use a combination of operations.
- Do *not* solve these problems.

1. Ilya ran 2.6 km and Mark ran 1.96 km.

How much farther did Ilya run?

____

2. Hassan drank a 0.3 L glass of milk and then a 0.4 L glass of milk.

How much milk did he drink?

____

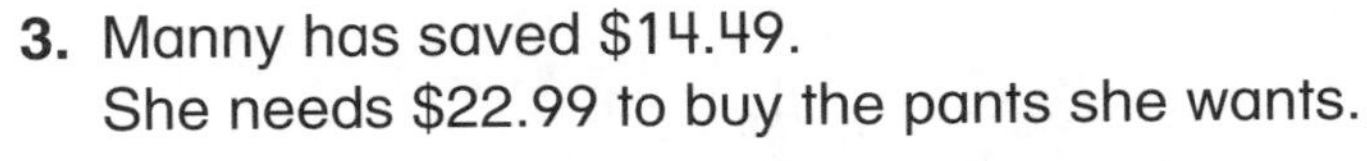

3. Manny has saved $14.49.
She needs $22.99 to buy the pants she wants.

How much more must she save?

____

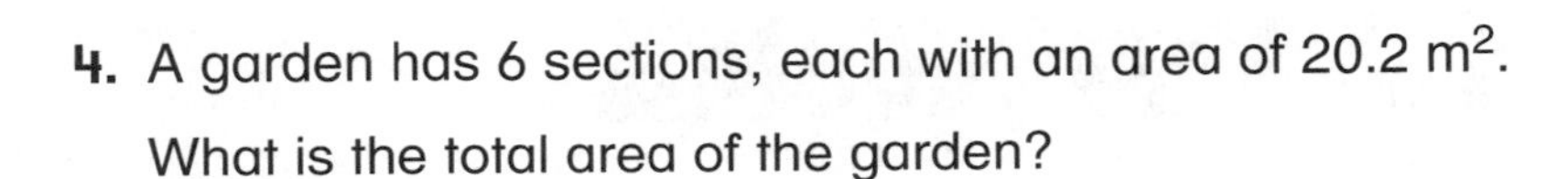

4. A garden has 6 sections, each with an area of 20.2 m^2.

What is the total area of the garden?

____

5. Angela's mother paid $47.94 for 6 pizzas for the class party.

How much did each pizza cost?

____

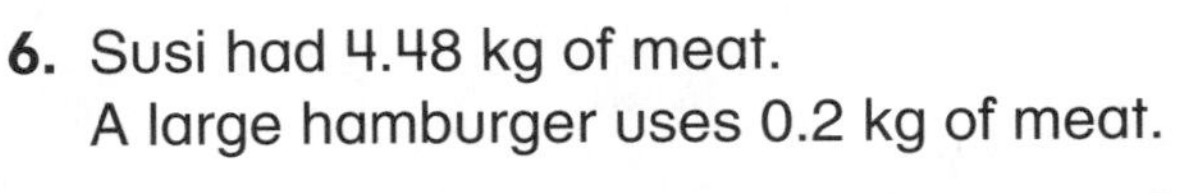

6. Susi had 4.48 kg of meat.
A large hamburger uses 0.2 kg of meat.

How many hamburgers of that size could she make?

____

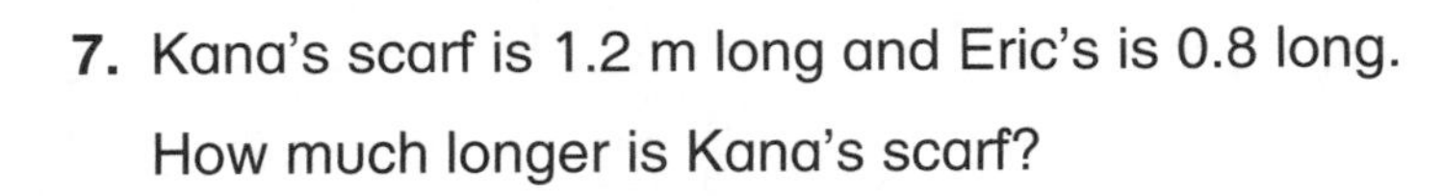

7. Kana's scarf is 1.2 m long and Eric's is 0.8 long.

How much longer is Kana's scarf?

____

Copyright © 2012 by Nelson Education Ltd.

Name:________________________________ Date:________________

8. Jake's backpack contains his jacket and textbook. It has a total mass of 4.2 kg. His jacket is 0.6 kg and his textbook is 1.1 kg.

   What is the mass of the backpack itself?

   ____

9. Rachel has 2.66 L of iced tea. Bru has twice as much iced tea as Rachel.

   How much iced tea does Bru have?

   ____

10. Michael would like to pour 0.3 L of juice into each of 8 glasses.

    How much juice does Michael need?

    ____

11. Jina bought 3 items. One cost $32.28, one cost $11.12, and one cost $8.43. Her parents will pay half of the total.

    How much will Jina need to pay?

    ____

12. It costs $7.99 a month to belong to an online e-zine.

    How much would it cost for a year?

    ____

13. Elke has 4.0 L of lemonade.

    How many glasses of 0.3 L of lemonade could she pour?

    ____

Copyright © 2012 by Nelson Education Ltd.

Solutions and Key to Pathways

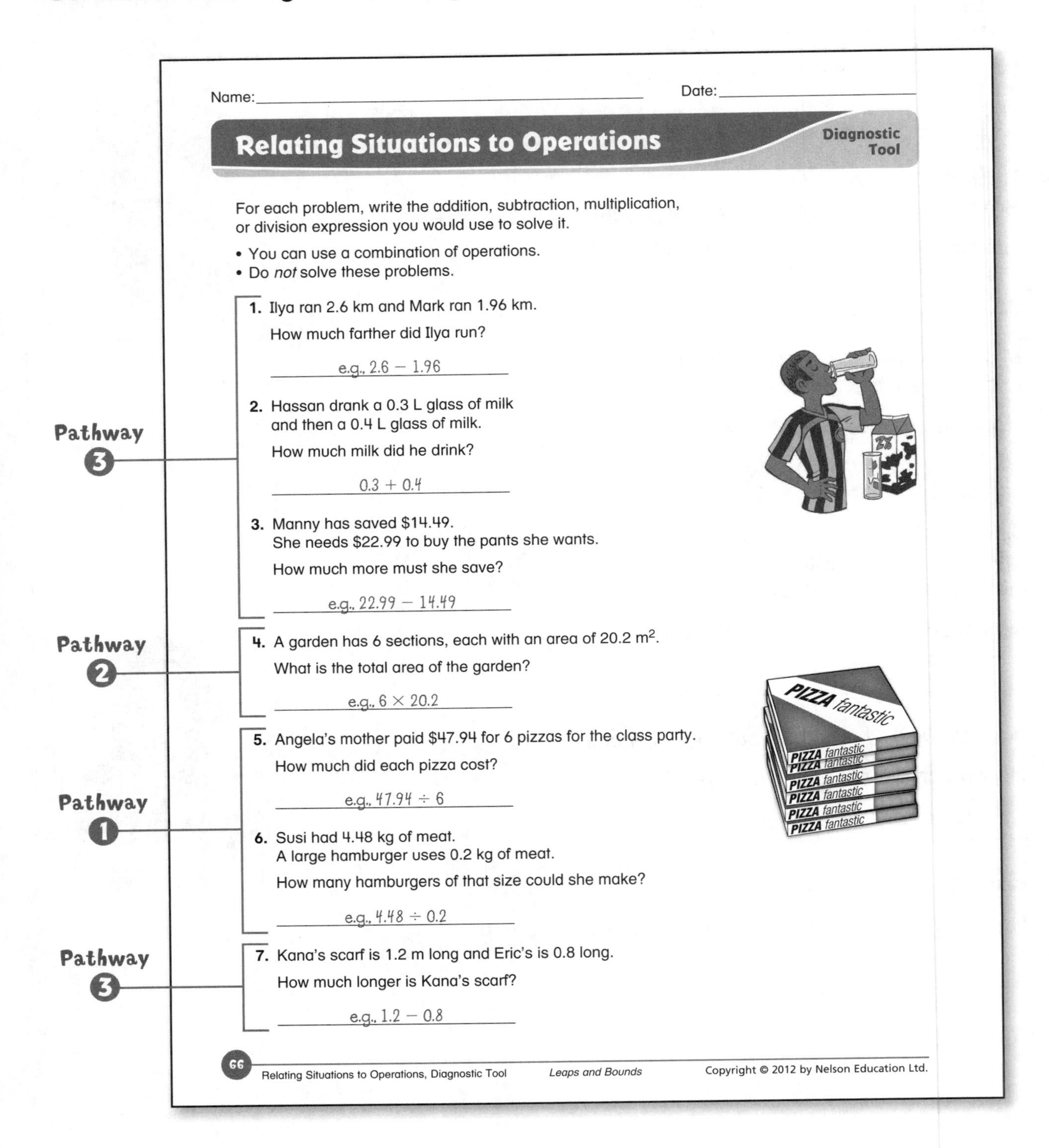

Name:_______________ Date:_______________

Relating Situations to Operations

Diagnostic Tool

For each problem, write the addition, subtraction, multiplication, or division expression you would use to solve it.

- You can use a combination of operations.
- Do *not* solve these problems.

Pathway 3

1. Ilya ran 2.6 km and Mark ran 1.96 km.
   How much farther did Ilya run?
   e.g., 2.6 − 1.96

2. Hassan drank a 0.3 L glass of milk and then a 0.4 L glass of milk.
   How much milk did he drink?
   0.3 + 0.4

3. Manny has saved $14.49.
   She needs $22.99 to buy the pants she wants.
   How much more must she save?
   e.g., 22.99 − 14.49

Pathway 2

4. A garden has 6 sections, each with an area of 20.2 m^2.
   What is the total area of the garden?
   e.g., 6 × 20.2

Pathway 1

5. Angela's mother paid $47.94 for 6 pizzas for the class party.
   How much did each pizza cost?
   e.g., 47.94 ÷ 6

6. Susi had 4.48 kg of meat.
   A large hamburger uses 0.2 kg of meat.
   How many hamburgers of that size could she make?
   e.g., 4.48 ÷ 0.2

Pathway 3

7. Kana's scarf is 1.2 m long and Eric's is 0.8 long.
   How much longer is Kana's scarf?
   e.g., 1.2 − 0.8

66 Relating Situations to Operations, Diagnostic Tool *Leaps and Bounds* Copyright © 2012 by Nelson Education Ltd.

Copyright © 2012 by Nelson Education Ltd.

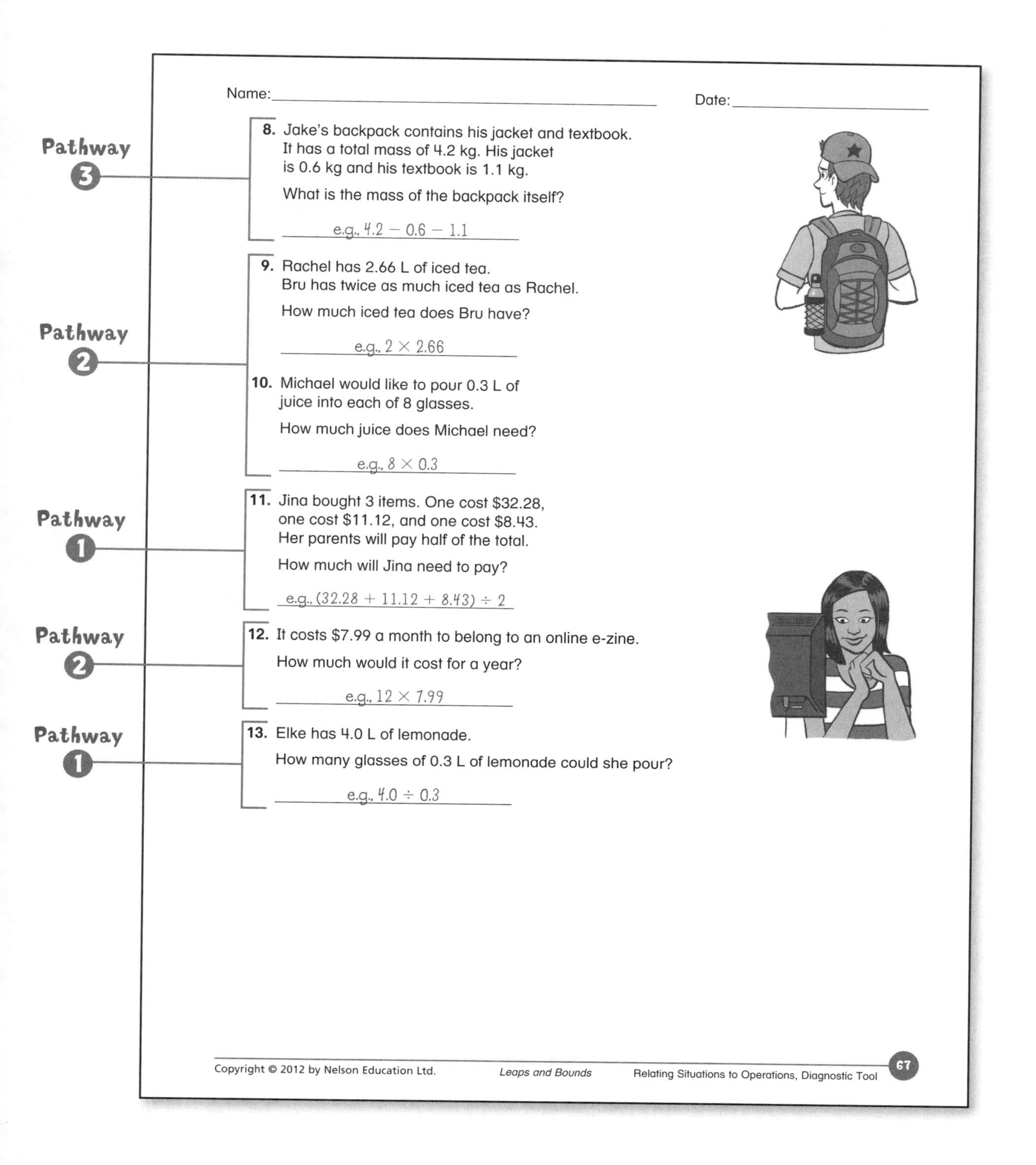

Name:_______________ Date:_______________

Pathway 3

8. Jake's backpack contains his jacket and textbook. It has a total mass of 4.2 kg. His jacket is 0.6 kg and his textbook is 1.1 kg.

   What is the mass of the backpack itself?

   e.g., $4.2 - 0.6 - 1.1$

Pathway 2

9. Rachel has 2.66 L of iced tea. Bru has twice as much iced tea as Rachel.

   How much iced tea does Bru have?

   e.g., 2×2.66

10. Michael would like to pour 0.3 L of juice into each of 8 glasses.

    How much juice does Michael need?

    e.g., 8×0.3

Pathway 1

11. Jina bought 3 items. One cost $32.28, one cost $11.12, and one cost $8.43. Her parents will pay half of the total.

    How much will Jina need to pay?

    e.g., $(32.28 + 11.12 + 8.43) \div 2$

Pathway 2

12. It costs $7.99 a month to belong to an online e-zine.

    How much would it cost for a year?

    e.g., 12×7.99

Pathway 1

13. Elke has 4.0 L of lemonade.

    How many glasses of 0.3 L of lemonade could she pour?

    e.g., $4.0 \div 0.3$

Pathway 1 OPEN-ENDED

Recognizing Division Situations

You will need

- 3 graduated cups with 1.5 L capacity (total) and a pourable material (optional)
- Student Resource pages 80–81

Open-Ended Intervention

Before Using the Open-Ended Intervention

Provide graduated cups and pourable material for students to use if they wish. Ask:

- ▸ Suppose you had 3 mugs, each holding 0.4 L of milk. How would you figure out how much milk there was altogether?
(e.g., *I'd multiply 3 by 0.4.*)
- ▸ There are 1.2 L altogether in 3 mugs, with the same amount in each. How would you figure out how much milk there was in each mug?
(e.g., *I'd divide 1.2 by 3.*)
- ▸ Suppose you had 3.0 L altogether and several 0.5 L mugs. How would you figure out how many mugs you would need? Why?
(e.g., *I'd divide, since there are equal groups.*)

Using the Open-Ended Intervention Student Resource pages 80–81

Read through the task on the student pages together. Ensure students know that the first picture shows the sharing meaning (How many are in each group?) and the one below shows the grouping meaning (How many groups are there?). Give students time to work, ideally in pairs.

Observe whether students

- recognize that division involves figuring out the number of equal groups or the amount in each equal group for a known total amount
- can create problems using both meanings of division (grouping and sharing)
- can relate division to multiplication
- can explain why a problem can be solved using division

Consolidating and Reflecting

Ensure understanding by asking questions such as these based on students' work:

- ▸ What made your first problem a division problem?
(e.g., *I knew the total and the number of equal groups, and I had to divide to figure out the size of each group.*)
- ▸ What made your last 2 problems different?
(e.g., *In Problem 2, I knew the number of equal groups but not the size of each. In Problem 3, I knew the size of each equal group but not the number of groups.*)
- ▸ Could you have solved all of your problems using multiplication? Explain.
(*Yes.* e.g., *You could just figure out what to multiply by to get the total.*)
- ▸ Why might someone say that a problem about people sharing money fairly and a problem about dividing an area into equal zones are similar?
(e.g., *Both are sharing problems.*)
- ▸ What is the main clue that helps you decide if a problem could be solved using division?
(e.g., *The problem gives the total, and it's about making equal groups.*)

 Copyright © 2012 by Nelson Education Ltd.

Recognizing Division Situations

Pathway 1 GUIDED

Guided Intervention

Before Using the Guided Intervention

Provide graduated cups and pourable material for students to use if they wish. Ask:

- Suppose you had 3 mugs, each holding 0.4 L of milk. How would you figure out how much milk there was altogether?
  (e.g., *I'd multiply 3 by 0.4.*)
- There are 1.2 L altogether in 3 mugs, with the same amount in each. How would you figure out how much milk there was in each mug?
  (e.g., *I'd divide 1.2 by 3.*)
- Suppose you had 3.0 L altogether and several 0.5 L mugs. How would you figure out how many mugs you would need? Why?
  (e.g., *I'd divide, since there are equal groups.*)

You will need

- 3 graduated cups with 1.5 L capacity (total) and a pourable material (optional)
- Student Resource pages 82–84

Using the Guided Intervention **Student Resource pages 82–84**

Work through the instructional section of the student pages together. Students may want to model the situations with materials.

Have students work through the **Try These** questions individually or in pairs.

Observe whether students

- distinguish division problems from other problems (Questions 1, 3, 6)
- recognize that "How many groups are there?" and "How many are in each group?" situations can be solved by dividing (Questions 1, 4, 5, 7)
- can explain why a problem can be solved using division (Questions 2, 3, 4, 5, 8)
- can tell what numbers to divide to solve a division problem (Questions 4, 5)
- realize that 2 problems with different division meanings can be created from the same context (Question 7)

Consolidating and Reflecting

Ensure understanding by asking questions such as these based on students' work:

- What made Question 1B a division problem?
  (e.g., *I knew the total and the number of equal groups, and I had to figure out the size of each group.*)
- What made Question 1C different from Question 1B?
  (e.g., *In C, I knew the size of each equal group, but in B, I knew how many equal groups there were.*)
- Why might someone say that a problem about people sharing money fairly and a problem about dividing an area into equal sections are similar?
  (e.g., *Both are sharing problems.*)
- Which of your division problems could you have solved by multiplying? Why?
  (e.g., *All of them; every division is a backwards multiplication.*)
- What main clue helps you decide if a problem could be solved by division?
  (e.g., *It gives the total, and it's about making equal groups.*)

Pathway 2 OPEN-ENDED

Recognizing Multiplication Situations

You will need

- play coins (loonies, dimes, and pennies)
- 1 cm Grid Paper (BLM 12)
- Student Resource pages 85–86

Open-Ended Intervention

Before Using the Open-Ended Intervention

Display $1.43 (1 loonie, 4 dimes, and 3 pennies). Ask:

- ▸ Suppose I told you that I had 2 times as much money in my hand. How would you model the amount I have?
  (e.g., *You'd have 2 groups of $1.43, so I'd show 2 sets of 1 loonie, 4 dimes, and 3 pennies.*)

Draw a 6.5 cm by 4 cm rectangle on grid paper, with 3 sides on grid lines.

- ▸ How would you figure out the area of this rectangle?
  (e.g., *I'd count the squares.*)
- ▸ Why might you also multiply 4 by 6.5?
  (e.g., *There are 4 rows, or groups, of 6.5 square centimetres.*)

Using the Open-Ended Intervention (Student Resource pages 85–86)

Read through the tasks on the student pages together. Ensure students realize that the first picture shows the equal groups meaning and the second picture shows a rate situation. Give students time to work, ideally in pairs.

Observe whether students

- recognize that multiplication can be used to solve problems involving equal groups, the area of a rectangle, and rates
- can create multiplication problems using different meanings of multiplication
- can explain why a problem can be solved using multiplication

Consolidating and Reflecting

Ensure understanding by asking questions such as these based on students' work:

- ▸ What made your first problem a multiplication problem?
  (e.g., *There were 6 equal lengths [each 3.24 m] end to end, so you can multiply one length by 6.*)
- ▸ Could you solve your problem using a different operation? How?
  (*Yes.* e.g., *You could add 3.24 six times.*)
- ▸ What made your first 2 problems different?
  (e.g., *One problem involves area and the other involves equal lengths, or groups.*)
- ▸ Why might someone say that a problem about 5 people spending $4.23 each and a problem about 2 melons each with a mass of 1.2 kg are the same sort of problem?
  (e.g., *Both involve rates or equal groups, so they can both be solved using multiplication.*)
- ▸ What made your Problem 3 a multiplication problem?
  (e.g., *There were 6 equal distances or groups [each 3.24 m] end to end, so to find the total length you can multiply one length by 6.*)

 Copyright © 2012 by Nelson Education Ltd.

Recognizing Multiplication Situations

Pathway 2 GUIDED

Guided Intervention

You will need

- play coins (loonies, dimes, and pennies)
- 1 cm Grid Paper (BLM 12)
- Student Resource pages 87–89

Before Using the Guided Intervention

Provide play coins. Ask:

- How could you model 2 × 1.43 with coins? Why? (e.g., *I'd show 2 sets of 1 loonie, 4 dimes, and 3 pennies because there are 2 equal groups.*)

Draw a 6.5 cm by 4 cm rectangle on grid paper, with 3 sides on grid lines.

- How would you figure out the area of this rectangle? (e.g., *I'd count the squares.*)
- Why might you multiply 4 by 6.5? (e.g., *There are 4 rows, or groups, of 6.5 square centimetres.*)

Using the Guided Intervention (Student Resource pages 87–89)

Work through the instructional section of the student pages together.

Have students work through the **Try These** questions individually or in pairs.

Observe whether students

- distinguish multiplication problems from those involving other operations (Questions 1, 3, 6)
- recognize that multiplication involves equal groups, the area of a rectangle, and rates (Questions 1, 3)
- can explain why a problem can be solved using multiplication (Questions 2, 3, 6, 8)
- can write a multiplication expression for a problem (Questions 1, 4)
- can create a multiplication problem (Question 7)
- realize that 2 problems with different multiplication meanings can be created from the same context (Question 5)

Consolidating and Reflecting

Ensure understanding by asking questions such as these based on students' work:

- How might you change Question 1D to make it a multiplication problem? (e.g., *Make all the sides of the triangle equal.*)
- What made Question 1E a multiplication problem? (e.g., *You multiply length and width to find the area of a rectangle.*)
- What made the problems in Question 5 different? (e.g., *One is about area, and the other is about equal groups.*)
- Why might someone say that a problem about 5 people spending $4.23 each and a problem about 2 melons with a mass of 1.2 kg each are the same sort of problem? (e.g., *Both involve rates or equal groups, so both can be solved using multiplication.*)
- Why can you solve a multiplication problem by adding? (e.g., *Multiplying is like adding the same number over and over.*)

Pathway 3 OPEN-ENDED

Recognizing Subtraction Situations

You will need

- yarn or string
- rulers
- scissors
- Student Resource pages 90–91

Open-Ended Intervention

Before Using the Open-Ended Intervention

Cut 2 pieces of yarn 3.1 m long. Show one of the pieces and ask:

- ▸ This yarn is 3.1 m long, and I want to cut off a piece 1.8 m long. How would you figure out how much yarn was left? (e.g., *I'd subtract 1.8 from 3.1.*)

Have students measure and cut to check. Then show the other 3.1 m piece of yarn and give the students the 1.8 m piece. Ask:

- ▸ My yarn is 3.1 m and yours is 1.8 m. How would you figure out how much longer mine is? (e.g., *I'd subtract 1.8 from 3.1.*)
- ▸ You have 1.8 m of yarn and you need 3.1 m. How would you figure out how much more you needed? (e.g., *I'd add up from 1.8 to get to 3.1.*)
- ▸ Why might you subtract instead? (e.g., *That would tell me the same thing—how much to add.*)

Using the Open-Ended Intervention (**Student Resource pages 90–91**)

Read through the tasks on the student pages together. Ensure students realize that the top picture shows the take-away meaning and the bottom one shows the comparison meaning. Give students time to work, ideally in pairs.

Observe whether students

- recognize that subtraction can be used to solve take-away situations, comparison situations, and missing addend situations
- can create subtraction problems using different meanings of subtraction
- can relate subtraction to addition
- can explain why a problem can be solved using subtraction

Consolidating and Reflecting

Ensure understanding by asking questions such as these based on students' work:

- ▸ What made your first problem a subtraction problem? (e.g., *When you figure out change, you can add up from the actual cost [$2.34] to what you paid [$4] to figure out how much more one is than the other.*)
- ▸ What made your first 2 problems similar? (e.g., *The whole was 4.0 and a part was 2.34.*)
- ▸ What made your first 2 problems different? (e.g., *In Problem 1, I thought about adding up to see how much more 4.0 was than 2.34, but in Problem 2, I thought of taking away 2.34 from 4.0.*)
- ▸ Why might someone say that a problem about how much more 4.2 km is than 1.7 km and a problem about how much more 2 L is than 1.4 L are the same kind of problem? (e.g., *Both are comparison problems, so both can be solved using subtraction.*)
- ▸ Could you have solved all of your problems using addition? How? (*Yes.* e.g., *You would have to figure out what to add to the known part to get to the known total.*)

Copyright © 2012 by Nelson Education Ltd.

Recognizing Subtraction Situations

Pathway 3 GUIDED

Guided Intervention

You will need

- yarn or string
- scissors
- rulers
- Student Resource pages 92–95

Before Using the Guided Intervention

Cut 2 pieces of yarn 3.1 m long. Show one of the pieces and ask:

- ▸ This yarn is 3.1 m long, and I want to cut off a piece 1.8 m long. How would you figure out how much yarn was left? (e.g., *I'd subtract 1.8 from 3.1.*)

Have students measure and cut to check. Then show the other 3.1 m piece of yarn and give the students the 1.8 m piece. Ask:

- ▸ My yarn is 3.1 m and yours is 1.8 m. How would you figure out how much longer mine is? (e.g., *I'd subtract 1.8 from 3.1.*)
- ▸ You have 1.8 m of yarn and you need 3.1 m. How would you figure out how much more you needed?
  (e.g., *I'd subtract 1.8 from 3.1, or I'd add up from 1.8 to get to 3.1.*)

Using the Guided Intervention (Student Resource pages 92–95)

Work through the instructional section of the student pages together.

Have students work through the **Try These** questions individually or in pairs.

Observe whether students

- distinguish subtraction problems from those involving other operations (Questions 1, 3, 6)
- recognize that subtraction can be used to solve take-away, comparison, and missing addend problem situations (Questions 1, 4, 5, 6, 7)
- can relate subtraction to addition (Question 2)
- can explain why a problem can be solved using subtraction (Questions 2, 4, 5, 8)
- realize that 2 problems with different subtraction meanings can be created from the same situation (Questions 6, 7)

Consolidating and Reflecting

Ensure understanding by asking questions such as these based on students' work:

- ▸ What made Question 1B a subtraction problem? (e.g., *When you figure out change, you can add up from the actual cost [$2.39] to what you paid [$5] to figure out how much more one is than the other, and that's subtracting.*)
- ▸ What made the problems in Question 7 similar? What made them different? (e.g., *Both involve 6.5 and subtraction. The first is a take-away problem, and the second involves figuring out how much more.*)
- ▸ Why might someone say that a problem about how much more 4.2 km is than 1.7 km and a problem about how much more 2 L is than 1.4 L are the same kind of problem?
  (e.g., *Both are comparison problems, so both can be solved using subtraction.*)
- ▸ Could you solve any subtraction problem using addition? (*Yes.* e.g., *you would have to figure out what to add to the known part to get to the known total.*)

Copyright © 2012 by Nelson Education Ltd.

Strand: Number

Comparing Fractions

Planning For This Topic

Materials for assisting students with comparing fractions consist of a diagnostic tool and 3 intervention pathways. Pathway 1 includes mixed numbers and improper fractions. Pathway 2 focuses exclusively on proper fractions. Pathway 3 concentrates on fraction equivalence with proper fractions.

Each pathway has an open-ended intervention and a guided intervention. Choose the type of intervention more suitable for your students' needs and your particular circumstances.

Curriculum Connections

Grades 5 to 8 curriculum connections for this topic are provided online. See www.nelson.com/leapsandbounds. All pathways are relevant to both curriculums.

Why might students struggle with comparing fractions?

Students might struggle with comparing fractions for any of the following reasons:

- They might not recognize that fraction comparisons, when using visual or concrete models, depend on the size of the wholes. Only fractions of the same size whole can be compared. (Note that, when comparing fractions symbolically, the assumption is that the wholes are the same size.)
- They might use whole-number reasoning rather than proportional reasoning (e.g., they mistakenly think that $\frac{1}{10}$ is greater than $\frac{1}{9}$, since $10 > 9$).
- They might be unaware that fractions with common numerators can be compared by just comparing the denominators (e.g., $\frac{2}{3}$ of a whole is greater than $\frac{2}{5}$ of the same whole because thirds are bigger pieces than fifths of the same whole, so 2 thirds is greater than 2 fifths).
- They might not be aware that they can use benchmarks to compare fractions (e.g., since $\frac{1}{3} < \frac{1}{2}$ and $\frac{7}{8} > \frac{1}{2}$, then $\frac{1}{3} < \frac{7}{8}$).
- They might not be confident about creating equivalent fractions to make fraction comparisons easier.
- They might use inappropriate strategies for comparison (e.g., they might calculate the difference between the numerator and denominator of each fraction and then compare the differences to tell which fraction is greater because this strategy "sometimes works").
- They might be uncomfortable with improper fractions and have difficulty comparing them.
- They might not understand the relationship between mixed numbers and the equivalent improper fractions.

Professional Learning Connections

PRIME: Number and Operations, Background and Strategies (Nelson Education Ltd., 2005), pages 108–111, 113–116
Making Math Meaningful to Canadian Students K–8 (Nelson Education Ltd., 2008), pages 203–210
Big Ideas from Dr. Small Grades 4–8 (Nelson Education Ltd., 2009), pages 48–50
Good Questions (dist. by Nelson Education Ltd., 2009), pages 20, 26, 28
More Good Questions (dist. by Nelson Education Ltd, 2010), page 69

Copyright © 2012 by Nelson Education Ltd.

Diagnostic Tool: Comparing Fractions

Use the diagnostic tool to determine the most suitable intervention pathway for comparing fractions. Provide Diagnostic Tool: Comparing Fractions, Teacher's Resource pages 78 and 79, and have students complete it in writing or orally. Provide fraction materials for modelling fractions, such as fraction rectangles, fraction strips (e.g., Fraction Strips, BLM 10), fraction circles (e.g., Fraction Circles, BLM 18) or Cuisenaire rods.

See solutions on Teacher's Resource pages 80 and 81.

Intervention Pathways

The purpose of the intervention pathways is to develop fraction number sense as well as to support later work with fraction operations.

There are 3 pathways:

- Pathway 1: Fractions and Mixed Numbers
- Pathway 2: Proper Fractions
- Pathway 3: Equivalent Fractions

Use the chart below (or the Key to Pathways on Teacher's Resource pages 80 and 81) to determine which pathway is most suitable for each student or group of students.

Diagnostic Tool Results	Intervention Pathway
If students struggle with Questions 9 to 11	use Pathway 1: Fractions and Mixed Numbers *Teacher's Resource pages 82–83* *Student Resource pages 96–101*
If students struggle with Questions 4 to 8	use Pathway 2: Proper Fractions *Teacher's Resource pages 84–85* *Student Resource pages 102–107*
If students struggle with Questions 1 to 3	use Pathway 3: Equivalent Fractions *Teacher's Resource pages 86–87* *Student Resource pages 108–113*

If students successfully complete Pathway 2, they may or may not need the additional intervention provided by Pathway 1. Either re-administer Pathway 1 questions from the diagnostic tool or encourage students to do a portion of the open-ended intervention for Pathway 1 to decide if more work in Pathway 1 would be beneficial.

Name: ______________________________ Date: ____________________

Comparing Fractions

Diagnostic Tool

You will need

- materials for modelling fractions (e.g., fraction rectangles, Fraction Strips (BLM 10), Fraction Circles (BLM 18), or coloured rods)

1. Circle the fractions that are equal or equivalent to $\frac{2}{3}$.

$\frac{3}{2}$ $\frac{4}{5}$ $\frac{4}{6}$ $\frac{10}{15}$

2. a) Sketch a picture to show $\frac{3}{5}$.

b) Use your picture to show an equivalent fraction for $\frac{3}{5}$. Write the fraction.

3. a) Create 3 fractions equivalent to $\frac{8}{12}$.

________ ________ ________

b) Show or explain how you created each fraction.

4. Show or explain how you know each statement is true.

a) $\frac{3}{8}$ is greater than $\frac{3}{10}$.

b) $\frac{2}{5}$ is less than $\frac{5}{6}$.

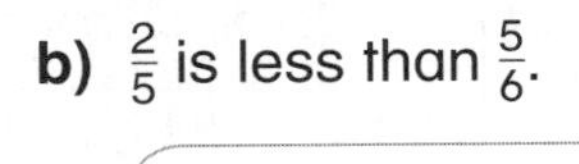

5. Circle the fraction that is *greater* in each pair.

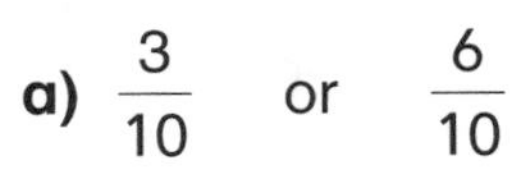

a) $\frac{3}{10}$ or $\frac{6}{10}$

b) $\frac{2}{3}$ or $\frac{2}{5}$

c) $\frac{1}{8}$ or $\frac{4}{5}$

d) $\frac{3}{8}$ or $\frac{1}{3}$

Copyright © 2012 by Nelson Education Ltd.

Name: ______________________________ Date: ______________

6. Show or explain how you know that $\frac{3}{8} < \frac{4}{10}$.

7. The fractions below are all less than 1. What value would make each comparison true? Write all possible solutions.

a) $\frac{3}{\blacksquare} > \frac{3}{5}$ ______________

b) $\frac{\blacksquare}{9} < \frac{7}{9}$ ______________

8. Circle the fractions and mixed numbers that are greater than $\frac{7}{4}$.

$\frac{13}{4}$ $\quad \frac{7}{5}$ $\quad 2\frac{1}{3}$ $\quad 1\frac{2}{3}$ $\quad \frac{11}{6}$

9. Write a mixed number for each improper fraction.

a) $\frac{8}{3} =$ ________ **b)** $\frac{13}{8} =$ ________

10. Write an improper fraction for each mixed number.

a) $2\frac{1}{2} =$ ________ **b)** $3\frac{2}{5} =$ ________

11. Show or explain how you know each comparison is true.

a) $4\frac{1}{2} > \frac{11}{3}$

b) $2\frac{2}{3} < \frac{11}{4}$

Copyright © 2012 by Nelson Education Ltd.

Solutions and Key to Pathways

Name:________________________ Date:________________

Comparing Fractions

Diagnostic Tool

You will need

- materials for modelling fractions (e.g., fraction rectangles, Fraction Strips (BLM 10), Fraction Circles (BLM 18), or coloured rods)

Pathway 3

1. Circle the fractions that are equal or equivalent to $\frac{2}{3}$.

   $\frac{3}{2}$ $\frac{4}{5}$ ($\frac{4}{6}$) ($\frac{10}{15}$)

2. a) Sketch a picture to show $\frac{3}{5}$.

   $\frac{6}{10}$

   b) Use your picture to show an equivalent fraction for $\frac{3}{5}$. Write the fraction.

3. a) Create 3 fractions equivalent to $\frac{8}{12}$.

   $\frac{2}{3}$ $\frac{4}{6}$ $\frac{16}{24}$

   b) Show or explain how you created each fraction.

   e.g., For $\frac{2}{3}$, I divided the numerator and denominator of $\frac{8}{12}$ by 4.
   For $\frac{4}{6}$, I multiplied the numerator and denominator of $\frac{2}{3}$ by 2.
   For $\frac{16}{24}$, I multiplied the numerator and denominator of $\frac{8}{12}$ by 2.

Pathway 2

4. Show or explain how you know each statement is true.

   a) $\frac{3}{8}$ is greater than $\frac{3}{10}$.

   e.g., Eighths are bigger than tenths, so 3 eighths is greater than 3 tenths.

   b) $\frac{2}{5}$ is less than $\frac{5}{6}$.

   e.g., $\frac{2}{5}$ is less than $\frac{1}{2}$, and $\frac{5}{6}$ is more than $\frac{1}{2}$, so $\frac{2}{5}$ is less than $\frac{5}{6}$.

5. Circle the fraction that is *greater* in each pair.

   a) $\frac{3}{10}$ or ($\frac{6}{10}$)

   b) ($\frac{2}{3}$) or $\frac{2}{5}$

   c) $\frac{1}{8}$ or ($\frac{4}{5}$)

   d) ($\frac{3}{8}$) or $\frac{1}{3}$

Copyright © 2012 by Nelson Education Ltd.

Copyright © 2012 by Nelson Education Ltd.

Name:______________________ Date:______________

Pathway 2

6. Show or explain how you know that $\frac{3}{8} < \frac{4}{10}$.

e.g., $\frac{3}{8}$ is $\frac{1}{8}$ below $\frac{1}{2}$ on a number line and $\frac{4}{10}$ is $\frac{1}{10}$ below $\frac{1}{2}$, so $\frac{4}{10}$ is closer to $\frac{1}{2}$ than $\frac{3}{8}$ is, which makes it greater.

7. The fractions below are all less than 1. What value would make each comparison true? Write all possible solutions.

a) $\frac{3}{■} > \frac{3}{5}$ ___4___

b) $\frac{■}{9} < \frac{7}{9}$ ___0, 1, 2, 3, 4, 5, 6___

8. Circle the fractions and mixed numbers that are greater than $\frac{7}{4}$.

$\frac{13}{4}$ (circled) $\frac{7}{5}$ $2\frac{1}{3}$ (circled) $1\frac{2}{3}$ $\frac{11}{6}$ (circled)

Pathway 1

9. Write a mixed number for each improper fraction.

a) $\frac{8}{3} =$ ___$2\frac{2}{3}$___ b) $\frac{13}{8} =$ ___$1\frac{5}{8}$___

10. Write an improper fraction for each mixed number.

a) $2\frac{1}{2} =$ ___$\frac{5}{2}$___ b) $3\frac{2}{5} =$ ___$\frac{17}{5}$___

11. Show or explain how you know each comparison is true.

a) $4\frac{1}{2} > \frac{11}{3}$

e.g., $4\frac{1}{2} > 4$ and $4 = \frac{12}{3}$.
$\frac{11}{3} < \frac{12}{3}$, so $\frac{11}{3} < 4\frac{1}{2}$.

b) $2\frac{2}{3} < \frac{11}{4}$

e.g., $\frac{11}{4} = 2\frac{3}{4}$ so I only have to explain why $\frac{2}{3} < \frac{3}{4}$.
$\frac{2}{3}$ is $\frac{1}{3}$ below 1 on a number line and $\frac{3}{4}$ is only $\frac{1}{4}$ below 1, which makes $\frac{3}{4}$ closer to 1 than $\frac{2}{3}$ is, so $\frac{2}{3} < \frac{3}{4}$.

Pathway 1 OPEN-ENDED

Fractions and Mixed Numbers

You will need

- materials for modelling fractions (e.g., Fraction Strips (BLM 10), Fraction Circles/Spinners (BLM 18), or commercial fraction pieces)
- pencils and paper clips
- Student Resource pages 96–97

Open-Ended Intervention

Before Using the Open-Ended Intervention

Provide fraction materials and use the fraction pieces to model $\frac{5}{4}$ (reading it aloud as "five fourths"). Ask:

- What do you notice?
  (e.g., *There are 5 pieces, and each is* $\frac{1}{4}$ *of a whole. I can make 1 whole with 4 pieces and then there's another* $\frac{1}{4}$. *It looks like* $\frac{5}{4}$ *is* $1\frac{1}{4}$.)
- Model a fraction that is greater than $\frac{5}{4}$.
  (e.g., $\frac{6}{4}$ *or* $1\frac{2}{4}$)
- How do you know it's greater?
  (e.g., *6 fourths is more than 5 fourths.*)
- Model $\frac{5}{3}$. Is it greater or less than $\frac{5}{4}$? How do you know?
  (*It's more.* e.g., *Both models have 5 pieces, but each third is bigger than each fourth.*)
- What mixed number name would you give $\frac{5}{3}$? Why?
  ($1\frac{2}{3}$. e.g., *It's 1 whole (3 thirds makes a whole) and* $\frac{2}{3}$ *of another whole.*)

Using the Open-Ended Intervention (Student Resource pages 96–97)

Read through the tasks on the student pages together. Provide fraction materials. Give students time to work, ideally in pairs.

Observe whether students

- can compare and order fractions and mixed numbers
- can explain their strategies for ordering a set of fractions and mixed numbers
- relate equivalent mixed numbers and improper fractions

Consolidating and Reflecting

Ensure understanding by asking questions such as these based on students' work:

- How could you have predicted that the middle fraction would be the least?
  (e.g., *It was the only fraction less than 1.*)
- Why was it easy to compare the mixed numbers?
  (e.g., *If the whole number parts are different, you can ignore the proper fraction parts and just compare the whole numbers. If the whole numbers are the same, you can just compare the proper fraction parts.*)
- How did you compare the improper fractions and mixed numbers?
  (e.g., *Sometimes I estimated what whole number the improper fraction was close to, and other times I used a model to help me change the improper fraction to a mixed number.*)
- How do you change an improper fraction to a mixed number?
  (e.g., *I think of how many wholes I can make by dividing the numerator by the denominator, and then I make a proper fraction from what is left over.*)

 Copyright © 2012 by Nelson Education Ltd.

Fractions and Mixed Numbers

Pathway 1
GUIDED

Guided Intervention

Before Using the Guided Intervention

Provide fraction materials and use the fraction pieces to model $\frac{5}{4}$ (reading it aloud as "five fourths"). Ask:

- What do you notice?
  (e.g., *There are 5 pieces, and each is* $\frac{1}{4}$ *of a whole. I can make 1 whole with 4 pieces and then there's another* $\frac{1}{4}$. *It looks like* $\frac{5}{4}$ *is* $1\frac{1}{4}$.)
- Model a fraction that is greater than $\frac{5}{4}$.
  (e.g., $\frac{6}{4}$ *or* $1\frac{2}{4}$)
- How do you know it's greater?
  (e.g., *6 fourths is more than 5 fourths.*)
- Model $\frac{5}{3}$. Is it greater or less than $\frac{5}{4}$? How do you know?
  (*It's more.* e.g., *Both models have 5 pieces, but each third is bigger than each fourth.*)
- What mixed number name would you give $\frac{5}{3}$? Why?
  ($1\frac{2}{3}$. e.g., *It's 1 whole (3 thirds makes a whole) and* $\frac{2}{3}$ *of another whole.*)

Using the Guided Intervention **Student Resource pages 98–101**

Work through the instructional section of the student pages together. Students could model each comparison with fraction materials.

Have students work through the **Try These** questions individually or in pairs.

Observe whether students

- can compare and order fractions and mixed numbers (Questions 1, 4, 5)
- can explain their strategies for comparing fractions (Questions 1, 2, 3, 4, 7, 8)
- can model a fraction comparison (Questions 2 and 3)
- can create fractions to meet criteria related to their size (Question 6)

Consolidating and Reflecting

Ensure understanding by asking questions such as these based on students' work:

- How could you have predicted the least fraction in Question 5?
  (e.g., *I knew it was* $\frac{7}{9}$ *because it was the only fraction less than 1.*)
- Why was it easy to compare the mixed numbers in Question 5?
  (e.g., *They had different whole number parts—3 and 1—and 3 is greater than 1.*)
- How did you compare improper fractions and mixed numbers?
  (e.g., *Sometimes I estimated what whole number the improper fraction was close to, and other times I used a model to help me change the improper fraction to a mixed number.*)
- How do you change an improper fraction to a mixed number?
  (e.g., *I think of how many wholes I could make by dividing the numerator by the denominator, and then I make a proper fraction from what is left over.*)

You will need

- materials for modelling fractions (e.g., Fraction Strips (BLM 10), Fraction Circles (BLM 18), or commercial fraction pieces)
- Student Resource pages 98–101

Copyright © 2012 by Nelson Education Ltd.

Pathway 2 OPEN-ENDED

Proper Fractions

You will need

- materials for modelling fractions (e.g., Fraction Strips (BLM 10), Fraction Circles/Spinners (BLM 18), or commercial fraction pieces)
- pencils and paper clips
- Student Resource pages 102–103

Open-Ended Intervention

Before Using the Open-Ended Intervention

Provide fraction materials and use the fraction pieces to model $\frac{3}{4}$ (reading it aloud as "three fourths"). Ask:

- What do you notice?
  (e.g., *It uses 3 pieces, and each piece is* $\frac{1}{4}$ *of a whole.*)
- Model a fraction that is greater than $\frac{3}{4}$.
  (e.g., $\frac{5}{6}$)
- How do you know it's greater?
  (e.g., $\frac{5}{6}$ *is just* $\frac{1}{6}$ *below 1 on a number line, while* $\frac{3}{4}$ *is* $\frac{1}{4}$ *below 1, and* $\frac{1}{4}$ *is more than* $\frac{1}{6}$.)
- Model $\frac{3}{5}$. Is it more or less than $\frac{3}{4}$? How do you know?
  (*It's less.* e.g., *Both have 3 pieces, but the pieces are smaller in* $\frac{3}{5}$ *because fifths are smaller than fourths.*)

Using the Open-Ended Intervention (Student Resource pages 102–103)

Read through the tasks on the student pages together. Provide fraction materials. Give students time to work, ideally in pairs.

Observe whether students

- can compare and order fractions
- can explain their strategies for ordering a set of fractions
- can create equivalent fractions

Consolidating and Reflecting

Ensure understanding by asking questions such as these based on students' work:

- Which fraction was easiest to figure out where to put in the order?
  (e.g., $\frac{10}{12}$ *was easiest because it was the only one that was more than half*.)
- Which fraction pair was really easy to compare?
  (e.g., *any pair with the same numerator or the same denominator*)
- What did you do when it was more difficult to compare 2 fractions?
  (e.g., *Sometimes I used fraction models; other times I used one or more equivalent fractions.*)
- How does using equivalent fractions help?
  (e.g., *You can make either the numerators or the denominators the same, and then you can just compare 2 numbers. If the numerators are the same, you compare the denominators. If the denominators are the same, you compare the numerators.*)

 Copyright © 2012 by Nelson Education Ltd.

Proper Fractions

Pathway 2 GUIDED

Guided Intervention

You will need

- materials for modelling fractions (e.g., Fraction Strips (BLM 10), Fraction Circles (BLM 18), or commercial fraction pieces)
- Student Resource pages 104–107

Before Using the Guided Intervention

Provide fraction materials and use the fraction pieces to model $\frac{3}{4}$ (reading it aloud as "three fourths"). Ask:

- What do you notice?
  (e.g., *It uses 3 pieces, and each piece is* $\frac{1}{4}$ *of a whole.*)
- Model a fraction that is greater than $\frac{3}{4}$. (e.g., $\frac{5}{6}$)
- How do you know it's greater?
  (e.g., $\frac{5}{6}$ *is just* $\frac{1}{6}$ *below 1 on a number line, while* $\frac{3}{4}$ *is* $\frac{1}{4}$ *below 1, and* $\frac{1}{4}$ *is more than* $\frac{1}{6}$.)
- Model $\frac{3}{5}$. Is it more or less than $\frac{3}{4}$? How do you know?
  (*It's less.* e.g., *Both have 3 pieces but the pieces are smaller in* $\frac{3}{5}$ *because fifths are smaller than fourths.*)

Using the Guided Intervention (Student Resource pages 104–107)

Work through the instructional section of the student pages together. Provide fraction materials for students to model the fraction comparisons.

Have students work through the **Try These** questions individually or in pairs.

Observe whether students

- can compare and order fractions (Questions 1, 2, 3, 4, 5, 8)
- can explain their strategies for comparing fractions (Questions 1, 2, 3, 7)
- can model a fraction comparison (Question 2)
- can create fractions to meet criteria related to size (Questions 4, 6, 8)

Consolidating and Reflecting

Ensure understanding by asking questions such as these based on students' work:

- Why did you use different strategies for the situations in Question 3?
  (e.g., *If the numerators were the same, I compared the denominator. If the fractions were far apart, I used a benchmark. If they were close, I used equivalent fractions.*)
- Why might Question 4c) be easier than 4d)?
  (e.g., *In 4c), the denominators were the same, so all I needed was a smaller numerator; in 4d), I had to think more about what to do. I decided to make equivalent fractions with a numerator of 12 to help me get started.*)
- How could you have predicted the greatest fraction in Question 5?
  (e.g., *I knew* $\frac{11}{12}$ *was almost 1. No other fraction except* $\frac{5}{6}$ *was even close to that. Since* $\frac{5}{6}$ *was* $\frac{1}{6}$ *away from 1 while* $\frac{11}{12}$ *was only* $\frac{1}{12}$ *away, I knew* $\frac{11}{12}$ *was greater.*)
- How does your work in Question 8 show why you can't just look at the size of the numerator and denominator to decide how big a fraction is?
  (e.g., *Sometimes a big numerator and denominator can be a smaller fraction than one with a small numerator and denominator, like* $\frac{20}{100}$ *and* $\frac{1}{4}$.)

Pathway 3 OPEN-ENDED

Equivalent Fractions

You will need

- materials for modelling fractions (e.g., Fraction Strips (BLM 10), Fraction Circles (BLM 18), or commercial fraction pieces)
- 1 cm Grid Paper (BLM 12)
- Student Resource pages 108–109

Open-Ended Intervention

Before Using the Open-Ended Intervention

Provide fraction materials and grid paper, and represent $\frac{4}{8}$ and $\frac{2}{4}$ using the same materials. Ask:

- What do you notice?
  (e.g., *They are both the same as a half. Half the circle is shaded each time.*)
- What other ways are there to show $\frac{1}{2}$ using fractions?
  (e.g., $\frac{3}{6}$)
- What name for $\frac{1}{2}$ involves tenths?
  ($\frac{5}{10}$)
- How could you show that?
  (e.g., *I'd make a 2-row-by-5-column grid and shade one row, or half of the grid, which is 5 out of 10 squares.*)

Using the Open-Ended Intervention (Student Resource pages 108–109)

Read through the tasks on the student pages together. Provide grid paper and fraction materials. Give students time to work, ideally in pairs.

Observe whether students can

- create equivalent fractions
- use a model to show fraction equivalence
- explain how they know 2 fractions are equivalent

Consolidating and Reflecting

Ensure understanding by asking questions such as these based on students' work:

- What fractions did you think of first? Why?
  (e.g., *I thought of $\frac{10}{20}$ and $\frac{20}{40}$ first, since $\frac{1}{2}$ is the easiest fraction.*)
- How did you prove that they were both equivalent to $\frac{1}{2}$?
  (e.g., *If you have one half, the numerator is half the denominator and that's what happens with $\frac{10}{20}$ and $\frac{20}{40}$.*)
- How does knowing $\frac{4}{20} = \frac{1}{5}$ help you create other fractions?
  (e.g., *I could use $\frac{4}{20} + \frac{4}{20} = \frac{8}{20}$ to get $\frac{8}{20} = \frac{2}{5}$, or $\frac{4}{20} + \frac{4}{20} + \frac{4}{20} = \frac{12}{20}$ to get $\frac{12}{20} = \frac{3}{5}$.*)
- Do you think there are more fractions than you found? Explain.
  (*Yes.* e.g., *I could have used bigger and bigger denominators for a numerator of 20, like 100, 200, 300, 400, 500, which could go on forever.*)
- If someone asked you for a rule for creating equivalent fractions, what rule would you give?
  (e.g., *Multiply or divide the numerator and denominator by the same number.*)

 Copyright © 2012 by Nelson Education Ltd.

Equivalent Fractions

Pathway 3 GUIDED

Guided Intervention

Before Using the Guided Intervention

Provide fraction materials and grid paper, and represent $\frac{4}{8}$ and $\frac{2}{4}$ using the same materials. Ask:

- What do you notice?
  (e.g., *They are both the same as a half. Half the circle is shaded each time.*)
- What other ways are there to show $\frac{1}{2}$ using fractions? (e.g., $\frac{3}{6}$)
- What name for $\frac{1}{2}$ involves tenths? ($\frac{5}{10}$)
- How could you show that?
  (e.g., *I'd make a 2-row-by-5-column grid and shade one row, or half of the grid, which is 5 out of 10 squares.*)

You will need

- materials for modelling fractions (e.g., Fraction Strips (BLM 10), Fraction Circles (BLM 18), or commercial fraction pieces)
- 1 cm Grid Paper (BLM 12)
- Student Resource pages 110–113

Using the Guided Intervention Student Resource pages 110–113

Work through the instructional section of the student pages together. Provide fraction materials and grids for students to model the equivalent fractions.

Have students work through the **Try These** questions individually or in pairs.

Observe whether students

- can recognize equivalent fractions from a model (Question 1)
- relate a model to the procedure for creating an equivalent fraction (Question 2)
- can model fraction equivalence (Questions 3, 4)
- can create equivalent fractions (Questions 5, 9, 10)
- can explain how they know fractions are equivalent (Questions 6, 7, 8, 11)

Consolidating and Reflecting

Ensure understanding by asking questions such as these based on students' work:

- Choose an equivalent fraction you created in Question 5. How can you show that the fractions are equivalent?
  (e.g., *For* $\frac{4}{5} = \frac{8}{10}$, *I'd shade 2 out of 5 rows in a 2-by-5 grid, which would be 8 out of 10 squares.*)
- In Question 6, you saw that you could divide the numerator and denominator by the same amount to get an equivalent fraction but could not subtract the same amount. Can you add the same amount? Explain.
  (*No.* e.g., $\frac{1}{3}$ *is not the same as* $\frac{2}{4}$.)
- In Question 8, Prableen's idea relates the numerator and denominator. Does the same idea work with fractions like $\frac{3}{4}$ and $\frac{6}{8}$? Why does that make sense?
  (*Yes.* e.g., *If you take* $\frac{6}{8}$ *and split each eighth into 2 equal parts, you have* $\frac{3}{4}$. *For both* $\frac{3}{4}$ *and* $\frac{6}{8}$, *there are* $\frac{3}{4}$ *the number of shaded parts as unshaded parts.*)
- If someone asked you for a rule for creating an equivalent fraction, what rule would you give?
  (e.g., *Multiply or divide the numerator and denominator by the same number.*)

Strand: Number

Fraction Operations

Planning For This Topic

Materials for assisting students with fraction operations consist of a diagnostic tool and 4 intervention pathways. Pathway 1 focuses on repeated addition of fractions in situations involving concepts of multiplication and division. Pathway 2 involves adding and subtracting mixed numbers. Pathway 3 focuses on subtracting proper and improper fractions. Pathway 4 involves addition with proper and improper fraction.

Each pathway has an open-ended intervention and a guided intervention. Choose the type of intervention more suitable for your students' needs and your particular circumstances.

Curriculum Connections

Grades 5 to 8 curriculum connections for this topic are provided online. See www.nelson.com/leapsandbounds. All 4 pathways are relevant to both the WNCP and Ontario curriculums.

Why might students struggle with fraction operations?

Students might struggle with fraction operations for any of the following reasons:

- They might not have a good sense of fraction size, so they cannot determine whether their answers make sense.
- They might add (or subtract) fractions incorrectly by adding (or subtracting) the numerators and adding (or subtracting) the denominators.
- They might not recognize subtraction situations. In particular, they might not recognize that comparing 2 fractions involves subtracting the fractions (and when representing it symbolically, subtracting the greater one from the lesser one).
- They might have learned only one strategy for adding and subtracting fractions that involves equivalent fractions with a common denominator (often the lowest common denominator). So, they might make careless errors when creating equivalent fractions, particularly without the support of manipulatives.
- They might have difficulty writing a mixed number as an improper fraction, or vice versa, without the support of manipulatives.
- They might not realize that, when adding mixed numbers, the whole number part of the sum can be 1 greater than the sum of the whole numbers in the addends due to the sum of the proper fraction parts.
- They might struggle with the subtraction of mixed numbers where the fraction part in the minuend is less than the fraction part in the subtrahend (e.g., $3\frac{5}{8} - 2\frac{7}{8}$).
- They might not see a division problem situation as a multiplication situation, and, as such, struggle with dividing a whole by a simple fraction instead of using repeated addition to solve the problem.

Professional Learning Connections

PRIME: Number and Operations, Background and Strategies (Nelson Education Ltd., 2005), pages 114–116

Making Math Meaningful to Canadian Students K–8 (Nelson Education Ltd., 2008), pages 210–215, 221

Big Ideas from Dr. Small Grades 4–8 (Nelson Education Ltd., 2009), pages 50–55

Good Questions (dist. by Nelson Education Ltd., 2009), pages 24, 49

More Good Questions (dist. by Nelson Education Ltd, 2010), page 71

 Copyright © 2012 by Nelson Education Ltd.

Diagnostic Tool: Fraction Operations

Use the diagnostic tool to determine the most suitable intervention pathway for fraction operations. Provide Diagnostic Tool: Fraction Operations, Teacher's Resource pages 90 to 93, and have students complete it in writing or orally. Provide fraction materials (e.g., fraction circles, fraction strips, and/or pattern blocks, which are all available commercially and as blackline masters), 2 cm Grid Paper (BLM 11), and coloured counters.

Pattern blocks are useful for modelling certain fractions (halves, thirds, and sixths).

See solutions on Teacher's Resource pages 94 to 97.

Intervention Pathways

The purpose of the intervention pathways is to help students determine sums and differences of fractions. Determining sums and differences of fractions will prepare students for later work with other operations and for working with algebraic expressions involving rational numbers.

There are 4 pathways:
- Pathway 1: Repeated Addition of Fractions
- Pathway 2: Adding and Subtracting Mixed Numbers
- Pathway 3: Subtracting Fractions
- Pathway 4: Adding Fractions

Use the chart below (or the Key to Pathways on Teacher's Resource pages 94 to 97) to determine which pathway is most suitable for each student or group of students.

Diagnostic Tool Results	Intervention Pathway
If students struggle with Questions 11 to 13	use Pathway 1: Repeated Addition of Fractions *Teacher's Resource pages 98–99* *Student Resource pages 114–119*
If students struggle with Questions 8 to 10	use Pathway 2: Adding and Subtracting Mixed Numbers *Teacher's Resource pages 100–101* *Student Resource pages 120–125*
If students struggle with Questions 5 to 7	use Pathway 3: Subtracting Fractions *Teacher's Resource pages 102–103* *Student Resource pages 126–131*
If students struggle with Questions 1 to 4	use Pathway 4: Adding Fractions *Teacher's Resource pages 104–105* *Student Resource pages 132–137*

Copyright © 2012 by Nelson Education Ltd.

Fraction Operations

Diagnostic Tool

You will need

- fraction materials (e.g., Fraction Strips (BLM 10), Fraction Circles (BLM 18), pattern blocks)
- 2 cm Grid Paper (BLM 11), counters

1. Name 2 equivalent fractions for each fraction.

a) $\frac{4}{9}$ ____ ____ **b)** $\frac{5}{6}$ ____ ____ **c)** $\frac{12}{18}$ ____ ____

2. Sketch a model to show that each equation is true.

a) $\frac{3}{5} + \frac{1}{3} = \frac{14}{15}$

b) $\frac{5}{6} + \frac{3}{4} = 1\frac{7}{12}$

3. Calculate. Show your work.

a) $\frac{2}{5} + \frac{1}{3} =$ ________

c) $\frac{5}{6} + \frac{3}{8} =$ ________

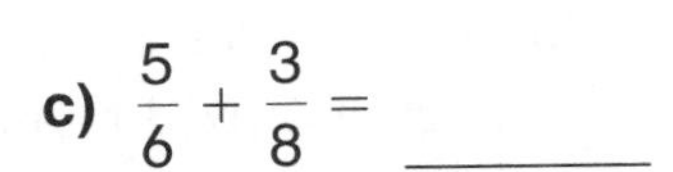

b) $\frac{4}{5} + \frac{1}{6} =$ ________

d) $\frac{6}{5} + \frac{4}{3} =$ ________

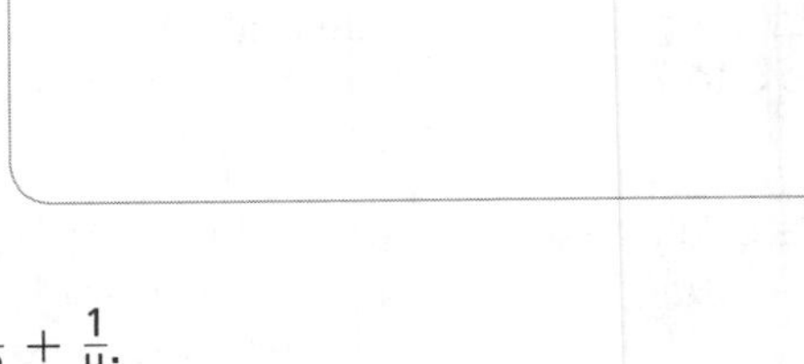

4. Create a problem that you could solve by adding $\frac{5}{12} + \frac{1}{4}$.

____

____

____

Copyright © 2012 by Nelson Education Ltd.

5. Sketch a model to show that each equation is true.
Explain one of your models.

a) $\frac{4}{5} - \frac{1}{3} = \frac{7}{15}$

b) 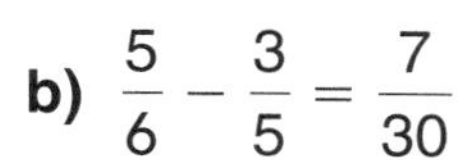 $\frac{5}{6} - \frac{3}{5} = \frac{7}{30}$

6. Calculate. Show your work.

a) $\frac{2}{5} - \frac{1}{3} =$ ________

c) 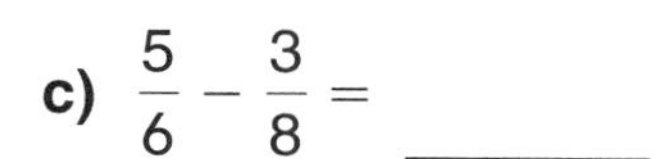 $\frac{5}{6} - \frac{3}{8} =$ ________

b) $\frac{4}{5} - \frac{1}{6} =$ ________

d) $\frac{4}{3} - \frac{3}{5} =$ ________

7. Create a problem that you could solve by calculating $\frac{3}{4} - \frac{5}{12}$.

____

____

____

Copyright © 2012 by Nelson Education Ltd.

8. Sketch a model to show each calculation.

a) $2\frac{1}{2} + 1\frac{3}{5} =$ ________

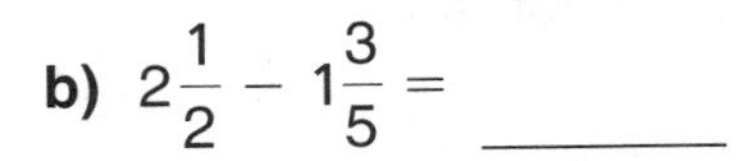

b) $2\frac{1}{2} - 1\frac{3}{5} =$ ________

9. Calculate. Show your work.

a) $1\frac{1}{5} + 2\frac{3}{10} =$ ________

b) $1\frac{2}{3} + 3\frac{5}{6} =$ ________

c) $3\frac{4}{5} + 5\frac{2}{3} =$ ________

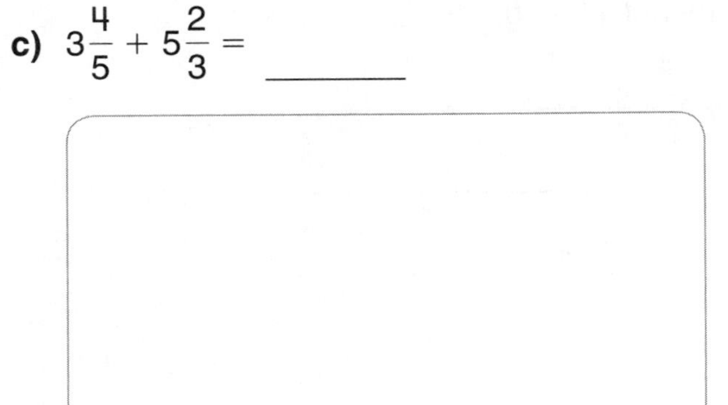

d) $2\frac{3}{10} - 1\frac{1}{5} =$ ________

e) $3\frac{5}{6} - 1\frac{2}{3} =$ ________

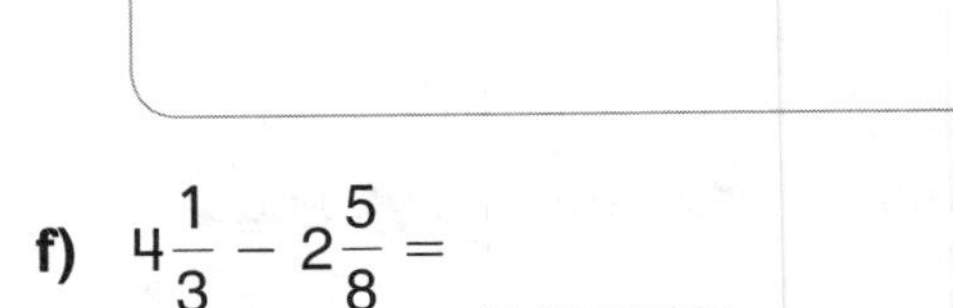

f) $4\frac{1}{3} - 2\frac{5}{8} =$ ________

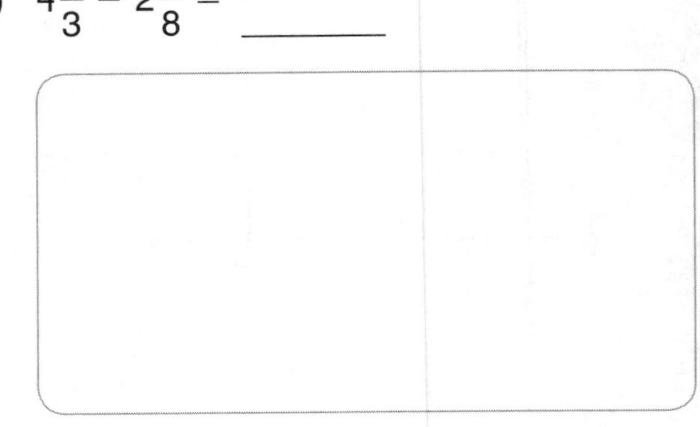

 Copyright © 2012 by Nelson Education Ltd.

10. Create a problem that you could solve by calculating $6 - 3\frac{2}{3}$.

____

____

____

11. Calculate. Show your work.

a) $\frac{2}{3} + \frac{2}{3} + \frac{2}{3} + \frac{2}{3} + \frac{2}{3} =$ ________

b) $6 \times \frac{3}{5} =$ ________

c) $2 \div \frac{1}{3} =$ ________

d) $3 \div \frac{3}{4} =$ ________

12. Amanda had only a $\frac{3}{4}$ cup measure. She filled it 8 times with flour and dumped the flour into a bowl. How many cups of flour did she put in the bowl? Show your work.

13. Jeff has 4 large full cans of paint and some small empty cans. Each small can holds $\frac{2}{3}$ as much as a large can. How many small cans hold the same amount as 4 large cans? Show your work.

Copyright © 2012 by Nelson Education Ltd.

Solutions and Key to Pathways

Fraction Operations

Diagnostic Tool

You will need

- fraction materials (e.g., Fraction Strips (BLM 10), Fraction Circles (BLM 18), pattern blocks)
- 2 cm Grid Paper (BLM 11), counters

Pathway 4

1. Name 2 equivalent fractions for each fraction.

a) $\frac{4}{9}$ $\frac{8}{18}$ $\frac{12}{27}$ **b)** $\frac{5}{6}$ $\frac{10}{12}$ $\frac{20}{24}$ **c)** $\frac{12}{18}$ $\frac{2}{3}$ $\frac{4}{6}$

2. Sketch a model to show that each equation is true.

a) $\frac{3}{5} + \frac{1}{3} = \frac{14}{15}$

e.g.,

b) $\frac{5}{6} + \frac{3}{4} = 1\frac{7}{12}$

e.g.,

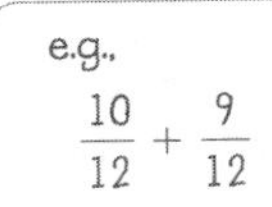

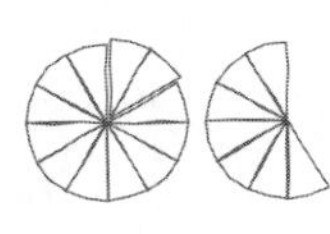

3. Calculate. Show your work.

a) $\frac{2}{5} + \frac{1}{3} = \frac{11}{15}$

e.g., $\frac{2}{5} + \frac{1}{3} = \frac{6}{15} + \frac{5}{15}$
$= \frac{11}{15}$

b) $\frac{4}{5} + \frac{1}{6} = \frac{29}{30}$

e.g., $\frac{4}{5} + \frac{1}{6} = \frac{24}{30} + \frac{5}{30}$
$= \frac{29}{30}$

c) $\frac{5}{6} + \frac{3}{8} = \frac{29}{24}$

e.g., $\frac{5}{6} + \frac{3}{8} = \frac{20}{24} + \frac{9}{24}$
$= \frac{29}{24}$

d) $\frac{6}{5} + \frac{4}{3} = \frac{38}{15}$

e.g., $\frac{6}{5} + \frac{4}{3} = \frac{18}{15} + \frac{20}{15}$
$= \frac{38}{15}$

4. Create a problem that you could solve by adding $\frac{5}{12} + \frac{1}{4}$.

e.g., I had 2 egg cartons. One was $\frac{5}{12}$ full and the other was $\frac{1}{4}$ full.
What fraction of an egg carton is that altogether?

Copyright © 2012 by Nelson Education Ltd.

Copyright © 2012 by Nelson Education Ltd.

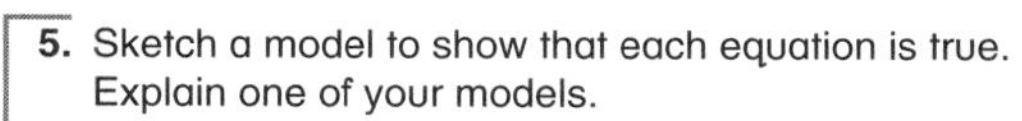

5. Sketch a model to show that each equation is true.
Explain one of your models.

a) $\frac{4}{5} - \frac{1}{3} = \frac{7}{15}$

e.g., I showed $\frac{4}{5}$ on a grid. I moved 1 counter to show $\frac{1}{3}$ in the bottom row and then took that row away. That left $\frac{7}{15}$.

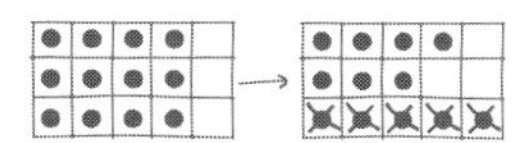

b) $\frac{5}{6} - \frac{3}{5} = \frac{7}{30}$

e.g.,

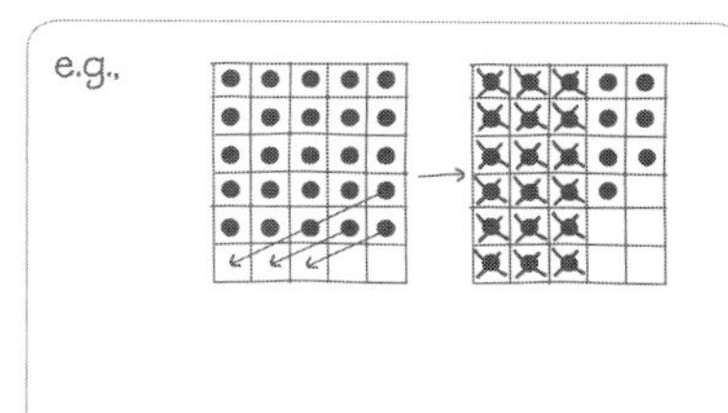

6. Calculate. Show your work.

a) $\frac{2}{5} - \frac{1}{3} = \frac{1}{15}$

e.g., $\frac{2}{5} - \frac{1}{3} = \frac{6}{15} - \frac{5}{15}$
$= \frac{1}{15}$

b) $\frac{4}{5} - \frac{1}{6} = \frac{19}{30}$

e.g., $\frac{4}{5} - \frac{1}{6} = \frac{24}{30} - \frac{5}{30}$
$= \frac{19}{30}$

c) $\frac{5}{6} - \frac{3}{8} = \frac{22}{48}$ or $\frac{11}{24}$

e.g., $\frac{5}{6} - \frac{3}{8} = \frac{20}{24} - \frac{9}{24}$
$= \frac{11}{24}$

d) $\frac{4}{3} - \frac{3}{5} = \frac{11}{15}$

e.g., $\frac{4}{3} - \frac{3}{5} = \frac{20}{15} - \frac{9}{15}$
$= \frac{11}{15}$

7. Create a problem that you could solve by calculating $\frac{3}{4} - \frac{5}{12}$.

e.g., I had $\frac{3}{4}$ cup of flour, and I used $\frac{5}{12}$ cup to bake some muffins.
What fraction of a cup is left?

Pathway 3

Copyright © 2012 by Nelson Education Ltd.

Copyright © 2012 by Nelson Education Ltd.

8. Sketch a model to show each calculation.

a) $2\frac{1}{2} + 1\frac{3}{5} =$ $4\frac{1}{10}$

e.g.,

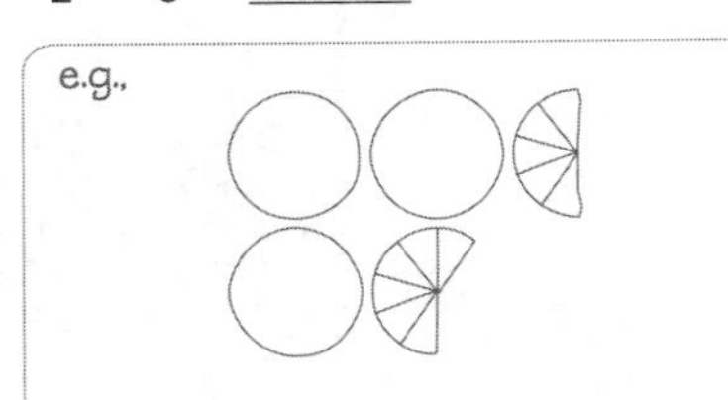

b) $2\frac{1}{2} - 1\frac{3}{5} =$ $\frac{9}{10}$

e.g.,

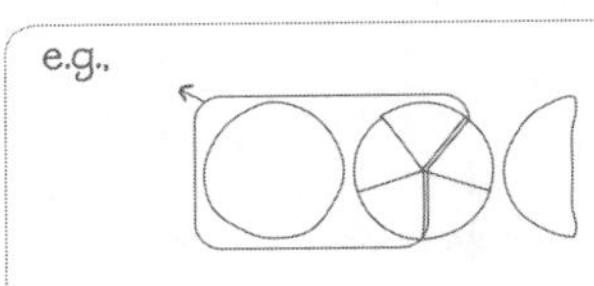

9. Calculate. Show your work.

a) $1\frac{1}{5} + 2\frac{3}{10} =$ $3\frac{5}{10}$ or $3\frac{1}{2}$

e.g., $1\frac{1}{5} + 2\frac{3}{10} = 1\frac{2}{10} + 2\frac{3}{10}$

$= 3\frac{5}{10}$

b) $1\frac{2}{3} + 3\frac{5}{6} =$ $5\frac{3}{6}$

e.g., $1\frac{2}{3} + 3\frac{5}{6} = 1 + 3 + \frac{2}{3} + \frac{5}{6}$

$= 4 + \frac{4}{6} + \frac{5}{6}$

$= 4\frac{9}{6}$ or $5\frac{3}{6}$

c) $3\frac{4}{5} + 5\frac{2}{3} =$ $9\frac{7}{15}$

e.g., $3\frac{4}{5} + 5\frac{2}{3} = 3 + 5 + \frac{4}{5} + \frac{2}{3}$

$= 8 + \frac{12}{15} + \frac{10}{15}$

$= 8 + \frac{22}{15} = 9\frac{7}{15}$

d) $2\frac{3}{10} - 1\frac{1}{5} =$ $1\frac{1}{10}$

e.g., $2\frac{3}{10} - 1\frac{2}{10} = 1\frac{1}{10}$

e) $3\frac{5}{6} - 1\frac{2}{3} =$ $2\frac{1}{6}$

e.g., $\frac{1}{3} + 1 + \frac{5}{6} = \frac{2}{6} + 1 + \frac{5}{6}$

$= 1\frac{7}{6}$ or $2\frac{1}{6}$

1 $1\frac{2}{3}$ 2 3 $3\frac{5}{6}$ 4

f) $4\frac{1}{3} - 2\frac{5}{8} =$ $1\frac{17}{24}$

e.g., $4\frac{1}{3} - 2\frac{5}{8} = 4\frac{8}{24} - 2\frac{15}{24}$

$= 3\frac{32}{24} - 2\frac{15}{24}$

$= 1\frac{17}{24}$

Pathway 2

92 Fraction Operations, Diagnostic Tool *Leaps and Bounds* Copyright © 2012 by Nelson Education Ltd.

Copyright © 2012 by Nelson Education Ltd.

Pathway 2

10. Create a problem that you could solve by calculating $6 - 3\frac{2}{3}$.

e.g., The school bought 6 pizzas, and $3\frac{2}{3}$ of them were eaten at lunch. How much pizza was left?

Pathway 1

11. Calculate. Show your work.

a) $\frac{2}{3} + \frac{2}{3} + \frac{2}{3} + \frac{2}{3} + \frac{2}{3} = \frac{10}{3}$

e.g., 5 groups of 2 thirds is 10 thirds.

c) $2 \div \frac{1}{3} = 6$

e.g.,

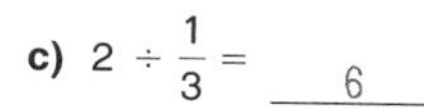

b) $6 \times \frac{3}{5} = \frac{18}{5}$

e.g., 6 groups of 3 fifths is 18 fifths.

d) $3 \div \frac{3}{4} = 4$

e.g., There are 4 groups of $\frac{3}{4}$ in 3 wholes.

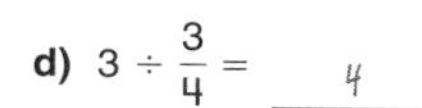

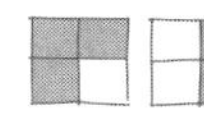

12. Amanda had only a $\frac{3}{4}$ cup measure. She filled it 8 times with flour and dumped the flour into a bowl. How many cups of flour did she put in the bowl? Show your work.

6 cups

e.g., 8 sets of 3 fourths is 24 fourths, or $\frac{24}{4}$.

$\frac{24}{4} = 6$

13. Jeff has 4 large full cans of paint and some small empty cans. Each small can holds $\frac{2}{3}$ as much as a large can. How many small cans hold the same amount as 4 large cans? Show your work.

6 cans; e.g., If he poured 1 large can into 1 small can, there would still be $\frac{1}{3}$ left in the large can. If he did it again with another large can, there would be another $\frac{1}{3}$ left; the $\frac{1}{3} + \frac{1}{3} = \frac{2}{3}$ cup would fill 1 more small can. If 2 large cans fill 3 small ones, then 4 large ones fill 6 small ones.

Copyright © 2012 by Nelson Education Ltd.

Copyright © 2012 by Nelson Education Ltd.

Pathway 1 OPEN-ENDED

Repeated Addition of Fractions

You will need

- fraction materials (e.g., Fraction Strips (BLM 10), Fraction Circles (BLM 18), pattern blocks, or other commercial materials)
- 2 cm Grid Paper (BLM 11)
- coloured counters
- Student Resource pages 114–115

Pattern blocks are useful for modelling halves, thirds, and sixths.

Note that the division of a whole number by a fraction (in Part B) is viewed through the lens of repeated addition of fractions.

Open-Ended Intervention

Before Using the Open-Ended Intervention

Provide pattern blocks and explain that you're going to name the hexagon as 1 whole. Ask:

- What blocks would you use to show $\frac{2}{3}$ of a hexagon?
  (e.g., *2 blue rhombuses*)
- How would you model $\frac{2}{3} + \frac{2}{3}$?
  (e.g., *I'd use 2 hexagons and cover each with 2 rhombuses.*)
- What fraction of a hexagon do the 4 rhombuses cover? How do you know?
  ($\frac{4}{3}$, *or* $1\frac{1}{3}$. e.g., *The rhombuses cover 1 whole hexagon and* $\frac{1}{3}$ *of another.*)
- Why might you write what you just did as $\frac{2}{3} + \frac{2}{3}$ or as $2 \times \frac{2}{3}$?
  (e.g., *When you put 2 things together, you're adding them or multiplying by 2.*)

Using the Open-Ended Intervention (Student Resource pages 114–115)

Read through the tasks on the student pages together. You might also model pouring water into 4 glasses until each is $\frac{2}{3}$ full, and then pouring everything into $2\frac{2}{3}$ glasses. Provide fraction materials. Give students time to work, ideally in pairs.

Observe whether students

- can model the repeated addition of fractions in a problem situation (multiplication and division)
- can estimate and determine sums of repeated fractions
- can explain their thinking about repeated addition of fractions

Consolidating and Reflecting

Ensure understanding by asking questions such as these based on students' work:

- Choose a number of glasses and a fraction you used in Part A. How did you estimate the number of cups?
  (e.g., *10 and* $\frac{3}{4}$*: for 2 glasses, I needed* $1\frac{1}{2}$ *cups. So for 4 glasses, I needed 3 cups. So for 10 glasses, I'd need 6 cups plus some more, so I estimated 7 cups.*)
- Choose a different number and fraction. How did you model the problem?
  (e.g., *8 and* $\frac{2}{3}$*: I used a hexagon pattern block to represent each glass. It also represents 1 cup of juice, since each glass holds 1 cup. I put 2 rhombuses on each hexagon to represent* $\frac{2}{3}$ *cup of juice. Then I rearranged the rhombuses to cover 5 hexagons, with 1 rhombus on a sixth hexagon, which is* $5\frac{1}{3}$*.*)
- Choose a situation in Part B. Explain how you modelled the problem.
  (e.g., *9 and* $\frac{2}{3}$ *cup: I used 9 hexagon pattern blocks to represent 9 glasses and then started to cover them with sets of 2 rhombus blocks (each set of 2 blocks represented the* $\frac{2}{3}$ *cup of juice in a ladle). I kept going until I covered all 9 blocks. I used 13 sets of 2 rhombuses and 1 more rhombus, which is half a set of 2 rhombuses. So I used* $13\frac{1}{2}$ *ladles altogether.*)

Copyright © 2012 by Nelson Education Ltd.

Repeated Addition of Fractions

Pathway 1 GUIDED

Guided Intervention

Before Using the Guided Intervention

Provide fraction materials (e.g., pattern blocks). Ask:

- Suppose the hexagon block is a whole. What blocks would you use to show $\frac{2}{3}$ of a hexagon block?
  (e.g., *2 blue rhombus blocks*)
- How would you model $\frac{2}{3} + \frac{2}{3}$?
  (e.g., *I'd put 4 rhombuses on 2 hexagon blocks.*)
- What fraction of a hexagon do the 4 rhombuses cover? How do you know?
  ($\frac{4}{3}$*, or* $1\frac{1}{3}$. e.g., *The rhombuses cover 1 whole hexagon and* $\frac{1}{3}$ *of another.*)
- Why might you write what you just did as $\frac{2}{3} + \frac{2}{3}$ or as $2 \times \frac{2}{3}$?
  (e.g., *When you put 2 things together you're adding them or multiplying by 2.*)

Using the Guided Intervention **Student Resource pages 116–119**

Work through the instructional section of the student pages together. Students should model each strategy with materials.

Have students work through the **Try These** questions individually or in pairs.

Observe whether students

- can model the repeated addition of fractions in a problem situation (multiplication and division) (Questions 1, 3)
- can estimate sums of repeated fractions (Question 2)
- can explain their thinking about repeated addition of fractions (Questions 2, 3, 4, 8)
- can determine sums of repeated fractions (Questions 4, 5, 6, 7)

Consolidating and Reflecting

Ensure understanding by asking questions such as these based on students' work:

- How do you know the statement in Question 2a) is true?
  (e.g., *I imagined modelling* $\frac{3}{5}$ *with the fraction circles 8 times, which is 24 fifths. If I put 5 together to make wholes, I could make 4 wholes and there would still be a bit left—that's why 8 groups of* $\frac{3}{5}$ *is more than 4.*)
- How did you or could you model ■ $\times \frac{7}{8} = 7$ in Question 5b)?
  (e.g., *I put out 7 fraction circle wholes. I kept putting sets of* $\frac{7}{8}$ *on them until they were all full. I used 8 sets of* $\frac{7}{8}$ *altogether.*)
- How could you have predicted what you noticed in Question 7c)?
  (e.g., *If I cover 3 wholes with sets of* $\frac{3}{5}$*, it's like covering them with 3 sets of* $\frac{1}{5}$*, so however many* $\frac{3}{5}$*s it takes, it takes 3 times as many* $\frac{1}{5}$*s.*)

You will need

- fraction materials (e.g., Fraction Strips (BLM 10), Fraction Circles (BLM 18), pattern blocks, or other commercial materials)
- 2 cm Grid Paper (BLM 11)
- coloured counters
- Student Resource pages 116–119

Pattern blocks are useful for modelling halves, thirds, and sixths.

Note that the division of a whole number by a fraction (as modelled at the bottom of Student Resource page 116) is viewed through the lens of repeated addition of fractions.

Copyright © 2012 by Nelson Education Ltd.

Pathway 2 OPEN-ENDED

Adding and Subtracting Mixed Numbers

You will need

- 2 cm Grid Paper (BLM 11)
- coloured counters (e.g., red and blue)
- Student Resource pages 120–121

Adding $\frac{1}{5} + \frac{2}{3}$

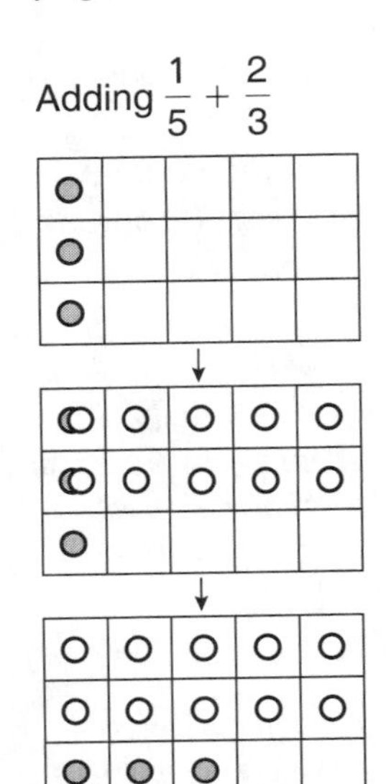

Open-Ended Intervention

Before Using the Open-Ended Intervention

Outline a 3-by-5 grid on large grid paper and fill 1 column with red counters (as shown in the margin). Ask:

- Why can you model both thirds and fifths on a 3-by-5 grid?
  (e.g., *You can show thirds by filling rows, and fifths by filling columns.*)
- How does this show $\frac{1}{5}$?
  (e.g., *1 of the 5 columns is filled.*)
- I want to add $\frac{2}{3}$ to my model for $\frac{1}{5}$ to show $\frac{1}{5} + \frac{2}{3}$. How do I do that?
  (e.g., *Fill 2 of the 3 rows with blue counters, move the 2 doubled-up counters to empty squares, and then see how much of the grid is filled. The sum is* $\frac{13}{15}$.)
- What would have happened if I had added $\frac{2}{5}$ and $\frac{2}{3}$ instead?
  (e.g., *All 15 squares in the grid would be filled and there'd be 1 extra counter to place in another 3-by-5 grid. The sum would be* $1\frac{1}{15}$.)

Using the Open-Ended Intervention (Student Resource pages 120–121)

Read through the tasks on the student pages together. Provide large grid paper and counters. Remind students that they can choose any 4 pairs of crops (of the 6 listed) in each chart. Give students time to work, ideally in pairs.

Observe whether students

- can model a fraction addition or subtraction
- can add and subtract mixed numbers
- can estimate sums and differences of mixed numbers
- relate the addition or subtraction of 2 mixed numbers to a problem situation

Consolidating and Reflecting

Ensure understanding by asking questions such as these based on students' work:

- Choose a crop pair. How did you estimate the sum of the rows?
  (e.g., $2\frac{1}{2} + 1\frac{3}{5}$*: 2 + 1 = 3 and* $\frac{3}{5}$ *is close to* $\frac{1}{2}$*, so* $\frac{1}{2} + \frac{1}{2}$ *is another 1, so about 4.*)
- Choose a different crop pair. How did you calculate the number of rows?
  (e.g., $3\frac{1}{4} + 1\frac{3}{5}$*: I knew 3 + 1 = 4 and used a 4-by-5 grid to add* $\frac{1}{4} + \frac{3}{5}$*. I filled 3 of 5 columns and 1 of 4 rows with counters and then moved the double counters to empty squares. The grid was* $\frac{17}{20}$ *full, so the sum was* $4\frac{17}{20}$.)
- Choose a crop pair. How did you estimate the difference?
  (e.g., $3\frac{1}{4} - 1\frac{3}{5}$*: I knew* $3\frac{3}{5} - 1\frac{3}{5} = 2$ *and, since* $3\frac{1}{4}$ *is less than* $3\frac{3}{5}$*, the difference was just less than 2, or about* $1\frac{3}{4}$.)
- Jessica noticed that when she used a grid model, she also created equivalent fractions with the same denominator. Do you agree? Explain.
  (*Yes.* e.g., *When you show fifths and thirds on the same 3-by-5 grid, you also show fifteenths.*)

Copyright © 2012 by Nelson Education Ltd.

Adding and Subtracting Mixed Numbers

Pathway 2 GUIDED

Guided Intervention

Before Using the Guided Intervention

Provide fraction circles and ask students to model $2\frac{1}{3}$. Ask:

- ▸ How does your model represent $2\frac{1}{3}$?
  (e.g., *It's 2 whole circles the same size and another $\frac{1}{3}$ of the same size circle.*)
- ▸ Why would it be easy to add 2 (wholes) to $2\frac{1}{3}$?
  (e.g., *You just add 2 and 2 to get 4 and keep the $\frac{1}{3}$, so you have $4\frac{1}{3}$ altogether.*)
- ▸ How would you take away $\frac{1}{6}$ from $2\frac{1}{3}$?
  (e.g., *I'd trade the 1 third piece for 2 sixth pieces, then I could take away 1 sixth.*)

Using the Guided Intervention (Student Resource pages 122–125)

Work through the instructional section of the student pages together. Provide large grid paper, counters, and other materials for modelling fractions.

Have students work through the **Try These** questions individually or in pairs.

Observe whether students

- relate a fraction model to an addition or subtraction (Questions 1, 2)
- can add and subtract mixed numbers (Questions 3, 4, 7)
- relate the addition or subtraction of mixed numbers to a problem situation (Question 5)
- can estimate the sum of mixed numbers (Question 6)
- recognize the advantage of using mixed numbers over improper fractions to estimate (Question 8)

Consolidating and Reflecting

Ensure understanding by asking questions such as these based on students' work:

- ▸ How did you use models for $2\frac{3}{5} + 3\frac{1}{8}$ in Question 3c)?
  (e.g., *I added 2 + 3 = 5 and made a 5-by-8 grid to model $\frac{3}{5} + \frac{1}{8}$. I filled 3 of 5 columns and 1 of 8 rows and then moved the double counters to empty squares. I counted how much of the grid was full and added it to 5.*)
- ▸ How did you decide on a problem for Question 5a)?
  (e.g., *I thought about situations where you take away fractions of things, like eating pieces of a pizza.*)
- ▸ How did you use estimation to answer Question 6?
  (e.g., *I knew it was going to be either 5 and a big proper fraction or 6 and a small proper fraction, so I looked for combinations of proper fractions that were close to 0 or close to 1.*)
- ▸ Jessica noticed that when she used a grid model, she also created equivalent fractions with the same denominator. Do you agree? Explain.
  (*Yes.* e.g., *When you show fifths and thirds on the same 3-by-5 grid, you also show fifteenths.*)

You will need

- fraction materials (e.g., Fraction Strips (BLM 10), Fraction Circles (BLM 18), pattern blocks, or other commercial materials)
- 2 cm Grid Paper (BLM 11)
- coloured counters
- Student Resource pages 122–125

Copyright © 2012 by Nelson Education Ltd.

Pathway 3 OPEN-ENDED

Subtracting Fractions

You will need

- fraction materials (e.g., Fraction Strips (BLM 10), Fraction Circles (BLM 18), or commercial materials)
- 2 cm Grid Paper (BLM 11)
- coloured counters (e.g., red and blue)
- Student Resource pages 126–127

Calculating $\frac{3}{5} - \frac{1}{3}$

○	○	○		
○	○	○		
○	○	○		

○	○	○	○	○

Open-Ended Intervention

Before Using the Open-Ended Intervention

Outline two 3-by-5 grids on large grid paper and fill 3 columns of one with red counters and 1 row of the other with blue counters (as shown in the margin). Ask:

- Why can you model both thirds and fifths on a 3-by-5 grid?
  (e.g., *You can show thirds by filling rows and fifths by filling columns.*)
- How does the grid with red counters show $\frac{3}{5}$?
  (e.g., *3 of 5 columns are full.*)
- How does the grid with blue counters show $\frac{1}{3}$?
  (e.g., *1 of 3 rows are full.*)
- How could you use these 2 grids to subtract $\frac{3}{5} - \frac{1}{3}$?
  (e.g., *I'd compare how many more squares are filled in the red-counter grid than the blue-counter grid. I can see that there are 4 more red counters than blue, so the difference is* $\frac{4}{15}$.)

Using the Open-Ended Intervention (Student Resource pages 126–127)

Read through the tasks on the student pages together. Provide fraction materials and large grid paper and counters. Note that the last task (at the bottom of Student Resource page 127) is asking students how much would be left if $\frac{1}{4}$ of the tank is used, not $\frac{1}{4}$ of what is in the tank. Give students time to work, ideally in pairs.

Observe whether students

- relate a fraction subtraction to a problem context
- can estimate fraction differences
- can model and solve a fraction subtraction
- can subtract proper and improper fractions

Consolidating and Reflecting

Ensure understanding by asking questions such as these based on students' work:

- Choose one of your fraction pairs. How did you estimate the difference?
  (e.g., $\frac{3}{8}$ *and* $\frac{1}{6}$: *I knew* $\frac{3}{8}$ *is about* $\frac{1}{3}$ *and* $\frac{1}{6}$ *is half of that, so the difference is about* $\frac{1}{6}$.)
- Choose a different pair of fractions. How did you use a model to determine the difference?
  (e.g., $\frac{2}{3}$ *and* $\frac{3}{4}$: *I modelled both using fraction circles and then figured out what to add to the* $\frac{2}{3}$ *model to make it as big as the* $\frac{3}{4}$ *model. When I added* $\frac{1}{12}$, *it worked.*)
- Why is subtracting when the denominators are different more complicated?
  (e.g., *If they're the same, you can just subtract the numerators, for example, 3 fifths − 1 fifth = 2 fifths. If they're different, the size of the fraction parts is different, so, for example, 3 fourths − 1 fifth is not 2 fourths or 2 fifths. It's hard to know what it is without using a model or equivalent fractions.*)

Copyright © 2012 by Nelson Education Ltd.

Subtracting Fractions

Pathway 3 GUIDED

Guided Intervention

Before Using the Guided Intervention

Provide fraction materials. Ask:

- How can you figure out how much more $\frac{3}{8}$ is than $\frac{1}{8}$, using a model? (e.g., *If you compare 3 eighth pieces and 1 eighth piece, you can see that $\frac{3}{8}$ is $\frac{2}{8}$ more.*)
- How might you do it without a model? (e.g., *I'd think "3 eighths − 2 eighths = 1 eighth."*)
- What about the subtraction $\frac{3}{8} - \frac{1}{3}$? (e.g., *I'd use a model, since I don't know how to compare fractions when they are different-sized parts of the whole.*)

Using the Guided Intervention (Student Resource pages 128–131)

Work through the instructional section of the student pages together. Provide fraction materials and large grid paper and counters for students to model the strategies. Note that Question 2 is asking students how much would be left if $\frac{1}{4}$ of a bowlful is eaten, not $\frac{1}{4}$ of what is in the bowl.

Have students work through the **Try These** questions individually or in pairs.

Observe whether students

- relate a model to a fraction subtraction (Questions 1, 2, 6)
- can subtract proper and improper fractions (Questions 2, 4, 6, 7, 10)
- can estimate fraction differences (Questions 3, 9)
- relate a fraction subtraction to a problem context (Questions 7, 8)
- recognize the efficiency of using a common denominator to subtract fractions (Question 9)

Consolidating and Reflecting

Ensure understanding by asking questions such as these based on students' work:

- How did you estimate $\frac{5}{6} - \frac{3}{4}$ in Question 3a)? (e.g., *I knew $\frac{3}{4}$ was close to $\frac{5}{6}$, since both are close to 1, so I knew the difference would be small.*)
- How did you calculate $\frac{13}{6} - \frac{3}{4}$ in Question 4c)? (e.g., *I knew $\frac{13}{6}$ is $2\frac{1}{6}$. I took $\frac{3}{4}$ from the 2 to get $1\frac{1}{4}$, so I had $1\frac{1}{4}$ and $\frac{1}{6}$ left. Since $\frac{1}{4} = \frac{3}{12}$ and $\frac{1}{6} = \frac{2}{12}$, $1\frac{1}{4} + \frac{1}{6} = 1\frac{5}{12}$.*)
- How could you use a model to check? (e.g., *$\frac{13}{6} - \frac{3}{4}$: I'd model $2\frac{1}{6}$ with fraction circles and then cover up $\frac{3}{4}$ of 1 circle. Then I'd see if $1\frac{5}{12}$ would describe the uncovered area.*)
- Why is subtracting when the denominators are different more complicated? (e.g., *If they're the same, you can just subtract the numerators, for example, 3 fifths − 1 fifth = 2 fifths. If they're different, the size of the fraction parts is different, so, for example, 3 fourths − 1 fifth is not 2 fourths or 2 fifths. It's hard to know what it is without using a model or equivalent fractions.*)

You will need

- fraction materials (e.g., Fraction Strips (BLM 10), Fraction Circles (BLM 18), or commercial materials)
- 2 cm Grid Paper (BLM 11)
- coloured counters
- Student Resource pages 128–131

Pathway 4 OPEN-ENDED

Adding Fractions

You will need

- fraction materials (e.g., Fraction Strips (BLM 10), Fraction Circles (BLM 18), or commercial materials)
- 2 cm Grid Paper (BLM 11)
- coloured counters (e.g., red and blue)
- Student Resource pages 132–133

Adding $\frac{1}{5} + \frac{2}{3}$

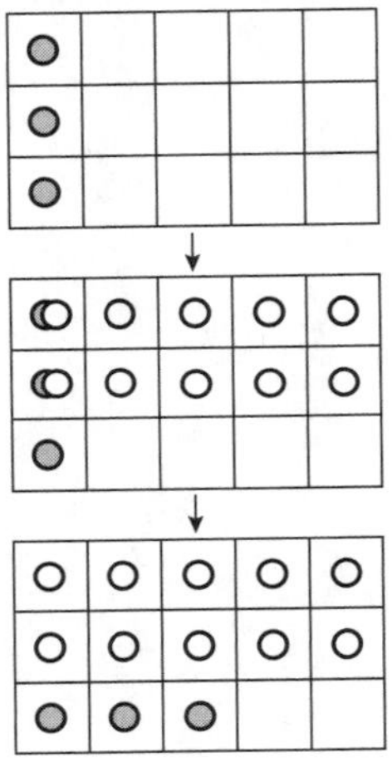

Open-Ended Intervention

Before Using the Open-Ended Intervention

Outline a 3-by-5 grid on large grid paper and fill 1 column with red counters (as shown in the margin). Ask:

- Why can you model both thirds and fifths on a 3-by-5 grid?
  (e.g., *You can show thirds by filling rows and fifths by filling columns.*)
- How does this show $\frac{1}{5}$? (e.g., *1 of the 5 columns is filled.*)
- I want to add $\frac{2}{3}$ to my model for $\frac{1}{5}$ to find the sum of $\frac{1}{5} + \frac{2}{3}$. How do I do that?
  (e.g., *Fill 2 of the 3 rows with blue counters, move the 2 doubled-up counters to empty squares, and then see how much of the grid is filled. The sum is* $\frac{13}{15}$.)
- What would have happened if I had added $\frac{2}{5} + \frac{2}{3}$ instead?
  (e.g., *All 15 squares in the grid would be filled and there'd be 1 extra counter to place in another 3-by-5 grid. The sum would be* $1\frac{1}{15}$.)

Using the Open-Ended Intervention (Student Resource pages 132–133)

Read through the tasks on the student pages together. Provide fraction materials, and grid paper and counters. Give students time to work, ideally in pairs.

Observe whether students

- relate a fraction addition to a problem context
- can estimate fraction sums
- can model and solve a fraction addition
- can add two fractions

Consolidating and Reflecting

Ensure understanding by asking questions such as these based on students' work:

- How did you decide if the sum was greater than 1?
  (e.g., *If both fractions were more than half, or if one was close to 1 and the other wasn't too small, I knew the sum would be greater.*)
- Choose one of your fraction pairs. Describe how you estimated the sum.
  (e.g., $\frac{3}{4}$ *and* $\frac{5}{9}$: $\frac{5}{9}$ *was a bit more than* $\frac{1}{2}$, *so I added* $\frac{3}{4}$ *and* $\frac{1}{2}$ *to get* $1\frac{1}{4}$ *and then added a bit more to get* $1\frac{1}{3}$.)
- Choose a different pair of fractions. Describe how you used a model to determine the sum.
  (e.g., $\frac{2}{3}$ *and* $\frac{3}{4}$: *I filled 2 of 3 rows and 3 of 4 columns of a 3-by-4 grid with counters. I moved the double counters to empty squares, but there were 5 extra counters. I put the 5 extra counters on a second empty 3-by-4 grid. Altogether, I filled 17 squares and, since each square is* $\frac{1}{12}$, *that's* $\frac{17}{12}$, *or* $1\frac{5}{12}$.)
- Jessica noticed that, when she used a grid model, she also created equivalent fractions with the same denominator. Do you agree? Explain.
  (*Yes.* e.g., *When fifths and thirds are on the same 3-by-5 grid, it shows fifteenths.*)

 Copyright © 2012 by Nelson Education Ltd.

Adding Fractions

Pathway 4 GUIDED

Guided Intervention

Before Using the Guided Intervention

Provide fraction materials. Ask:

- Why might someone think it's easy to add $\frac{3}{8} + \frac{3}{8}$?
  (e.g., *3 eighths + 3 eighths = 6 eighths*)
- How would you add $\frac{3}{8}$ and $\frac{3}{4}$, using a model?
  (e.g., *I'd use fraction circles. I'd put 3 eighth pieces together with 3 fourth pieces to make 1 whole circle and 1 more eighth.*)
- Why could you have predicted that $\frac{3}{8} + \frac{3}{4}$ would be more than 1?
  (e.g., *If you start with* $\frac{3}{4}$*, you only need* $\frac{1}{4}$ *to make 1, and* $\frac{3}{8}$ *is more than* $\frac{1}{4}$*.*)

Using the Guided Intervention (Student Resource pages 134–137)

Work through the instructional section of the student pages together. Provide fraction materials, and grid paper and counters for students to follow along.

Have students work through the **Try These** questions individually or in pairs.

Observe whether students

- relate a model to a fraction addition (Questions 1, 2, 3)
- relate a fraction addition to a problem situation (Questions 1, 9)
- can estimate fraction sums (Questions 4, 5, 10)
- can add fractions (Questions 6, 7, 8)
- recognize the efficiency of using a common denominator to add fractions (Question 11)

Consolidating and Reflecting

Ensure understanding by asking questions such as these based on students' work:

- How did you decide if the sum was greater than 1 for $\frac{5}{6} + \frac{1}{3}$ in Question 4b)?
  (e.g., *If you add* $\frac{1}{6}$ *to* $\frac{5}{6}$*, you get 1, but* $\frac{1}{3}$ *is more than* $\frac{1}{6}$*, so* $\frac{5}{6} + \frac{1}{3}$ *is more than 1.*)
- Jessica noticed that, when she used a grid model in Question 6d), she also created equivalent fractions with the same denominator. Do you agree? Explain.
  (*Yes;* e.g., *When you show fifths and thirds on the same 3-by-5 grid, you also show fifteenths.*)
- Explain how you did or could use a model for $\frac{3}{8} + \frac{3}{4}$ in Question 7a).
  (e.g., *I modelled eighths and fourths with fraction strips. I put the* $\frac{3}{8}$ *strip and* $\frac{3}{4}$ *strip end to end, and then I compared the length to 2 whole strips in eighths. The* $\frac{3}{8}$ *strip and* $\frac{3}{4}$ *strip end to end were the same length as 1 whole strip and another* $\frac{1}{8}$*, or* $1\frac{1}{8}$*.*)
- How did you decide what to say for Question 10b)?
  (e.g., *If you add* $\frac{1}{3}$ *to 1, you get* $1\frac{1}{3}$*. Since* $\frac{7}{8}$ *is less than 1, then* $\frac{7}{8} + \frac{1}{3}$ *is less than* $1\frac{1}{3}$*, or about* $1\frac{1}{4}$*.*)

You will need

- fraction materials (e.g., Fraction Strips (BLM 10), Fraction Circles (BLM 18), pattern blocks, or other commercial materials)
- 2 cm Grid Paper (BLM 11)
- coloured counters
- Student Resource pages 134–137

Strand: Number

Rates, Percents, and Ratios

Planning For This Topic

Materials for assisting students with rates, ratios, and percents consist of a diagnostic tool and 3 intervention pathways. Pathway 1 focuses on rates, Pathway 2 on percents, and Pathway 3 on ratios.

Note that some mathematicians do not distinguish between ratios and rates. Others do, and the distinction is made in these materials. There might be debates over whether something like number of cups of sugar per cookie is a rate or not. If students raise the question, be aware that there are arguments both ways.

Each pathway has an open-ended intervention and a guided intervention. Choose the type of intervention more suitable for your students' needs and your particular circumstances.

Curriculum Connections

Grades 5 to 8 curriculum connections for this topic are provided online. See www.nelson.com/leapsandbounds. Pathway 1 may not be required by those following the WNCP curriculum, since rate is a Grade 8 topic. Pathways 2 and 3 are relevant for the WNCP and the Ontario curriculums.

Why might students struggle?

Students might struggle with rate, percent, and ratio for any of the following reasons:

- They might be more comfortable comparing numbers additively than multiplicatively (e.g., they think of 12 as 8 more than 4, rather than as 3×4).
- They might have difficulty creating equivalent ratios, particularly when the respective terms are not whole-number multiples (e.g., they might be comfortable with 2:3 and 4:6 being equivalent but not with 4:6 and 6:9).
- They might not realize that, to solve a ratio, percent, or rate problem, they often need to create an equivalent ratio or rate.
- They might have difficulty if the multiplicative relationship is not a whole number relationship (e.g., calculating the price of 8 items if the price of 3 is known).
- They might struggle with fractions, making it difficult to solve many rate and ratio problems.
- They might be more successful calculating a percent of a whole than calculating the whole when the percent is known (e.g., they are more comfortable calculating 23% of 80 than calculating the whole if 18.4 is 23%).

Professional Learning Connections

PRIME: Number and Operations, Background and Strategies (Nelson Education Ltd., 2005), pages 131–132

Making Math Meaningful to Canadian Students K–8 (Nelson Education Ltd., 2008), pages 249–264

Big Ideas from Dr. Small Grades 4–8 (Nelson Education Ltd., 2009), pages 74–83

Good Questions (dist. by Nelson Education Ltd., 2009), pages 32, 34, 37, 52–53, 57

More Good Questions (dist. by Nelson Education Ltd., 2010), page 78

Copyright © 2012 by Nelson Education Ltd.

Diagnostic Tool: Rates, Percents, and Ratios

Use the diagnostic tool to determine the most suitable intervention pathway for rates, percents, and ratios. Provide Diagnostic Tool: Rates, Percents, and Ratios, Teacher's Resource pages 108 and 109, and have students complete it in writing or orally. Provide calculators for students to use if they wish.

See solutions on Teacher's Resource pages 110 and 111.

Intervention Pathways

The purpose of the intervention pathways is to help students solve proportion problems. Not only is this important for everyday life, but it supports further mathematics study in linear relations.

There are 3 pathways:

- Pathway 1: Using Rates
- Pathway 2: Using Percents
- Pathway 3: Using Ratios

Use the chart below (or the Key to Pathways on Teacher's Resource pages 110 and 111) to determine which pathway is most suitable for each student or group of students.

Diagnostic Tool Results	Intervention Pathway
If students struggle with Questions 8 to 11	use Pathway 1: Using Rates *Teacher's Resource pages 112–113* *Student Resource pages 138–143*
If students struggle with Questions 4 to 7	use Pathway 2: Using Percents *Teacher's Resource pages 114–115* *Student Resource pages 144–149*
If students struggle with Questions 1 to 3	use Pathway 3: Using Ratios *Teacher's Resource pages 116–117* *Student Resource pages 150–155*

If students successfully complete Pathway 3 (or 2), they may or may not need the additional intervention provided by Pathway 2 (or 1). Either re-administer Pathway 2 (or 1) questions from the diagnostic tool or encourage students to do a portion of the open-ended intervention for Pathway 2 (or 1) to decide if more work in that pathway would be beneficial.

Name:__________ Date:__________

Rates, Percents, and Ratios

Diagnostic Tool

You will need

- a calculator (optional)

1. **a)** What is the ratio of the number of triangles to the number of circles? ___:___

   b) What is the ratio of the number of triangles to the number of shapes? ___:___

2. The ratio of the number of adults to the number of students on a class trip is 14 to 122.

   a) What is the ratio of the number of students to the number of people in the group? _____:_____

   b) Why is 7 : 61 another way to write the ratio 14 : 122?

   ____

   ____

   c) On another class trip, there were the same number of students but each adult supervised more students. Would the ratio of students to the people in the group more likely be 12 : 122 or 16 : 122? Explain your thinking.

   ____

   ____

3. Fill in the blanks to make equivalent ratios.

   a) 3:8 = 9: _____

   b) 4:12 = 7: _____

   c) 7:9 = _____ :36

   d) 12:30 = 10: _____

4. What percent of each grid is shaded?

   a) _____

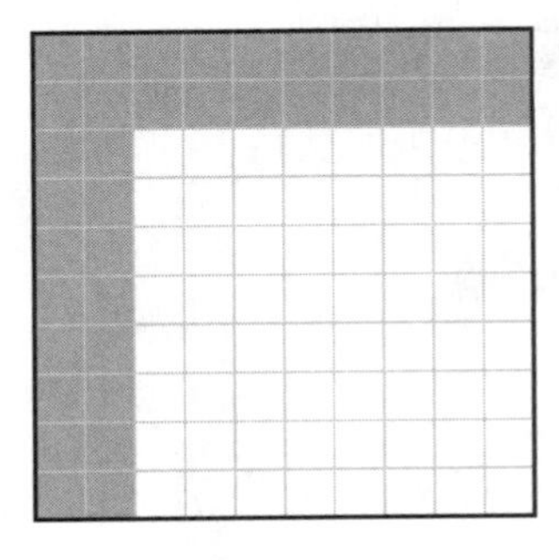

   b) _____

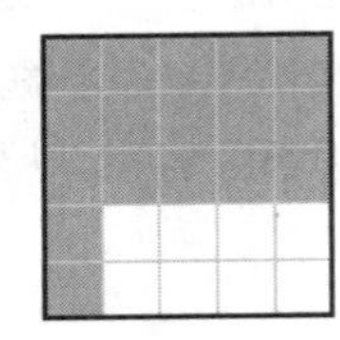

Copyright © 2012 by Nelson Education Ltd.

Name:______________________________ Date:______________

5. Complete each statement *without* using a calculator.

a) 25% of 200 = ________.

b) 50% of 480 = ________.

c) 75% of 800 = ________.

d) 10% of 48 = ________.

6. Mette's meal cost $20 and he left a 15% tip.
How much was the tip?

7. Ellie withdrew $60 from her bank account. It was 40% of what she had in the account. How much did she have in the account to start with?

8. If 5 hamburgers cost $9.45, how much would 3 hamburgers cost?

9. Which ice cream is more expensive per litre? How do you know?

3 L of ice cream for $10 or 4 L of ice cream for $12

10. Complete each statement to make an equivalent rate.

a) 3 boxes for $8 = 6 boxes for $ ____

b) 72 km in 1 h = ____ km in 1.5 h

c) 28 points in 4 games = ____ points in 3 games

11. Marie can walk 2 km in 21 minutes.

a) How long would it take her to walk 1 km? ______________

b) About how far could she walk in 1 min? ______________

Copyright © 2012 by Nelson Education Ltd.

Solutions and Key to Pathways

Name:____________________ Date:____________________

Rates, Percents, and Ratios

Diagnostic Tool

You will need

- a calculator (optional)

Pathway 3

1. a) What is the ratio of the number of triangles to the number of circles? 4 : 5

   b) What is the ratio of the number of triangles to the number of shapes? 4 : 9

2. The ratio of the number of adults to the number of students on a class trip is 14 to 122.

   a) What is the ratio of the number of students to the number of people in the group? 122 : 136

   b) Why is 7:61 another way to write the ratio 14:122?

   e.g., If there are 14 adults for 122 students, then half the adults would supervise half the students and that's 7:61.

   c) On another class trip, there were the same number of students but each adult supervised more students. Would the ratio of students to the people in the group more likely be 12:122 or 16:122? Explain your thinking.

   12:122. e.g., There would be fewer adults for the same number of students.

3. Fill in the blanks to make equivalent ratios.

   a) 3:8 = 9: 24

   b) 4:12 = 7: 21

   c) 7:9 = 28 :36

   d) 12:30 = 10: 25

Pathway 2

4. What percent of each grid is shaded?

   a) 36%

   b) 68%

 Copyright © 2012 by Nelson Education Ltd.

 Copyright © 2012 by Nelson Education Ltd.

Pathway 2

Pathway 1

Name:______________________ Date:______________

5. Complete each statement *without* using a calculator.

a) 25% of 200 = 50.

b) 50% of 480 = 240.

c) 75% of 800 = 600.

d) 10% of 48 = 4.8.

6. Mette's meal cost $20 and he left a 15% tip. How much was the tip?

$3
e.g., $0.15 \times 20 = 3$

7. Ellie withdrew $60 from her bank account. It was 40% of what she had in the account. How much did she have in the account to start with?

$150
e.g., 40% of ? = $60
$60 \div 0.4 = 150$

8. If 5 hamburgers cost $9.45, how much would 3 hamburgers cost?

$5.67
e.g., $9.45 ÷ 5 = $1.89
$1.89 × 3 = $5.67

9. Which ice cream is more expensive per litre? How do you know?

3 L of ice cream for $10 or 4 L of ice cream for $12

3 L for $10. e.g., 3 L for $10 is about $1 L for $3.33, which is more than 4 L for $12, which is 1 L for $3. With 4 L for $12, you get 1 extra litre for just $2.

10. Complete each statement to make an equivalent rate.

a) 3 boxes for $8 = 6 boxes for $ 16

b) 72 km in 1 h = 108 km in 1.5 h

c) 28 points in 4 games = 21 points in 3 games

11. Marie can walk 2 km in 21 minutes.

a) How long would it take her to walk 1 km? 10.5 min

b) About how far could she walk in 1 min? almost 0.1 km

Copyright © 2012 by Nelson Education Ltd.

Copyright © 2012 by Nelson Education Ltd.

Pathway 1 OPEN-ENDED

Using Rates

You will need

- calculators (optional)
- Student Resource pages 138–139

Open-Ended Intervention

Before Using the Open-Ended Intervention

Tell students that there is a world record set by someone who ran 100 m in a little more than 15 s while skipping rope. Ask:

- Suppose the runner kept the same pace the whole time. How long do you think it took to run 50 m?
  (e.g., *around 7.5 s*)
- Why might someone call 50 m in 7.5 s an equivalent rate to 100 m in 15 s?
  (e.g., *The rates say the same thing—if you double the distance, you double the time.*)
- Do you think there is enough information to estimate how far the runner could go in other numbers of seconds (assuming the pace stays the same)?
  (*Yes.* e.g., *You could figure out how far the runner went in 1 s and then use that information for other numbers of seconds.*)

Using the Open-Ended Intervention **Student Resource pages 138–139**

Read through the tasks on the student pages together. Allow calculators. Give students time to work, ideally in pairs.

Observe whether students

- can create equivalent rates
- relate real-life situations to the use of rates
- can solve rate problems
- can explain their thinking when solving rate problems

Consolidating and Reflecting

Ensure understanding by asking questions such as these based on students' work:

- Choose one of the rate situations. What makes it a rate?
  (e.g., *Telling how much ice cream costs for a certain amount of money is a rate because it compares an amount of ice cream to an amount of money.*)
- How did you create equivalent rates?
  (e.g., *I usually multiplied both numbers by 2, but if I needed a unit rate, I had to divide to get a 1 for one of the units.*)
- Choose a problem. How did you use equivalent rates to solve it?
  (e.g., *How long would it take for the elephant's heart to beat 10 000 times? The given rate was 140 times in 4 min, so I figured out that there would be 9800 beats in 280 min, which is an equivalent rate.*)
- Think of your own examples and the ones that were given. What do you think are the most common uses for rate?
  (e.g., *speed and prices*)

 Copyright © 2012 by Nelson Education Ltd.

Using Rates

Pathway 1 GUIDED

Guided Intervention

You will need

- calculators (optional)
- Student Resource pages 140–143

Before Using the Guided Intervention

Tell students that there is a world record set by someone who ran 100 m in a little more than 15 s while skipping rope. Ask:

- Suppose the runner kept the same pace the whole time. How long do you think it took to run 50 m? (e.g., *around 7.5 s*)
- Why might someone call 50 m in 7.5 s an equivalent rate to 100 m in 15 s? (e.g., *The rates say the same thing—if you double the distance, you double the time.*)
- Do you think there is enough information to estimate how far the runner could go in other numbers of seconds (assuming the pace stays the same)? (*Yes.* e.g., *You could figure out how far the runner went in 1 s and then use that information for other numbers of seconds.*)

Using the Guided Intervention **Student Resource pages 140–143**

Work through the instructional section of the student pages together. Have students create each equivalent rate and work through the strategies. Allow calculators. In the section on solving rate problems, if students have difficulty working with 18 bars, suggest that they start by solving an easier problem using 16 bars.

Have students work through the **Try These** questions individually or in pairs.

Observe whether students

- recognize and can create equivalent rates (Questions 1, 2)
- relate real-life situations to the use of rates (Questions 1 to 10)
- can solve rate problems (Questions 3 to 9)
- can explain their thinking when solving rate problems (Questions 5 to 8)

Consolidating and Reflecting

Ensure understanding by asking questions such as these based on students' work:

- How did you create an equivalent rate in Question 2c)? (e.g., *If there were 81 grams for 30 cm³, there would be 27 grams for 10 cm³ [an equivalent rate] and 54 grams for 20 cm³ [another equivalent rate]*).
- Why is the information in Question 3 a rate? (e.g., *It's about how much of something happens in a certain amount of time, so it's a speed rate.*)
- How did you solve Question 5? (e.g., *I divided 1000 by 280, since that would tell me how many groups of 280 there are in 1000. That's what I needed to know to figure out the number of minutes.*)
- Think of the examples of rates that were given and others you might think of yourself. What do you think are the most common uses for rates? (e.g., *speed and prices*)

Using Percents

You will need

- Hundredths Grids (BLM 8)
- calculators (optional)
- Student Resource pages 144–145

Open-Ended Intervention

Before Using the Open-Ended Intervention

Show students a hundredths grid. Ask:

- What does 50% mean? How do you know?
  (e.g., *half; It means 50 out of 100, and 50 is half of 100.*)
- How many squares are in 20% of the grid? How do you know?
  (*20.* e.g., *20% means 20 out of 100.*)
- What percent would be most of the grid?
  (e.g., *a percent in the 90s*)
- Suppose the whole grid was worth 200. What would 10% be worth? Explain.
  (*20.* e.g., *Each square has a value of 2, so 10%, or 10 squares, would be worth 20.*)
- How would you use the grid to figure out 80% of 300?
  (e.g., *If the whole grid is worth 300, each square is worth 3, so 80 squares are worth 240.*)

Using the Open-Ended Intervention **Student Resource pages 144–145**

Read through the tasks on the student pages together. Provide hundredths grids and allow calculators. Give students time to work, ideally in pairs.

Observe whether students

- relate percents, fractions, and decimals
- can create real-life situations that use percent
- can solve percent problems by estimating or calculating
- can explain their thinking when solving percent problems

Consolidating and Reflecting

Ensure understanding by asking questions such as these based on students' work:

- Choose one of the statements you used. How did you decide if it was true?
  (e.g., *For the first one, I changed fractions to percents.*)
- Is there another way you could have done it?
  (e.g., *I could have changed 35% to a fraction to see if* $\frac{1}{4} + \frac{1}{10} = \frac{35}{100}$, *or I could have tried some examples.*)
- Why can you calculate 85% of a number by multiplying by 0.85?
  (e.g., $85\% = \frac{85}{100} = 0.85$, *and you are taking 0.85 of the whole, so you multiply.*)
- What would be different if you were figuring out what number 170 is 85% of?
  (e.g., *I wouldn't multiply by 0.85, since I don't know the number I am supposed to be multiplying.*)
- Why was your untrue statement not true?
  (e.g., *My untrue statement was about taking 50% off a price that is already 50% off, meaning that it is free. You are taking 50% of different numbers if you do it in parts, but you are taking percents of the same number if you do it all at once.*)

 Copyright © 2012 by Nelson Education Ltd.

Using Percents

Pathway 2 GUIDED

Guided Intervention

Before Using the Guided Intervention

Show students a hundredths grid. Ask:

- What does 50% mean? How do you know?
  (e.g., *half; It means 50 out of 100, and 50 is half of 100.*)
- How many squares are in 20% of the grid? How do you know?
  (*20.* e.g., *20% means 20 out of 100.*)
- What percent would be most of the grid?
  (e.g., *a percent in the 90s*)
- Suppose the whole grid was worth 200. What would 10% be worth? Explain.
  (*20.* e.g., *Each square has a value of 2, so 10%, or 10 squares, would be worth 20.*)

You will need

- Hundredths Grids (BLM 8)
- calculators (optional)
- Student Resource pages 146–149

Using the Guided Intervention **Student Resource pages 146–149**

Work through the instructional section of the student pages together. Provide hundredths grids and have students model and/or work through each strategy.

Have students work through the **Try These** questions individually or in pairs. Allow calculators.

Observe whether students

- can interpret models of percents (Question 1)
- relate percents to real-life contexts (Question 2)
- can explain their thinking when solving percent problems (Questions 2, 3, 6, 7, 8, 9)
- relate percents to equivalent ratios, fractions, and decimals (Questions 3, 4, 5)
- can solve percent problems (Questions 6, 7, 8)

Consolidating and Reflecting

Ensure understanding by asking questions such as these based on students' work:

- In Question 3b), which benchmark did you use to estimate 30% of 418? Why?
  (e.g., *I used 10% and then I multiplied by 3.*)
- Is there another benchmark you could have used?
  (e.g., *I could have used 25%, since 25 and 30 are close.*)
- Why is it easy to calculate 10%?
  (e.g., *It's easy to divide by 10—you just move the digits over. 10% of 418 is 41.8.*)
- How did you solve Question 7a)?
  (e.g., *I knew 60% was 3 h. That means 20% is 1 h. Since 100% = 5 × 20%, it's 5 h.*)
- How can you use a fraction to check if your answer to Question 8 makes sense?
  (e.g., *I know that 35% is about $\frac{1}{3}$, and $\frac{1}{3}$ of 24 h is 8 h.*)

Pathway 3
OPEN-ENDED

Using Ratios

You will need

- red and green pattern blocks
- Student Resource pages 150–151

Open-Ended Intervention

Before Using the Open-Ended Intervention

Provide red and green pattern blocks and ask students to make a pattern or design that uses 3 red blocks for every 2 green ones. Ask:

- How many red and green blocks could you use?
  (e.g., *3 and 2, 6 and 4, 9 and 6*)
- How do you know 6:4 is equivalent to 3:2?
  (e.g., *If there are 6 of one item for every 4 of another, you can match 3 of the first items with 2 of the other items.*)
- What did you notice about those numbers?
  (e.g., *They were all groups of 3 for the red blocks and groups of 2 for the green ones.*)
- What is the ratio of red blocks to total blocks?
  (e.g., *9 to 15*)

Using the Open-Ended Intervention Student Resource pages 150–151

Read through the tasks on the student pages together.

Give students time to work, ideally in pairs.

Observe whether students
- can create equivalent ratios
- relate ratios to real-life situations
- recognize that there are different ratios that describe the same situation
- can solve ratio problems
- can explain their thinking when solving ratio problems

Consolidating and Reflecting

Ensure understanding by asking questions such as these based on students' work:

- Choose one of the ratio situations. What makes it a ratio?
  (e.g., *If you are comparing the number of boys to girls who are born, it is a comparison and the units are the same: number of people.*)
- How did you create equivalent ratios?
  (e.g., *I usually multiplied both numbers by a number like 2 or 3.*)
- Choose a problem you created. How did you use equivalent ratios to solve it?
  (e.g., *For my second paint problem, I used equivalent ratios when I changed 2:1.5 to 4:3 and then to 16:12.*)
- Think of your own examples and the ones that were given. What do you think are the most common uses of ratios?
  (e.g., *They are used to compare boys to girls or seniors to young people or in other ways to sort people. They are also used a lot in recipes.*)

Copyright © 2012 by Nelson Education Ltd.

Using Ratios

Pathway 3 GUIDED

Guided Intervention

You will need

- red and green pattern blocks
- coloured tiles or counters
- calculators (optional)
- coloured pencils
- Student Resource pages 152–155

Before Using the Guided Intervention

Provide red and green pattern blocks and ask students to make a pattern or design that uses 3 red blocks for every 2 green ones. Ask:

- How many red and green blocks could you use? (e.g., *3 and 2, 6 and 4, 9 and 6*)
- How do you know 6 : 4 is equivalent to 3 : 2?
  (e.g., *If there are 6 of one item for every 4 of another, you can match 3 of the first items with 2 of the other items.*)
- What did you notice about those numbers?
  (e.g., *They were all groups of 3 for the red blocks and groups of 2 for the green ones.*)
- What is the ratio of red blocks to total blocks? (e.g., *9 to 15*)

Using the Guided Intervention **Student Resource pages 152–155**

Work through the instructional section of the student pages together. Provide tiles or counters for students to model each ratio.

Have students work through the **Try These** questions individually or in pairs. Provide coloured pencils and have calculators available.

Observe whether students

- can model a given ratio (Question 1)
- recognize that there are different ratios that describe the same situation (Questions 2, 9)
- can compare ratios (Question 3)
- relate real-life situations to the use of ratios (Questions 3, 6, 7)
- can create equivalent ratios (Questions 4, 5)
- can solve ratio problems (Questions 6, 7, 8)

Consolidating and Reflecting

Ensure understanding by asking questions such as these based on students' work:

- How did you create an equivalent ratio in Question 5d)?
  (e.g., *I know 6 is $\frac{3}{4}$ of 8, so I knew that I needed a number that was $\frac{3}{4}$ of 36.*)
- How did you solve Question 6?
  (e.g., *6 blue to 4 yellow is the same ratio as 3 blue to 2 yellow. Since there were 18 yellow cans used and 18 is nine 2s, I knew I needed nine 3s, or 27 blue cans.*)
- Why is the information in Question 8 a ratio?
  (e.g., *It compares 2 units or measurements that are both lengths, so they are the same type.*)
- Think of the examples of ratios that were given and others you might think of yourself. What do you think is the most common use of ratios?
  (e.g., *They are used to compare boys to girls or seniors to young people or in other ways to sort people. They are also used a lot in recipes.*)

Copyright © 2012 by Nelson Education Ltd.

Multiplicative Relationships

Planning For This Topic

Materials for assisting students with multiplicative relationships consist of a diagnostic tool and 3 intervention pathways. Pathway 1 focuses on divisibility rules. Pathway 2 involves classifying numbers as primes, composites, squares, or square roots. Pathway 3 focuses on factors and multiples.

Each pathway has an open-ended intervention and a guided intervention. Choose the type of intervention more suitable for your students' needs and your particular circumstances.

Curriculum Connections

Grades 5 to 8 curriculum connections for this topic are provided online. See www.nelson.com/leapsandbounds. The topic of divisibility rules is not required in the Ontario curriculum, so Ontario teachers may choose not to use Pathway 1. The topic of squares and square roots is not required in the WNCP curriculum until Grade 8, so using Pathway 2 for remediation applies to Ontario students only. Pathway 3 is pertinent to both curriculums.

Why might students struggle with multiplicative relationships?

Students might struggle with multiplicative relationships for any of the following reasons:

- They might not know their multiplication facts.
- They might mix up the terms *factor* and *multiple*.
- They might not consider that one number can be a factor of, or a multiple of, a lot of different numbers.
- They might not realize that any statement about one number being a factor of another number automatically implies a statement about one number being a multiple of another (e.g., 3 is a factor of 12, so 12 is a multiple of 3).
- They might not recognize the relationship between various factors and multiples (e.g., if 6 is a factor of a number, so is 3; if 20 is a multiple of a number, so is 40).
- They might assume that all numbers ending in 1, 3, 7, or 9 are prime.
- They might not realize that you cannot always be sure a number is prime until you check for divisibility by quite a few factors. However, you can tell whether it is a composite if it has just one factor other than 1 and the number itself.
- They might be unaware that there are quick ways to check for divisibility.
- They might assume that a divisibility rule that works for one factor is similar to one that works for all factors (e.g., they might assume that, as a number is divisible by 3 if the sum of its digits is divisible by 3, then a number is divisible by 5 if the sum of its digits is divisible by 5).

Professional Learning Connections

PRIME: Number and Operations, Background and Strategies (Nelson Education Ltd., 2005), pages 98–102

Making Math Meaningful to Canadian Students K–8 (Nelson Education Ltd., 2008), pages 148–156

Big Ideas from Dr. Small Grades 4–8 (Nelson Education Ltd., 2009), pages 20–24

Good Questions (dist. by Nelson Education Ltd., 2009), pages 33, 35, 53, 55

Copyright © 2012 by Nelson Education Ltd.

Diagnostic Tool: Multiplicative Relationships

Use the diagnostic tool to determine the most suitable intervention pathway for working with multiplicative relationships. Provide Diagnostic Tool: Multiplicative Relationships, Teacher's Resource pages 120 and 121, and have students complete it in writing or orally. Provide square tiles or counters.

See solutions on Teacher's Resource pages 122 and 123.

Intervention Pathways

The purpose of the intervention pathways is to help students recognize what factors, multiples, square numbers, and square roots are, in part to prepare them for later work with algebraic expressions and also to help with simplifying fractions. The topics of prime numbers, composite numbers, and divisibility rules are also introduced, even though those particular topics are not further developed in curriculum outcomes until later grades.

There are 3 pathways:

- Pathway 1: Divisibility Rules
- Pathway 2: Prime Numbers and Perfect Squares
- Pathway 3: Factors and Multiples

Use the chart below (or the Key to Pathways on Teacher's Resource pages 122 and 123) to determine which pathway is most suitable for each student or group of students.

Diagnostic Tool Results	Intervention Pathway
If students struggle with Questions 8 to 9	use Pathway 1: Divisibility Rules *Teacher's Resource pages 124–125* *Student Resource pages 156–161*
If students struggle with Questions 5 to 7	use Pathway 2: Prime Numbers and Perfect Squares *Teacher's Resource pages 126–127* *Student Resource pages 162–166*
If students struggle with Questions 1 to 4	use Pathway 3: Factors and Multiples *Teacher's Resource pages 128–129* *Student Resource pages 167–172*

Name: ______________________ Date: ______________

Multiplicative Relationships

Diagnostic Tool

You will need

- square tiles or counters

1. Create a model using counters or tiles to show that 12 is a multiple of 3. Sketch and explain your model.

2. List 4 factors of each number.

a) 80 ____ ____ ____ ____

b) 189 ____ ____ ____ ____

c) 492 ____ ____ ____ ____

d) 392 ____ ____ ____ ____

3. List 4 multiples of each number.

a) 2 ____ ____ ____ ____

b) 5 ____ ____ ____ ____

c) 21 ____ ____ ____ ____

d) 13 ____ ____ ____ ____

4. A number is a multiple of 8. Describe 2 or more other things you know about the number.

____

____

5. Show or explain how you know that 7 is prime but 8 is not.

Copyright © 2012 by Nelson Education Ltd.

Name:______________________________ Date:__________________

6. Create a model that shows that 10 is *not* a prime number. Sketch and explain your model.

7. Show or explain how you know that 36 is a perfect square.

8. How can you tell that each statement is true *without dividing*?

a) 418 is divisible by 2.

____

____

____

b) 513 is divisible by 3 and 9, but not 6.

____

____

____

9. Show or explain why a number that is divisible by 6 is also divisible by 2.

Copyright © 2012 by Nelson Education Ltd.

Solutions and Key to Pathways

Name: ______________________ Date: ______________

Multiplicative Relationships

Diagnostic Tool

You will need

- square tiles or counters

Pathway 3

1. Create a model using counters or tiles to show that 12 is a multiple of 3. Sketch and explain your model.

   e.g., 12 makes 4 groups of 3 with 0 extra counters, so 12 is a multiple of 3.

2. List 4 factors of each number.

a) 80	10	5	2	1
b) 189	9	3	1	189
c) 492	2	4	1	492
d) 392	1	4	2	196

3. List 4 multiples of each number.

a) 2	2	4	6	8
b) 5	10	20	40	80
c) 21	42	63	210	420
d) 13	26	39	52	520

4. A number is a multiple of 8. Describe 2 or more other things you know about the number.

   e.g., It is even, a multiple of 4, and has 2 as a factor.

Pathway 2

5. Show or explain how you know that 7 is prime but 8 is not.

   e.g., 7 has only 2 factors (7 and 1), but 8 has 4 (8, 1, 2 and 4).

120 Multiplicative Relationships, Diagnostic Tool *Leaps and Bounds* Copyright © 2012 by Nelson Education Ltd.

 Copyright © 2012 by Nelson Education Ltd.

Name:______________________ Date:______________

Pathway 2

6. Create a model that shows that 10 is *not* a prime number. Sketch and explain your model.

e.g., I can make more than 1 rectangle with 10 tiles. The rectangles have different lengths and widths:

7. Show or explain how you know that 36 is a perfect square.

e.g., I'd make a 6 × 6 square out of 36 tiles.

Pathway 1

8. How can you tell that each statement is true *without dividing*?

a) 418 is divisible by 2.

e.g., 418 has a ones digit of 8, which is even.

b) 513 is divisible by 3 and 9, but not 6.

e.g., 5 + 1 + 3 = 9, and 9 ÷ 3 = 3, so 513 is a multiple of 3 and 9.

513 is not even, so it's not a multiple of 6.

9. Show or explain why a number that is divisible by 6 is also divisible by 2.

e.g., If a number can be grouped in 6s, then you can split each group into 3 groups of 2, which makes it divisible by 2.

Pathway 1 OPEN-ENDED

Divisibility Rules

You will need

- counters
- base ten blocks (optional)
- calculators
- Student Resource pages 156–157

Open-Ended Intervention

Before Using the Open-Ended Intervention

Write the numeral 34 281 and cover the 4281 part. Ask:

- Is this five-digit number odd?
  (e.g., *I can't tell, since I can't see the ones digit.*)
- The first digit, 3, is odd. Why doesn't that matter?
  (e.g., *First of all, that 3 means 30 000, which is even, but it's only the last digit that matters when deciding about even or odd.*)
- What do I mean if I say a number is divisible by 3?
  (e.g., *3 divides into that number, with no remainder.*)

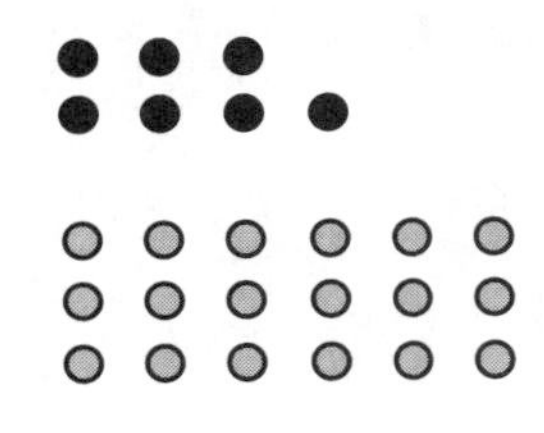

Arrange 7 counters and 18 counters as shown in the margin. Ask:

- How does the black counter arrangement show that 7 is not a multiple of 2?
  (e.g., *There is a leftover counter after you make pairs.*)
- How does the grey counter arrangement show that 3 is a factor of 18?
  (e.g., *It shows that* $18 = 6 \times 3$.)

Using the Open-Ended Intervention (Student Resource pages 156–157)

Read through the information about divisibility and the tasks on the student pages together. Provide base ten blocks if available and allow calculators. Give students time to work, ideally in pairs.

Observe whether students

- can develop divisibility rules for the factors 2, 3, 5, 6, 9, and 10
- can explain some divisibility rules

Consolidating and Reflecting

Ensure understanding by asking questions such as these based on students' work:

- How many digits do you care about, and which ones, when deciding if a number is divisible by 5? Why?
  (*Only the ones.* e.g., *The tens, hundreds, thousands, etc., are all divisible by 5, so it matters only what is left in the ones place.*)
- 45■8 is divisible by 3. What has to be true about the missing digit? Why?
  (*It has to be 1, 4, or 7 so that the sum of the digits is divisible by 3.*)
- Which divisibility rules consider all the digits in the number?
  (*the rules for 3, 6, and 9*)
- Which divisibility rules consider one or more but not all of the digits?
  (*the rules for 2, 5, and 10*)
- Why is it reasonable that the rules for 2, 5, and 10 are similar and the rules for 3, 6, and 9 are similar?
  (e.g., *10 is a multiple of 2 and 5; 6 and 9 are multiples of 3.*)

Copyright © 2012 by Nelson Education Ltd.

Divisibility Rules

Pathway 1 GUIDED

Guided Intervention

Before Using the Guided Intervention

Write the numeral 34 281 and cover the 4281 part. Ask:

- Is this five-digit number odd? (e.g., *I can't tell, since I can't see the ones digit.*)
- The first digit, 3, is odd. Why doesn't that matter?
  (e.g., *First of all, that 3 means 30 000, which is even, but it's only the last digit that matters when deciding about even or odd.*)
- What do I mean if I say a number is divisible by 3?
  (e.g., *3 divides into that number, with no remainder.*)

Arrange 7 counters and 18 counters as shown in the margin. Ask:

- How does the black counter arrangement show that 7 is not a multiple of 2?
  (e.g., *There is a leftover counter after you make pairs.*)
- How does the grey counter arrangement show that 3 is a factor of 18?
  (e.g., *It shows that* $18 = 6 \times 3$.)

Using the Guided Intervention **Student Resource pages 158–161**

Work through the instructional section of the student pages together and have students try out each divisibility rule on a variety of numbers.

Have students work through the **Try These** questions individually or in pairs.

Observe whether students

- can apply divisibility rules (Questions 1, 3)
- relate divisibility by one number to divisibility by another (Questions 2, 7, 8)
- recognize which digits of a number are relevant when determining divisibility (Question 3)
- can explain some divisibility rules (Questions 4, 5, 6)
- recognize the usefulness of divisibility rules (Question 9)

Consolidating and Reflecting

Ensure understanding by asking questions such as these based on students' work:

- How did you answer Question 3b)?
  (e.g., *The sum of the digits has to be a multiple of 3, so I added digits to go with 2 and 8 to make multiples of 3. The other 2 digits could be 10, 13, 16,...*)
- Explain the divisibility rule you chose in Question 5. (e.g., *The rule for 2; Every second number is another group of 2, so you skip over the 1s, 3s, 5s, 7s, and 9s.*)
- How did you answer Question 8? (e.g., *If a number is divisible by 3 and 5, it has both 3 and 5 as factors, so it must have 15 as a factor, too.*)
- Which divisibility rules consider all the digits in the number?
  (*the rules for 3, 6, and 9*)
- Which divisibility rules consider one or more but not all the digits?
  (*the rules for 2, 5, and 10*)

You will need

- counters
- Student Resource pages 158–161

Copyright © 2012 by Nelson Education Ltd.

Pathway 2 OPEN-ENDED

Prime Numbers and Perfect Squares

You will need

- square tiles
- calculators
- Student Resource pages 162–163

Open-Ended Intervention

Before Using the Open-Ended Intervention

Provide square tiles. Ask:

- ▸ How many different rectangles can you make with 10 tiles? (*2*)
- ▸ Why might some people say there are 2 rectangles, while others would say there are 4?
  (e.g., *Some people think the 1-by-10 rectangle is different from the 10-by-1 rectangle, but it's not.*)
- ▸ Why does showing 10 as a 2-by-5 rectangle show that 2 and 5 are factors of 10?
  (e.g., *A factor is a number you multiply. You multiply the length and width to get the area. The number of tiles is the area.*)
- ▸ How many factors can you find in making rectangles with 2 or 3 tiles? (*2*)
- ▸ How many factors can you find in making rectangles with 1 tile? (*only 1*)

Using the Open-Ended Intervention **Student Resource pages 162–163**

Read through the introduction and the tasks on the student pages together. Provide square tiles and calculators. Give students time to work, ideally in pairs.

Observe whether students

- can determine whether a number is prime
- can determine whether a number is a perfect square
- explore conjectures related to primes and perfect squares

Consolidating and Reflecting

Ensure understanding by asking questions such as these based on students' work:

- ▸ Why does every number except 1 have 2 or more factors?
  (e.g., *You can always make a long rectangle with a width of 1, and the width and length are factors.*)
- ▸ Think of a big number that is not prime.
  (e.g., *1 000 000*)
- ▸ How do you know it's not prime?
  (e.g., *It's even, so 2 is a factor.*)
- ▸ Why can you always write a composite number as a product of primes?
  (e.g., *If a number has factors, you can keep breaking up the factors until you can't anymore,* e.g., $28 = 2 \times 14$, *then* $28 = 2 \times 2 \times 7$. *Both 2 and 7 are primes.*)
- ▸ Why do perfect squares alternate between even and odd?
  (e.g., *You multiply numbers by themselves to get perfect squares, and numbers alternate between even and odd. Evens × evens are even, and odds × odds are odd.*)
- ▸ How would you determine whether 529 is a perfect square?
  (e.g., *I'd estimate to figure out what number to start with. I know that* $20 \times 20 = 400$ *and* $30 \times 30 = 900$, *so I would try* 21×21, 23×23, 27×27, *and* 29×29 *to see if they work.*)

 Copyright © 2012 by Nelson Education Ltd.

Prime Numbers and Perfect Squares

Pathway 2
GUIDED

Guided Intervention

Before Using the Guided Intervention

Provide square tiles. Ask:

- How many different rectangles can you make with 10 tiles? (*2*)
- Why might some people say there are 4 rectangles?
  (e.g., *Some people think the 1-by-10 rectangle is different from the 10-by-1 rectangle, but it's not.*)
- Why does showing 10 as a 2-by-5 rectangle show that 2 and 5 are factors of 10?
  (e.g., *A factor is a number you multiply. You multiply the length and width to get the area. The number of tiles is the area.*)

You will need

- square tiles
- calculators
- Student Resource pages 164–166

Using the Guided Intervention (Student Resource pages 164–166)

Work through the instructional section of the student pages together. Provide square tiles and have students model the rectangles.

Have students work through the **Try These** questions individually or in pairs. Allow calculators.

Observe whether students

- can determine whether a number is prime (Questions 1, 2, 3, 5, 6, 10)
- can write a composite as a product of primes (Question 4)
- explore conjectures related to primes and perfect squares (Questions 6, 9)
- can determine whether a number is a perfect square (Questions 7, 8, 10)

Consolidating and Reflecting

Ensure understanding by asking questions such as these based on students' work:

- Why does every number except 1 have 2 or more factors?
  (e.g., *You can always make a long rectangle with a width of 1, and the width and length are factors.*)
- What number over 100 000 would be a lot of work to check to see if it's prime? Why? (e.g., *100 001; You have to try a lot of factors to see if they work.*)
- In Question 4, how did you write a composite number as a product of primes?
  (e.g., *If the number has factors, you can keep breaking up the factors until you can't anymore,* e.g., $28 = 2 \times 14$, *then* $28 = 2 \times 2 \times 7$. *Both 2 and 7 are primes.*)
- Why do perfect squares alternate between being even and odd?
  (e.g., *You multiply numbers by themselves to get perfect squares, and numbers alternate between even and odd. Evens × evens are even, and odds × odds are odd.*)
- How can you use the idea in Question 8 to check whether 529 is a perfect square?
  (e.g., *I'd figure out a perfect square close to, but less than, 529 by multiplying different numbers, and then test to see if the square of the next whole number was more than 529.*)

Copyright © 2012 by Nelson Education Ltd.

Pathway 3 OPEN-ENDED

Factors and Multiples

You will need

- square tiles
- calculators
- Student Resource pages 167–168

Open-Ended Intervention

Before Using the Open-Ended Intervention

Print the words *factor* and *multiple* and ask students to use both words in one sentence. (e.g., *Since 2 is a factor of 8, then 8 is a multiple of 2.*) Ask:

- Is a factor of a number usually more, less, or the same as the number? (e.g., *usually less, but some are the same. A factor can't be more, since you multiply by another whole number to get the number. Factors of 200 are 1, 2, 4, 5, 10, 20,…, which are all less than the number. 200 is also a factor of itself, so a factor can be the same as the number.*)
- Is a multiple of a number always or usually greater than the number? (*Usually*. e.g., *20 is a multiple of 20, so it isn't always greater. However, you can multiply the number by any other whole number, starting with 2, to get a multiple that will be greater.*)

Using the Open-Ended Intervention (Student Resource pages 167–168)

Read through the introduction on the student pages and have students use square tiles to model rectangles for 72. Read through the tasks together. Give students time to work, ideally in pairs. Allow calculators.

Observe whether students

- relate the dimensions of rectangles to the notion of factoring
- can check if one number is a factor of another number and list the factors
- realize you can multiply by a whole number to get a multiple
- relate factors and multiples of numbers to one another

Consolidating and Reflecting

Ensure understanding by asking questions such as these based on students' work:

- When is using rectangles a good way to figure out factors? (e.g., *for smaller numbers; you need too many tiles for large numbers, and it would take too long*)
- Why is 1 a factor of every number? (e.g., *Every number is 1 times itself.*)
- How do you know 180 has a lot of factors? (e.g., *5, 4, and 9 are factors, so 20 [5 × 4], 36 [4 × 9], and 45 [5 × 9] are also factors, and since 4 is 2 × 2 and 9 is 3 × 3, there are even more.*)
- Why does a number always have more multiples than factors? (e.g., *For multiples, you can keep multiplying the number by bigger numbers. For factors, you keep dividing but, once you hit the number, there are no more factors.*)
- If 4 is a factor of a number, so is 2. How do you know this is true? (e.g., *4 = 2 × 2*)
- If 600 is a multiple of a number, so is 1200. How do you know this is true? (e.g., *If 600 = ■ × N, then 1200 = ■ × 2 × N, so ■ is a factor of 1200.*)

 Copyright © 2012 by Nelson Education Ltd.

Factors and Multiples

Pathway 3
GUIDED

Guided Intervention

You will need

- square tiles
- calculators
- Student Resource pages 169–172

Before Using the Guided Intervention

Print the words *factor* and *multiple* and ask students to use both words in one sentence. (e.g., *Since 2 is a factor of 8, then 8 is a multiple of 2.*) Ask:

- Is a factor of a number usually more, less, or the same as the number? (e.g., *usually less, but some are the same; A factor can't be more, since you multiply by another whole number to get the number. Factors of 200 are 1, 2, 4, 5, 10, 20,…, which are all less than the number. 200 is also a factor of itself, so a factor can be the same as the number.*)
- Is a multiple of a number always or usually greater than the number? (*Usually.* e.g., *20 is a multiple of 20, so it isn't always greater. However, you can multiply the number by any other whole number, starting with 2, to get a multiple that will be greater.*)

Using the Guided Intervention (Student Resource pages 169–172)

Work through the instructional section of the student pages together. Provide square tiles and have students model the numbers, using rectangles. Have students work through the **Try These** questions individually or in pairs. Allow calculators.

Observe whether students

- relate the dimensions of rectangles to the notion of factoring (Question 1)
- can check whether one number is a factor or multiple of another, and list the factors or multiples (Questions 2, 3, 7)
- relate factors and multiples of a number to one another (Questions 4, 6)
- realize you can multiply by a whole number to get a multiple (Questions 5, 7, 8)
- solve problems related to factors and multiples (Question 8)

Consolidating and Reflecting

Ensure understanding by asking questions such as these based on students' work:

- How does Question 1 help you relate rectangles and factors? (e.g., *The lengths and the widths of the rectangles are factors.*)
- How do you know that a number like 120 has a lot of factors (Question 3b))? (e.g., *When you start making the factor tree, the numbers you first break 120 into, 10×12, have a lot of factors.*)
- Think about Questions 4 and 6. Why, if you know one thing about a factor or a multiple, might you know other things as well? (e.g., *If the factor has factors, that means there are smaller factors. Any multiple of a multiple of a number is a multiple of the number, so you can multiply multiples to get more multiples.*)
- How is Question 8b) a factor and multiple problem? (e.g., *Each book's thickness has to be a factor of 96, and 96 is a multiple of the book's thickness.*)

Copyright © 2012 by Nelson Education Ltd.

Strand: Number

Integers

Planning For This Topic

Materials for assisting students with integers consist of a diagnostic tool and 3 intervention pathways. Pathway 1 focuses on subtracting integers. Pathway 2 is about adding integers. Pathway 3 involves representing and comparing integers.

Integer models using counters and number lines are both presented; students can choose which they prefer.

Each pathway has an open-ended intervention and a guided intervention. Choose the type of intervention more suitable for your students' needs and your particular circumstances.

Professional Learning Connections

PRIME: Number and Operations, Background and Strategies (Nelson Education Ltd., 2005), pages 133–134

Making Math Meaningful to Canadian Students K–8 (Nelson Education Ltd., 2008), pages 267–275, 279–281

Big Ideas from Dr. Small Grades 4–8 (Nelson Education Ltd., 2009), pages 84–89

Good Questions (dist. by Nelson Education Ltd., 2009), pages 38, 51–52

More Good Questions (dist. by Nelson Education Ltd, 2010), page 70

Curriculum Connections

Grades 5 to 8 curriculum connections for this topic are provided online. See www.nelson.com/leapsandbounds. All 3 pathways are relevant to WNCP and Ontario curriculums.

Why might students struggle with integers?

Students might struggle with integers for any of the following reasons:

- They might be confused when comparing integers (e.g., students might think $-40 > -3$ since $40 > 3$).
- They might have difficulties with whole number operations that affect their ability to operate with integers.
- They might struggle with the use of the – sign for both negative integers and subtraction, and the use of the + sign for both positive integers and addition.
- They might consider subtraction only in terms of take-away, making it difficult to model the subtraction of a negative number on a number line.
- They might confuse how to handle the signs for subtracting than for adding because the "rules" are different (e.g., for $5 + (-3)$, you use the sign of the integer farthest from 0 for the answer, but for $3 - (-5)$, you don't).
- They might over-generalize some relationships with signs (e.g., since $4 - (-3)$ is the same as the positive amount $4 + 3$, students might think that 2 negatives make a positive when adding -4 and -3).

 Copyright © 2012 by Nelson Education Ltd.

Diagnostic Tool: Integers

Use the diagnostic tool to determine the most suitable intervention pathway for working with integers. Provide Diagnostic Tool: Integers, Teacher's Resource pages 132 and 133, and have students complete it in writing or orally. Provide integer tiles, or counters in 2 colours.

See solutions on Teacher's Resource pages 134 and 135.

Intervention Pathways

The purpose of the intervention pathways is to help students determine sums and differences of integers, and to represent and compare them, to prepare students for later work with other integer operations and algebraic expressions involving signed numbers.

There are 3 pathways:

- Pathway 1: Subtracting Integers
- Pathway 2: Adding Integers
- Pathway 3: Representing and Comparing Integers

Use the chart below (or the Key to Pathways on Teacher's Resource pages 134 and 135) to determine which pathway is most suitable for each student or group of students.

Diagnostic Tool Results	Intervention Pathway
If students struggle with Questions 8 to 10	use Pathway 1: Subtracting Integers *Teacher's Resource pages 136–137* *Student Resource pages 173–178*
If students struggle with Questions 5 to 7	use Pathway 2: Adding Integers *Teacher's Resource pages 138–139* *Student Resource pages 179–184*
If students struggle with Questions 1 to 4	use Pathway 3: Representing and Comparing Integers *Teacher's Resource pages 140–141* *Student Resource pages 185–189*

Copyright © 2012 by Nelson Education Ltd.

Name:_______________________ Date:_______________

Integers

Diagnostic Tool

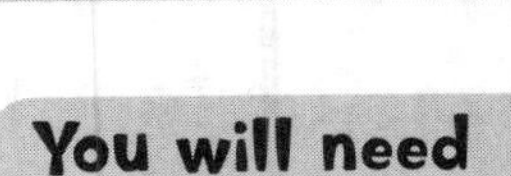

You will need

- integer tiles, or counters (2 colours)

1. a) Mark these integers on the number line: −3, −7, +2, −10.

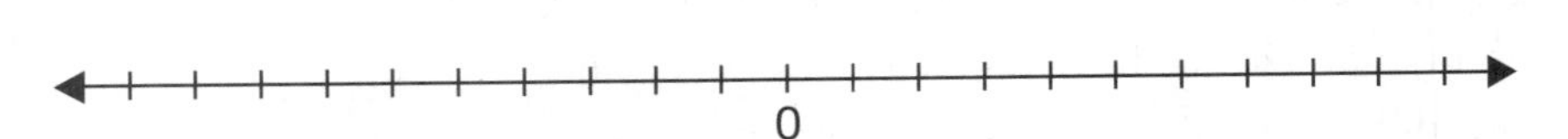

b) Order the integers from least to greatest.

_______ _______ _______ _______

2. The counter models below represent +4 and −3.

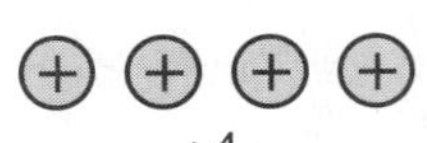

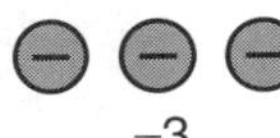

a) Use counter models to represent +2 and −2.
Sketch each model.

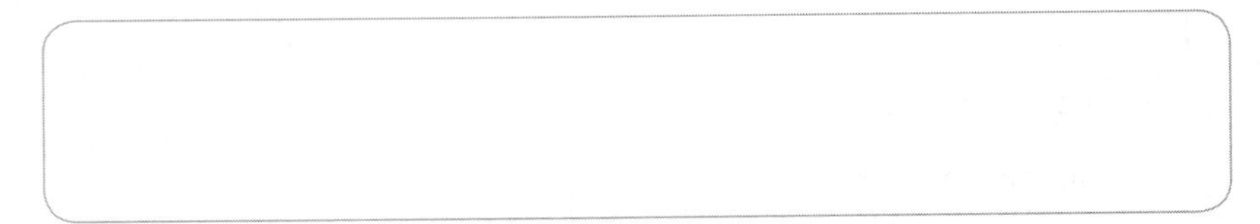

b) Explain how +2 and −2 are the same and different.

same: ______________________________

____

different: ______________________________

____

3. a) Name 2 opposite integers that are close together.

_______ _______

b) Name 2 opposite integers that are far apart.

_______ _______

4. Describe 3 situations that the integer −5 might describe.

____

____

____

 Copyright © 2012 by Nelson Education Ltd.

Name: ______________________________ Date: ____________________

5. Calculate.

a) $+5 + (-2) =$ ________

b) $-4 + (-7) =$ ________

c) $-10 + (+9) =$ ________

d) $+5 + (-6) =$ ________

6. Sketch a model that would explain your answer to Question 5a).

7. You add an integer to -5 and the sum is negative.

$-5 + ■ = -■$

What do you know about the integer you added?

____

8. Calculate.

a) $-4 - (-3) =$ ________

b) $+5 - (-3) =$ ________

c) $-3 - (+8) =$ ________

d) $+5 - (+7) =$ ________

9. Sketch a model that would explain your answer to Question 8a).

10. Create a problem you could solve by calculating $-4 - (-1)$.

____

____

____

Copyright © 2012 by Nelson Education Ltd.

Solutions and Key to Pathways

Name: ______ Date: ______

Integers

Diagnostic Tool

You will need

- integer tiles, or counters (2 colours)

1. a) Mark these integers on the number line: −3, −7, +2, −10.

−10 −7 −3 0 +2

b) Order the integers from least to greatest.

−10 −7 −3 +2

2. The counter models below represent +4 and −3.

(+)(+)(+)(+) +4 (−)(−)(−) −3

a) Use counter models to represent +2 and −2.
Sketch each model.

(+)(+) +2 (−)(−) −2

b) Explain how +2 and −2 are the same and different.

same: e.g., Each integer is represented by 2 counters.

different: e.g., colours; +2 uses lighter counters and −2 uses darker counters

3. a) Name 2 opposite integers that are close together.

+1 −1

b) Name 2 opposite integers that are far apart.

+10 −10

4. Describe 3 situations that the integer −5 might describe.

e.g., You owe $5.

Your score is 5 below average.

5 °C below 0

Pathway 3

 Copyright © 2012 by Nelson Education Ltd.

 Copyright © 2012 by Nelson Education Ltd.

Name:______________________ Date:______________

5. Calculate.

a) $+5 + (-2) =$ __+3__

b) $-4 + (-7) =$ __−11__

c) $-10 + (+9) =$ __−1__

d) $+5 + (-6) =$ __−1__

Pathway 2

6. Sketch a model that would explain your answer to Question 5a).

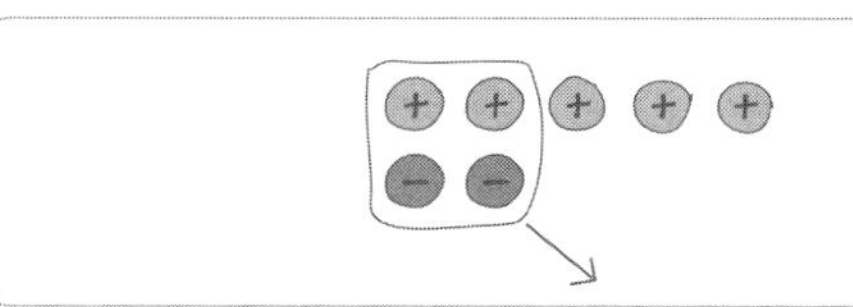

7. You add an integer to -5 and the sum is negative.

$-5 + ■ = -■$

What do you know about the integer you added?

It has to be less than +5.

8. Calculate.

a) $-4 - (-3) =$ __−1__

b) $+5 - (-3) =$ __+8__

c) $-3 - (+8) =$ __−11__

d) $+5 - (+7) =$ __−2__

Pathway 1

9. Sketch a model that would explain your answer to Question 8a).

e.g.,

10. Create a problem you could solve by calculating $-4 - (-1)$.

e.g., The temperature went from −1 °C to −4 °C. How many degrees did it change by, and did it get warmer or colder?

Pathway 1 OPEN-ENDED

Subtracting Integers

You will need

- integer tiles or counters (2 colours)
- Number Lines (BLM 25, optional)
- Student Resource pages 173–174

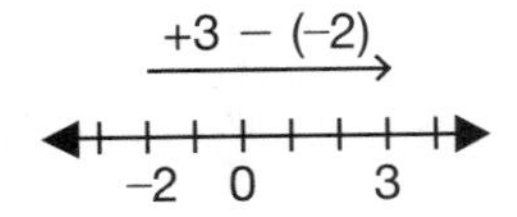

Open-Ended Intervention

Before Using the Open-Ended Intervention

Provide integer tiles or counters. Ask:

- In what real-life situations do you subtract? (e.g., *if you take something away, or if you want to know how much more one number is than another*)
- What might $-3 - (-2)$ mean? (e.g., *Take* -2 *from* -3 *so you have* -1 *left.*)
- How could you show $-3 - (-2)$ with a counter model? (e.g., *I'd use 3 negative counters and take away 2 of them so there's 1 negative counter left.*)
- Why does adding 2 positive and 2 negative counters to -3 not change its value? (e.g., *You are really just adding 0.*)
- How does going from -2 to $+3$ on a number line represent $+3 - (-2)$? (e.g., *Subtraction is the opposite of addition, so you can figure out what to add to* -2 *to get to* $+3$ *to solve* $+3 - (-2)$.)

Using the Open-Ended Intervention **(Student Resource pages 173–174)**

Read through the tasks on the student pages together. Provide integer tiles or counters. You might provide Number Lines (BLM 25), or have students sketch their own. Talk about what the student in the illustration is doing and why $-6 - (-2) = ■$ can be solved using $-2 + ■ = -6$. Give students time to work, ideally in pairs.

Observe whether students

- interpret subtraction as take-away, comparison, and figuring out what to add
- can subtract integers
- can use and explain models to subtract integers
- can describe real-life integer subtraction situations

Consolidating and Reflecting

Ensure understanding by asking questions such as these based on students' work:

- Choose a subtraction from Part 2. Describe your model. (e.g., *For* $-40 - (-11)$, *I imagined 40 negative counters and took away 11 of them, leaving* -29.)
- Why is $-4 - (-10)$ different from $-10 - (-4)$? (e.g., *You go in opposite directions on a number line, so* $-4 - (-10)$ *is positive and* $-10 - (-4)$ *is negative.*)
- Why is $4 - (-2)$ the same as $4 + 2$? (e.g., *If you subtract* $4 - (-2)$ *as* $-2 + ■ = 4$ *on a number line, you go 2 to get from* -2 *to 0 and then another 4 to get to 4, a total of 6.*)

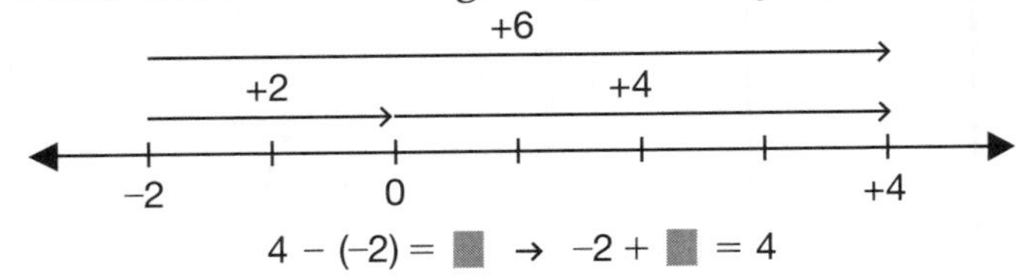

 Copyright © 2012 by Nelson Education Ltd.

Subtracting Integers

Pathway 1 GUIDED

Guided Intervention

Before Using the Guided Intervention

Provide integer tiles or counters. Ask:

- In what real-life situations do you subtract?
  (e.g., *if you take something away, or if you want to know how much more one number is than another*)
- What might $-3 - (-2)$ mean?
  (e.g., *Take* -2 *from* -3 *so you have* -1 *left.*)
- What else could $-3 - (-2)$ mean?
  (e.g., *how much more* -3 *is than* -2)

You will need

- integer tiles or counters (2 colours)
- Number Lines (BLM 25, optional)
- Student Resource pages 175–178

Using the Guided Intervention (Student Resource pages 175–178)

Work through the instructional section of the student pages together. Students should model with integer tiles or counters and/or number lines.

Have students work through the **Try These** questions individually or in pairs.

Observe whether students

- can use and explain models to subtract integers (Questions 1, 2, 3, 4, 9)
- interpret subtraction as take-away, comparison, and figuring out what to add (Questions 1, 2, 3, 4, 5, 7, 10)
- can subtract integers (Questions 4, 8, 9, 10)
- can describe real-life integer subtraction situations (Question 6)

Consolidating and Reflecting

Ensure understanding by asking questions such as these based on students' work:

- Describe the model you used to subtract $+4 - (-7)$ in Question 4a).
  (e.g., *I showed 4 positive counters and then added 7 pairs of positive and negative counters and then took away 7 negative counters.*)
- I noticed in Question 5b) that $+7 - (-3)$ is the same as $7 + 3$. Why does that make sense?
  (e.g., *On a number line, if you go from* -3 *to* $+7$*, you go right 3 to get to 0 and right another 7 to get to* $+7$*, a total of 10*).
- In Question 9, why is the answer to $+2 - (+5)$ different from the answer to $+2 - (-5)$, even though you started with and took away the same number of counters?
  (e.g., *For both, I started with 2 positive counters and added 3 pairs of negative and positive counters. But for* $+2 - (+5)$*, I took away 5 positive counters and was left with 3 negative counters, while for* $+2 - (-5)$*, I took away 5 negative counters and was left with 7 positive counters.*)

Pathway 2
OPEN-ENDED

Adding Integers

You will need

- integer tiles or counters (2 colours)
- Number Lines (BLM 25, optional)
- Student Resource pages 179–180

Open-Ended Intervention

Before Using the Open-Ended Intervention

Provide integer tiles or counters. Ask:

- What might $-3 + (-2)$ mean?
  (e.g., *You owe someone $3 and then you borrow another $2, so you owe $5 altogether.*)
- How could you show $-3 + (-2)$ with a counter model?
  (e.g., *I'd combine 3 negative counters with 2 negative counters to get 5 negative counters altogether.*)
- How could you show $-3 + (-2)$ with a number-line model?
  (e.g., *I'd start at* -3 *and then go left 2 units to add* -2.)
- Why do you think $+1 + (-1) = 0$?
  (e.g., $+1$ *and* -1 *are opposites, so they cancel each other out. If you had $1 but you owed $1, it's as if you don't have any money.*)

Using the Open-Ended Intervention Student Resource pages 179–180

Read through the tasks on the student pages together. Provide integer tiles or counters. You might provide Number Lines (BLM 25), or students could sketch their own. Talk about what the student in the illustration is doing, and what the result will be and why $(+6 + (-5) = +1)$.

Give students time to work, ideally in pairs.

Observe whether students

- can add integers
- can use and explain models to add integers
- can describe real-life integer addition situations

Consolidating and Reflecting

Ensure understanding by asking questions such as these based on students' work:

- How did you predict the least sum in Part 2?
  (e.g., *I looked for the 2 negative integers farthest from 0 so their sum would be the farthest left.*)
- Describe the model you used for your first addition in Part 3.
  (e.g., *For* $-30 + (+18)$, *I combined 30 negative counters with 18 positive counters, matched 18 negatives with 18 positives to get 0, and there were 12 negatives left, or* -12.)
- Describe an instance when you used the zero principle and why.
  (e.g., *For* $-24 + (+18)$, *I had both positive and negative counters and wanted to combine 18 of each so I could remove them to see what was left over.*)

 Copyright © 2012 by Nelson Education Ltd.

Adding Integers

Pathway 2
GUIDED

Guided Intervention

Before Using the Guided Intervention

Provide integer tiles or counters. Ask:

- What might $-3 + (-2)$ mean?
  (e.g., *You owe someone $3 and then you borrow another $2, so you owe $5 altogether.*)
- How could you show $-3 + (-2)$ with a counter model?
  (e.g., *I'd combine 3 negative counters with 2 negative counters to get 5 negative counters altogether.*)
- How could you show $-3 + (-2)$ with a number-line model?
  (e.g., *I'd start at* -3 *and then go left 2 units to add* -2.)

You will need

- integer tiles or counters (2 colours)
- Number Lines (BLM 25, optional)
- Student Resource pages 181–184

Using the Guided Intervention **Student Resource pages 181–184**

Work through the instructional section of the student pages together. Provide integer tiles or counters for modelling the addition. You might provide Number Lines (BLM 25), or have students sketch their own.

Have students work through the **Try These** questions individually or in pairs.

Observe whether students

- can use and explain models to add integers (Questions 1, 2, 3, 4, 10, 11)
- can add integers (Questions 4, 6, 7, 8, 9)
- can describe real-life integer addition situations (Question 5)
- recognize that addition always involves combining (Question 10)

Consolidating and Reflecting

Ensure understanding by asking questions such as these based on students' work:

- Describe the model you used for Question 4a).
  (e.g., *For* $-4 + (-7)$, *I combined 4 negative counters with 7 negative counters to get 11 negative counters altogether.*)
- Describe an instance when you used the zero principle and why.
  (e.g., *For Question 4b),* $+6 + (-2)$, *I had both positive and negative counters and wanted to combine 2 of each so I could remove them to see what was left over.*)
- Why were there 2 possible numbers of counters (3 or 13) for the possible sums in Question 9?
  (e.g., *It depends on whether you use the zero principle or not. If you are adding integers with the same sign, the sum is 13 or* -13, *so you use 13 counters. If you're adding integers with opposite signs, you use the zero principle to remove counters and end up with 3 counters.*)

Copyright © 2012 by Nelson Education Ltd.

Representing and Comparing Integers

You will need

- integer tiles or counters (2 colours)
- Student Resource pages 185–186

Open-Ended Intervention

Before Using the Open-Ended Intervention

Ask:

- ▸ When have you seen negative numbers used?
  (e.g., *for temperatures*)

Draw a vertical number line from −5 to 5. Ask:

- ▸ Where would you place +2 °C?
  (*2 spaces above 0*)
- ▸ Where would you place −3 °C?
  (*3 spaces below 0*)
- ▸ Which number do you think is greater? How do you know?
  (*+2.* e.g., *It is farther up the number line.*)

Using the Open-Ended Intervention (Student Resource pages 185–186)

Read through the tasks on the student pages together. Provide integer tiles or counters. Talk about who has the higher score (Haroon or Corey) and how they know.

Give students time to work, ideally in pairs.

Observe whether students

- can model integers
- can identify the opposite of an integer
- can compare and order integers
- can describe real-life situations where negative numbers are used

Consolidating and Reflecting

Ensure understanding by asking questions such as these based on students' work:

- ▸ Why was there only one possibility for the first blank?
  (*−30 was the only integer less than −10.*)
- ▸ Were there any blanks with only a few possibilities?
  (e.g., *There were not a lot of integers less than −5.*)
- ▸ Did you find it easier to use counters or a number line to represent an integer? Why?
  (e.g., *I prefer counters since it's easy to see very quickly how many there are and what sign the number is.*)
- ▸ Why is a number line a good model for ordering integers?
  (e.g., *Once you put them on the number line, they are already in order from least to greatest.*)
- ▸ Why is it always easy to compare a positive integer with a negative integer?
  (e.g., *Any positive integer is always greater than any negative integer.*)

 Copyright © 2012 by Nelson Education Ltd.

Representing and Comparing Integers

Pathway 3 GUIDED

Guided Intervention

You will need

- integer tiles or counters (2 colours)
- Student Resource pages 187–189

Before Using the Guided Intervention

Ask:

- ▸ When have you seen negative numbers used?
  (e.g., *for temperatures*)

Draw a vertical number line from −5 to 5. Ask:

- ▸ Where would you place +2 °C?
  (*2 spaces above 0*)
- ▸ Where would you place −3 °C?
  (*3 spaces below 0*)
- ▸ Which number do you think is greater? How do you know?
  (*+2.* e.g., *It is farther up the number line.*)

Using the Guided Intervention Student Resource pages 187–189

Work through the instructional section of the student pages together. Provide integer tiles or counters for modelling the integers.

Have students work through the **Try These** questions individually or in pairs.

Observe whether students

- can model integers (Questions 1, 3)
- can identify the opposite of an integer (Questions 1, 3, 5)
- can distinguish integers from non-integers (Question 2)
- can compare and order integers (Questions 4, 7, 8, 9, 10)
- can describe real-life situations where negative numbers are used (Question 6)

Consolidating and Reflecting

Ensure understanding by asking questions such as these based on students' work:

- ▸ How did you choose an integer closer to −2 than +4 in Question 4e)?
  (e.g., *I knew anything left of −2 would work, and so would −1 and 0.*)
- ▸ How would models for −15 and +15 compare (Question 5)?
  (e.g., *Both use 15 counters but of different colours. On a number line, they're both 15 units from 0, but on opposite sides.*)
- ▸ Why is −5 > −6, even though +5 < +6, in Question 7a)?
  (e.g., *If +5 < +6, it means +5 is closer to 0 than +6 is. So the opposite of +5, −5, is also closer to 0 than the opposite of 6, −6; this makes −5 greater than −6*).
- ▸ Why is it always easy to compare a positive integer with a negative integer (Question 9)?
  (e.g., *Any positive integer is greater than any negative integer.*)

Strand: Patterns and Algebra

Patterns and Algebra Overview

How were the patterns and algebra topics chosen?

This resource provides materials for assisting students with 2 patterns and algebra topics. These topics are drawn from curriculum outcomes across the country for Grades 5 to 7. Topics are divided into levels, called pathways, that address gaps in students' prerequisite skills and knowledge.

How were the pathways determined?

At the Grade 7/8 level, there is increasing attention to algebra, while continuing to work on patterns. As well, the connections among the topics of patterns and algebra are developed.

The first topic in this strand focuses on patterns. The most rudimentary pathway addresses how students describe, create, and extend simple recursive patterns. These are patterns where each term is determined by operating on the previous term or terms. The next pathway focuses on a variety of representations for pattern rules—concrete or pictorial models, tables of values, graphs, and recursive rules and relationship rules—involving one operation (e.g., patterns such as 23, 24, 25, 26,… or 8, 16, 24, 32,…). The most advanced pathway focuses on linear relations, which includes representing linear relations using tables of values, graphs, or algebraic equations, and using the representations to solve problems.

The second topic focuses on algebra. The simplest pathway introduces the use of letter variables to represent unknown values in algebraic expressions and equations, and the use of algebraic equations (involving variables) to represent relationships. The next pathway involves preserving equality while solving simple equations using various models, such as a length model or a pan balance model. The most advanced pathway is about solving problems by first representing them with equations and then solving them algebraically (with the help of models) and/or graphically.

What patterns and algebra topics have been omitted?

The focus in this topic is on "algebraic thinking," so the kinds of repeating visual patterns created by performing translations and rotations are not covered, since they do not support the algebraic ideas being developed.

Although the equations and expressions used in these materials involve both variables and constants, there is no focus on the vocabulary term *constant*.

Copyright © 2012 by Nelson Education Ltd.

Materials

The materials listed below for assisting students struggling with patterns and algebra are likely in the classroom or easily accessible. Blackline masters are provided at the back of this resource. Many of these materials are optional and are therefore listed only in the Teacher's Resource.

- square tiles, linking cubes, or counters in 2 colours
- coloured pencils
- toothpicks
- play coins
- base ten blocks
- pan balance
- paper bags
- algebra tiles: x-tiles and 1-tiles
- calculators

BLM 12: 1 cm Grid Paper
BLM 13: 0.5 cm Grid Paper

Patterns and Algebra Topics and Pathways

Topics and pathways in this strand are shown below.
Each pathway has an open-ended intervention and a guided intervention.

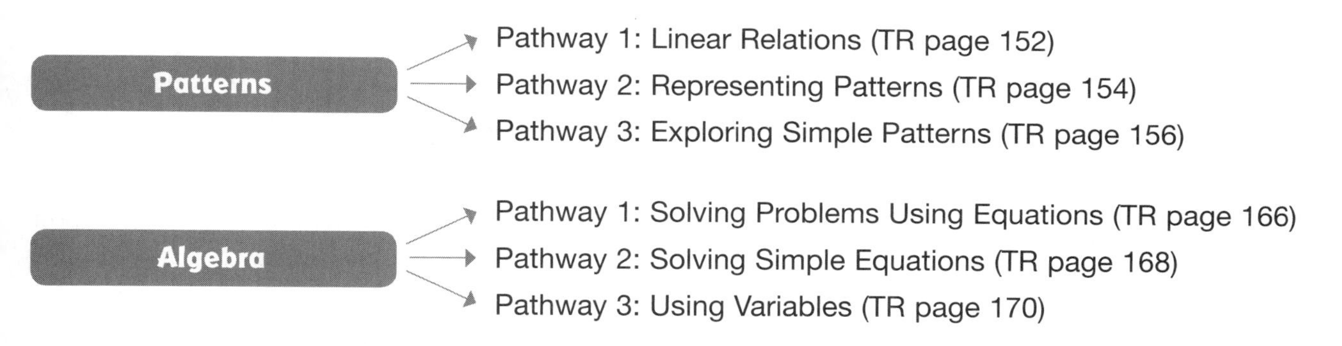

Strand: Patterns and Algebra

Patterns

Planning For This Topic

Materials for assisting students with patterns consist of a diagnostic tool and 3 intervention pathways. All 3 pathways involve number patterns only. Pathway 1 focuses on making predictions about linear relations where the relationship between the term number and the term value involves 2 operations (e.g., a pattern such as 4, 7, 10, 13,… that can be represented by the rule $v = 3 \times n + 1$). Pathway 2 focuses on various representations for a pattern where the relationship rule can be represented using a single operation (as $v = n + ■$ or $v = ■ \times n$). Pathway 3 involves determining recursive patterns and pattern rules (e.g., "Start at 3 and add 2 each time"), as well as creating and extending simple patterns. Recursive patterns are those where each term is determined by operating on the previous term or terms.

Each pathway has an open-ended intervention and a guided intervention. Choose the type of intervention more suitable for your students' needs and your particular circumstances.

Curriculum Connections

Grades 5 to 8 curriculum connections for this topic are provided online. See www.nelson.com/leapsandbounds. Pathway 1 may be most suitable for students working with the WNCP curriculum. If Pathway 1 is used for Ontario students, some preparatory work in graphing may be required. The other 2 pathways are appropriate for students working with either curriculum.

Professional Learning Connections

PRIME: Patterns and Algebra, Background and Strategies (Nelson Education Ltd., 2005), pages 42–55, 62–63
Making Math Meaningful to Canadian Students K–8 (Nelson Education Ltd., 2008), pages 570–573, 584–585
Big Ideas from Dr. Small Grades 4–8 (Nelson Education Ltd., 2009), pages 2–5, 9–10
Good Questions (dist. by Nelson Education Ltd., 2009), pages 131–133, 142–145
More Good Questions (dist. by Nelson Education Ltd., 2010), pages 19–21, 24–28

Why might students struggle with patterns?

Students might struggle with patterns for any of the following reasons:

- They might be weak in number sense and not recognize, for example, common differences between numbers in a growing or shrinking pattern.
- They might not realize that there are many ways to continue a pattern when only some of the terms are given (e.g., they might think that, if a pattern begins 4, 8, 12, 16,…, the next number *has to be* 20 instead of *could be* 20, since it could be a repeating pattern [4, 8, 12, 16, 4, 8, 12, 16, ...], or it could increase by 4 each time for the first 4 terms and then by 5 for the next 4 terms, and so on [4, 8, 12, 16, 21, 26, 31, 36, ...]).
- They might be able to identify patterns when the first few terms are given, but not when other terms are given, such as only the 3rd, 4th, and 10th terms.
- They might not recognize the difference between pattern rules that relate a term number to a term value, often called relationship rules (e.g., $v = 3n + 1$ indicates that you multiply the term number by 3 and add 1 to get the term value), and recursive pattern rules that describe how terms change from prior terms (e.g., "Start at 4 and keep adding 3"). Note that we will refer to recursive pattern rules as just pattern rules in this resource.

Copyright © 2012 by Nelson Education Ltd.

- They might identify a pattern or relationship rule that works only for the 1st and 2nd terms.
- They might have difficulty creating a relationship that relates a term value to a term number.
- They might have difficulty graphing a pattern from a table of values.
- They might not recognize that the same pattern can be represented in different ways to highlight different aspects of the pattern.

Diagnostic Tool: Patterns

Use the diagnostic tool to determine the most suitable intervention pathway for patterns. Provide Diagnostic Tool: Patterns, Teacher's Resource pages 146 to 148, and have students complete it in writing or orally. Provide calculators for students who want them. Empty grids are provided on the student page for creating graphs—students will need to use a scale for the *y*-axis that fits the grid. Alternatively, they could use grid paper.

See solutions on Teacher's Resource pages 149 to 151.

Intervention Pathways

The purpose of the intervention pathways is to help students recognize how various representations of patterns give us insight into those patterns. These concepts will be extended, particularly as students work with linear relations.

There are three pathways:
- Pathway 1: Linear Relations
- Pathway 2: Representing Patterns
- Pathway 3: Exploring Simple Patterns

Use the chart below (or the Key to Pathways on Teacher's Resource pages 149 to 151) to determine which pathway is most suitable for each student or group of students.

Diagnostic Tool Results	Intervention Pathway
If students struggle with Questions 7 to 9	use Pathway 1: Linear Relations *Teacher's Resource pages 152–153* *Student Resource pages 190–196*
If students struggle with Questions 5 and 6	use Pathway 2: Representing Patterns *Teacher's Resource pages 154–155* *Student Resource pages 197–203*
If students struggle with Questions 1 to 4	use Pathway 3: Exploring Simple Patterns *Teacher's Resource pages 156–157* *Student Resource pages 204–209*

Copyright © 2012 by Nelson Education Ltd.

Name:______________________ Date:______________________

Patterns

Diagnostic Tool

1. Use the pattern rule to list the first 5 terms in each pattern.

   a) Start at 8 and keep adding 4.

   8, _______, _______, _______, _______,…

   b) Start at 56 and keep subtracting 2.

   56, _______, _______, _______, _______,…

2. Each model represents the first 4 figures of a pattern. Write the first 5 terms of its number pattern.

   a)   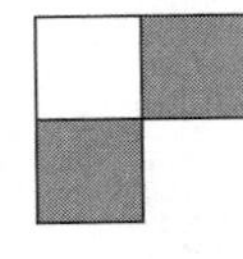 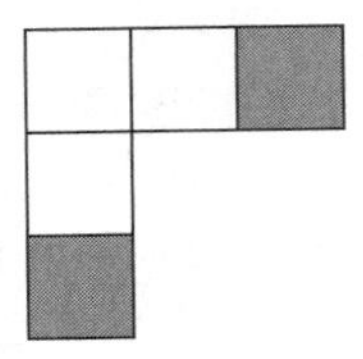 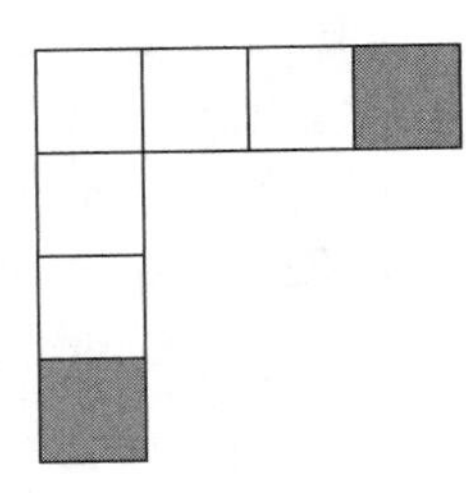

   1, _______, _______, _______, _______,…

   b) 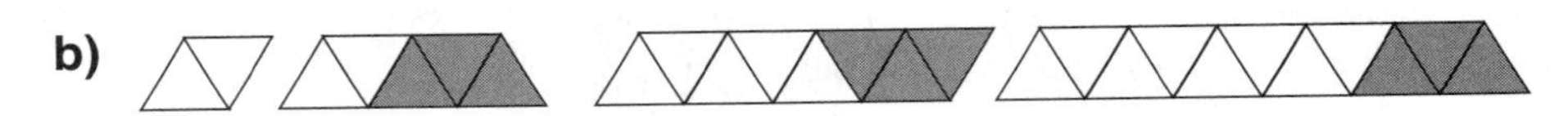

   2, _______, _______, _______, _______,…

3. What is the 10th number in each pattern in Question 2.

   a) _______ **b)** _______

4. Write a pattern rule for each number pattern.

   a) 40, 43, 46, 49, 52,…

   ____

   b) 40, 37, 34, 31, 28,…

   ____

   c) 3, 6, 12, 24, 48,…

   ____

   d) 5000, 1000, 200, 40, 8,…

   ____

 Copyright © 2012 by Nelson Education Ltd.

Name:____________________ Date:____________________

5. Use the number pattern 6, 12, 18, 24,....

a) Sketch a model to represent the first 4 numbers in the pattern. Describe how your model shows the number pattern.

b) Complete this table of values.

Term number (x)	**Term value (y)**
1	
2	
3	
4	

c) Graph the table of values.

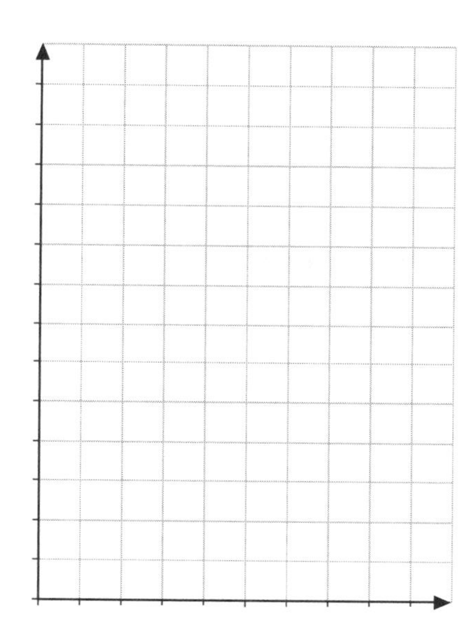

d) Write a relationship rule that describes how the term value relates to the term number for the pattern.

____

6. To figure out the 100th term in the pattern in Question 5, why might you use the relationship rule instead of the table of values or graph?

____

____

____

Name: ____________________ Date: ____________________

7. The graph of a linear relation forms a straight line.
Which tables of values below represent a linear relation? ____________

Table A

Term number (x)	Term value (y)
1	4
2	7
3	10
4	13

Table B

Term number (x)	Term value (y)
1	4
2	9
3	16
4	25

Table C

Term number (x)	Term value (y)
1	4
2	9
3	14
4	19

8. A plumber charges $35 for a home visit, plus an additional $75 for each hour worked.

a) Complete the table of values to show how much the plumber would charge for different numbers of hours worked.

Plumber Charges

Number of hours	Charge ($)
1	
2	
3	
4	

b) Graph the table of values.

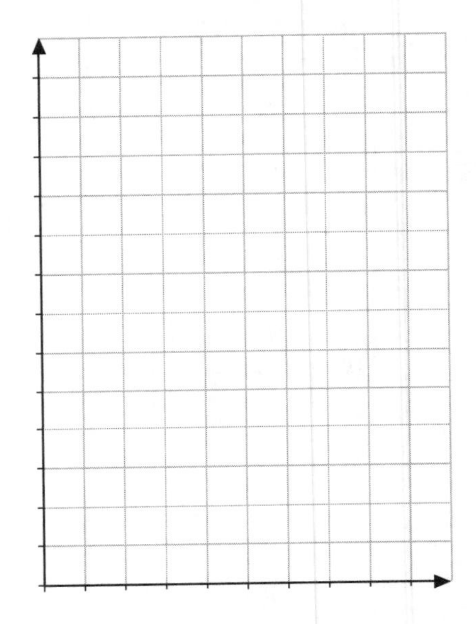

c) Write a relationship rule that could be used to calculate the charge if you know the number of hours worked.

____

9. How would the graph in Question 8 change if the plumber charged $60 an hour plus the $35 for the home visit? Why?

____

____

____

 Copyright © 2012 by Nelson Education Ltd.

Solutions and Key to Pathways

Name:______________________ Date:______________

Patterns

Diagnostic Tool

Pathway 3

1. Use the pattern rule to list the first 5 terms in each pattern.
   a) Start at 8 and keep adding 4.

      8, 12, 16, 20, 24,...
   b) Start at 56 and keep subtracting 2.

      56, 54, 52, 50, 48,...
2. Each model represents the first 4 figures of a pattern. Write the first 5 terms of its number pattern.
   a)

      1, 3, 5, 7, 9,...
   b)

      2, 5, 8, 11, 14,...
3. What is the 10th number in each pattern in Question 2.

   a) 19 b) 29
4. Write a pattern rule for each number pattern.
   a) 40, 43, 46, 49, 52,...

      Start at 40 and add 3 each time.
   b) 40, 37, 34, 31, 28,...

      Start at 40 and subtract 3 each time.
   c) 3, 6, 12, 24, 48,...

      Start at 3 and double each time.
   d) 5000, 1000, 200, 40, 8,...

      Start at 5000 and divide by 5 each time.

146 Patterns, Diagnostic Tool *Leaps and Bounds* Copyright © 2012 by Nelson Education Ltd.

Name:______________________ Date:______________

5. Use the number pattern 6, 12, 18, 24,....

a) Sketch a model to represent the first 4 numbers in the pattern. Describe how your model shows the number pattern.

e.g., It starts with a row of 6 and then a row of 6 is added each time.

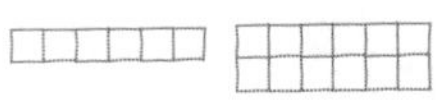
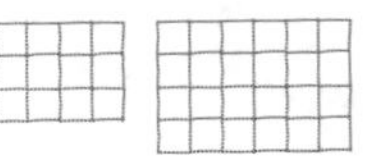

b) Complete this table of values.

Term number (x)	Term value (y)
1	6
2	12
3	18
4	24

c) Graph the table of values.

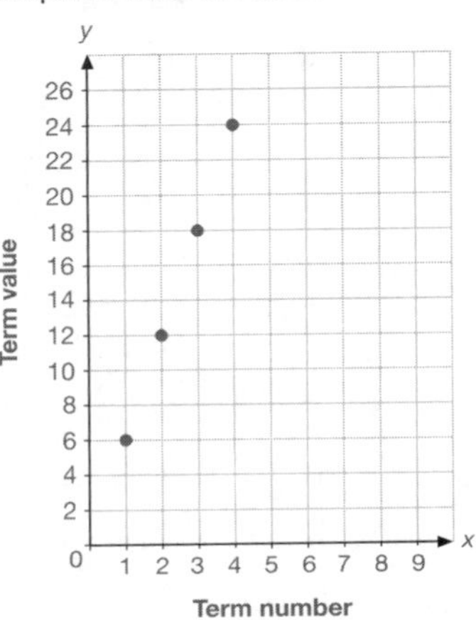

d) Write a relationship rule that describes how the term value relates to the term number for the pattern.

e.g., Term value = 6 × Term number, or $y = 6x$

6. To figure out the 100th term in the pattern in Question 5, why might you use the relationship rule instead of the table of values or graph?

e.g., I can just multiply 6 x 100, which is easy compared to continuing the table, or making the graph that wide and high.

Pathway 2

Copyright © 2012 by Nelson Education Ltd.

Copyright © 2012 by Nelson Education Ltd.

Name:______________________________ Date:__________________

7. The graph of a linear relation forms a straight line.
Which tables of values below represent a linear relation? A and C

Table A

Term number (x)	Term value (y)
1	4
2	7
3	10
4	13

Table B

Term number (x)	Term value (y)
1	4
2	9
3	16
4	25

Table C

Term number (x)	Term value (y)
1	4
2	9
3	14
4	19

8. A plumber charges $35 for a home visit, plus an additional $75 for each hour worked.

a) Complete the table of values to show how much the plumber would charge for different numbers of hours worked.

Plumber Charges

Number of hours	Charge ($)
1	110
2	185
3	260
4	335

b) Graph the table of values.

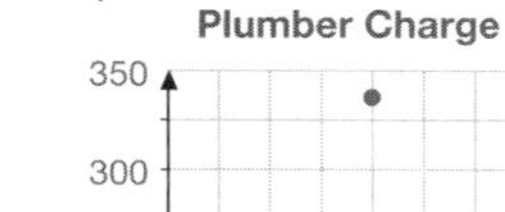

Plumber Charge

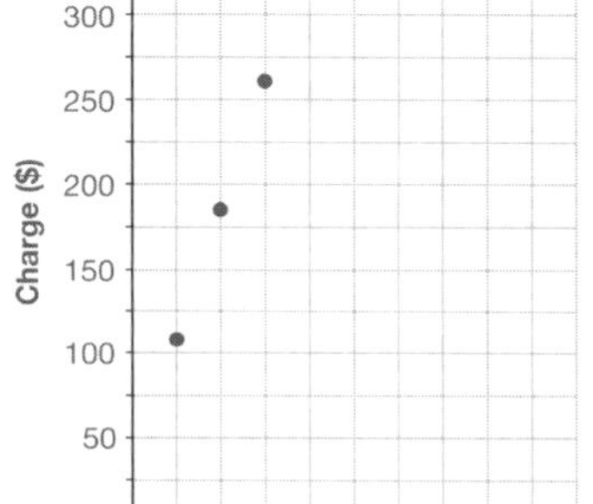

c) Write a relationship rule that could be used to calculate the charge if you know the number of hours worked.

e.g., Charge = 75 × number of hours + 35, or $c = 75h + 35$

9. How would the graph in Question 8 change if the plumber charged $60 an hour plus the $35 for the home visit? Why?

e.g., It would not be as steep, since you wouldn't go up the graph as much each time he worked another hour.

Pathway 1

 Copyright © 2012 by Nelson Education Ltd.

Copyright © 2012 by Nelson Education Ltd.

Linear Relations

You will need

- square tiles, linking cubes, or counters in 2 colours
- coloured pencils
- 1 cm Grid Paper (BLM 12) or 0.5 cm Grid Paper (BLM 13)
- calculators (optional)
- Student Resource pages 190–191

Term number (x)	Term value (y)

Open-Ended Intervention

Before Using the Open-Ended Intervention

Provide square tiles, linking cubes, or counters and ask students to model the pattern 2, 4, 6, 8,.... Then show them an empty table of values (as shown in the margin). Ask:

- How can you complete a table of values that relates the term value to the term number in the pattern?
  (e.g., *The first column would have the term numbers 1, 2, 3, 4 and the second column would have the term values 2, 4, 6, 8.*)
- Each pair of numbers in the table, for example, 1 and 2 or 2 and 4, is an ordered pair. How would you graph each pair?
  (e.g., *The term number is the x-coordinate, and the term value is the y-coordinate. For each, I'd start at (0, 0) and go right the first amount (the term number) and up the second amount (the term value).*)

Using the Open-Ended Intervention Student Resource pages 190–191

Read through the tasks on the student pages together. Provide grid paper and allow calculators. Give students time to work, ideally in pairs.

Observe whether students

- can create a growing linear number pattern
- can model a number pattern in a way that makes the growth apparent
- can create a table of values from a number pattern
- can graph a table of values
- can represent a linear number pattern with an equation
- recognize what makes a relationship linear

Consolidating and Reflecting

Ensure understanding by asking questions such as these based on students' work:

- Choose one of your patterns. What was easiest—modelling, making a table of values, or graphing? Why?
  (e.g., *The table was easiest, since you can just copy the numbers in the pattern.*)
- How do the table, graph, and rule all represent the same pattern?
  (e.g., *They all show how the pattern changes but in different ways.*)
- What would the table of values be useful for? (e.g., *continuing the pattern*)
- What would the graph be useful for?
  (e.g., *to see if the pattern is a linear relation*)
- What would the equation be useful for?
  (e.g., *to figure out the value of a term, such as the 50th term*)
- How are your graphs the same? different?
  (e.g., *They all go up the same amount each time, so they form straight lines. They have different slants.*)

 Copyright © 2012 by Nelson Education Ltd.

Guided Intervention

Before Using the Guided Intervention

Provide square tiles, linking cubes, or counters and ask students to model the pattern 6, 10, 14, 18,…. Ask:

- How can you explain your model? (e.g., *I used 6 cubes, then 4 more, then another 4, then another 4,….*)
- How can you use your model to predict the 10th number? (e.g., *Since adding 4 nine more times is* 9×4*, the 10th number would be* $6 + 36 = 42$*.*)
- What numbers in the 40s will appear in the pattern? How do you know? (e.g., *42 and 46 will be in the pattern since you add 4 to continue the pattern. Nothing else works because you can't subtract 4 from 42 or add 4 to 46 to get another number in the 40s.*)

You will need

- square tiles, linking cubes, or counters in 2 colours
- 1 cm Grid Paper (BLM 12) or 0.5 cm Grid Paper (BLM 13)
- calculators (optional)
- toothpicks (optional)
- Student Resource pages 192–196

Using the Guided Intervention Student Resource pages 192–196

Work through the instructional section of the student pages together. Have students use the table, the graph, and the equation to figure out the number of hours of volunteering after 10 weeks.

Have students work through the **Try These** questions individually or in pairs. Provide grid paper, calculators, and toothpicks for Question 6.

Observe whether students

- can relate a table of values and an equation to a graph (Question 1)
- can graph a table of values (Questions 2, 3, 5, 6)
- can create a table of values to represent a situation (Questions 3, 5, 6)
- can represent a linear situation with an equation (Questions 3, 5, 6, 7)
- can compare graphs or situations that describe linear relations (Questions 4, 8)
- recognize what makes a relationship linear (Question 5)

Consolidating and Reflecting

Ensure understanding by asking questions such as these based on students' work:

- In Question 1, how did the steepness of the points help you choose the table? (e.g., *If the numbers went up by a lot in a table, I knew to look for a steep graph.*)
- What makes the situation in Question 3 a linear relation? (e.g., *The cost went up by the same amount each hour.*)
- What was easiest for you in Question 6—modelling, making a table of values, or graphing? Why? (e.g., *the table, since you just copy the numbers in the pattern*)
- What would the table of values be useful for? (e.g., *continuing the pattern*)
- What would the graph be useful for? (e.g., *to see if the pattern is a linear relation*)
- Why might the equation in Question 7 be useful? (e.g., *to figure out the value of a term, such as the 50th term*)

Copyright © 2012 by Nelson Education Ltd.

Representing Patterns

You will need

- linking cubes, square tiles or counters in 2 colours
- 1 cm Grid Paper (BLM 12) or 0.5 cm Grid Paper (BLM 13)
- calculators (optional)
- Student Resource pages 197–198

Term number (x)	Term value (y)
1	4
2	8
3	12
4	16

Open-Ended Intervention

Before Using the Open-Ended Intervention

Model the pattern 4, 8, 12, 16,… using stacks of 4 linking cubes. Show 1 stack, then 2 stacks side by side, then 3 stacks side by side, and so on. Ask:

- What is the number pattern and pattern rule for this pattern? (e.g., *4, 8, 12, 16,…; Start at 4 and add 4 each time.*)
- Is the pattern familiar? (e.g., *Yes, it's the 4 times table.*)

Create a table of values for the cube pattern (as shown in the margin). Ask:

- How does the table help you see the relationship rule? (e.g., *The number on the right is always 4 times the number on the left.*)
- How would you graph the pattern? (e.g., *For each point, I'd use the number in the first column of the table as the x-coordinate, and the one in the second column as the y-coordinate.*)

Using the Open-Ended Intervention **Student Resource pages 197–198**

Read through the tasks on the student pages together. Give students time to work, ideally in pairs. Provide linking cubes, square tiles, or counters, and calculators for students who want them.

Observe whether students

- can model patterns
- can create a table of values for a pattern
- can create a graph from a table of values
- can write and use pattern rules (recursive) and relationship rules
- can compare representations of patterns in a meaningful way

Consolidating and Reflecting

Ensure understanding by asking questions such as these based on students' work:

- What was easiest for you—modelling, making a table of values, or graphing? Why? (e.g., *the table, since you just copy the numbers in the pattern*)
- What would the table of values be useful for? (e.g., *continuing the pattern*)
- What would the graph be useful for? (e.g., *seeing that the pattern is predictable since it forms a line*)
- Why might a relationship rule be more useful than a pattern rule? (e.g., *It would be easier to figure out the value of a term, such as the 50th term.*)
- Which pattern models are the most alike? Why? (e.g., *the second and third, since they are just times tables*)
- How were their graphs alike? (e.g., *Their points both formed a line that would go through (0, 0) if I extended them.*)

 Copyright © 2012 by Nelson Education Ltd.

Representing Patterns

Pathway 2
GUIDED

Guided Intervention

You will need

- linking cubes in 2 colours
- square tiles or counters in 2 colours
- coloured pencils
- calculators (optional)
- Student Resource pages 199–203

Before Using the Guided Intervention

Model the pattern 4, 8, 12, 16,… using stacks of 4 linking cubes. Show 1 stack, then 2 stacks side by side, then 3 stacks side by side, and so on. Ask:

- ▸ What is the number pattern and pattern rule for this pattern?
  (e.g., *4, 8, 12, 16,…; Start at 4 and add 4 each time.*)
- ▸ Is the pattern familiar?
  (e.g., *Yes, it's the 4 times table.*)
- ▸ What is another number pattern like this one?
  (e.g., *5, 10, 15, 20,…, since it's also a times table*)

Using the Guided Intervention (Student Resource pages 199–203)

Work through the instructional section of the student pages together. Provide linking cubes, square tiles, or counters, and allow calculator use.

Have students work through the **Try These** questions individually or in pairs.

Observe whether students

- can create a table of values for a pattern (Questions 1, 8)
- can create a graph from a table of values (Questions 2, 8)
- can model patterns and use a model to describe a pattern (Questions 3, 4, 5, 8)
- recognize that there are 2 types of rules, and can write and use pattern rules (recursive) and relationship rules to determine term values (Questions 6, 7)
- can compare representations of patterns in a meaningful way (Question 8)

Consolidating and Reflecting

Ensure understanding by asking questions such as these based on students' work:

- ▸ How were your tables in Question 1 different?
  (e.g., *For the first pattern, the numbers increased by 1, but for the second, they increased by 8.*)
- ▸ What did you do to your model in Question 3a) to make the pattern obvious?
  (e.g., *I added a row of 4 squares each time.*)
- ▸ Which rule in Question 6 is more useful? Why?
  (e.g. *Liam's is more useful, since it is easy to use to figure out any term; with Sindy's, you'd have to do a lot of adding.*)
- ▸ In Question 8, what was easiest—modelling, making a table of values, or graphing? Why?
  (e.g., *The table was easiest, since you just copy the numbers in the pattern.*)
- ▸ What would the table of values be useful for?
  (e.g., *continuing the pattern*)
- ▸ What would the graph be useful for?
  (e.g., *seeing that the pattern is predictable, since it forms a line*)

Copyright © 2012 by Nelson Education Ltd.

Pathway 3
OPEN-ENDED

Exploring Simple Patterns

You will need

- play coins (optional)
- square tiles, linking cubes, or counters in 2 colours
- coloured pencils
- base ten blocks (optional)
- Student Resource pages 204–205

Open-Ended Intervention

Before Using the Open-Ended Intervention

Ask students to skip count by 25s to 200. Ask:

- How is that a pattern?
  (e.g., *It goes up by the same amount every time: 25.*)
- How would you continue the pattern?
  (e.g., *I'd keep adding 25.*)
- How could you model the pattern using coins?
  (e.g., *1 quarter, then 2 quarters, then 3 quarters, and so on*)
- How would you figure out the 20th term?
  (e.g., *It takes 4 numbers to get to 100, so it would take 4 more to get to 200, 4 more to get to 300, 4 more to get to 400, and then 4 more to get to 500. That's 20 altogether, so the 20th term is 500.*)
- What would the pattern rule be if you had skip counted by 25s starting at 100 and went to 300?
  (e.g., *Start at 100 and add 25 each time.*)

Using the Open-Ended Intervention **Student Resource pages 204–205**

Read through the tasks on the student page together. Give students time to work, ideally in pairs. Provide square tiles, linking cubes, or counters, and provide play coins and base ten blocks for students who want them.

Observe whether students

- can create a pattern to meet a condition
- can create a table of values for a pattern
- can create a concrete or pictorial model for a pattern
- can write a pattern rule
- can extend a pattern to determine a term

Consolidating and Reflecting

Ensure understanding by asking questions such as these based on students' work:

- Choose one of your patterns. How did your model show how the pattern changed?
  (e.g., *I used a different colour square to show what I added each time, so what you add is really easy to see.*)
- How did you decide what model to use?
  (e.g., *If the numbers were big, I used base ten blocks. If the numbers were money values, like 5s, 10s, or 25s, I used coins. If they were small, I used tiles.*)
- How did you figure out where 20 was in your shrinking pattern?
  (e.g., *I started with 800 and went down by 20s, so I knew I had to go down by 780 altogether. That's 39 twenties, so I knew it was the 40th term.*)

 Copyright © 2012 by Nelson Education Ltd.

Exploring Simple Patterns

Pathway 3
GUIDED

Guided Intervention

Before Using the Guided Intervention

Have students skip count by 25s to 200. Ask:

- ▸ How is that a pattern?
(e.g., *It goes up by the same amount every time: 25.*)
- ▸ How would you continue the pattern?
(e.g., *I'd keep adding 25.*)
- ▸ How would you model the pattern using coins?
(e.g., *1 quarter, then 2 quarters, then 3 quarters, and so on*)

You will need

- play quarters
- square tiles, linking cubes, or counters in 2 colours
- coloured pencils
- base ten blocks
- Student Resource pages 206–209

Using the Guided Intervention Student Resource pages 206–209

Work through the instructional section of the student pages together. Provide square tiles, linking cubes, or counters and base ten blocks. Have students model the growing and shrinking patterns.

Have students work through the **Try These** questions individually or in pairs.

Observe whether students

- can write a pattern, given the pattern rule (Question 1)
- can extend a pattern (Questions 2, 3)
- can model a pattern (Question 3)
- can determine a pattern rule (Questions 4, 6, 9)
- can compare patterns (Question 5)
- can create a pattern to meet a condition (Questions 6, 7, 9)
- recognize that different patterns can start the same way (Question 8)

Consolidating and Reflecting

Ensure understanding by asking questions such as these based on students' work:

- ▸ How did you answer Question 2a)?
(e.g., *I kept the pattern going for 11 terms.*)
- ▸ How were your models in Question 3 different?
(e.g., *One showed increases of 5, while the second showed increases of 3.*)
- ▸ In Question 6, you created a pattern that grew quickly. How did you make that happen?
(e.g., *I used multiplication instead of addition, so it got bigger faster.*)
- ▸ What made your pattern in Question 9 unusual?
(e.g., *I added a different amount each time, but it's still a pattern because it's predictable.*)

Strand: Patterns and Algebra

Algebra

Planning For This Topic

Materials for assisting students with working with algebra consist of a diagnostic tool and 3 intervention pathways. Pathway 1 focuses on solving problems algebraically and using graphs. Pathway 2 involves preserving equality and solving simple equations. Pathway 3 explores the uses of variables.

Each pathway has an open-ended intervention and a guided intervention. Choose the type of intervention more suitable for your students' needs and your particular circumstances.

Curriculum Connections

Grades 5 to 8 curriculum connections for this topic are provided online. See www.nelson.com/leapsandbounds. All pathways are suitable for both the Ontario and WNCP curriculums, although solving problems using graphs is optional for Ontario students (part of Pathway 1).

Why might students struggle with algebra?

Students might struggle with using variables, modelling problems algebraically, and solving equations for any of the following reasons:

- They might confuse certain variables with numbers or other symbols (e.g., the letter o with the number 0, the letter x with the multiplication sign $\times$, and the letter l with the number 1).
- They might think that the variable is an abbreviation for a word rather than a representation of a varying quantity, since we sometimes do use a variable that reflects the context (e.g., c for cost).
- They might misinterpret an expression like $3m$ as a two-digit number in the 30s, not realizing that $3m$ means $3 \times m$, and not 30, 31, 32,…, or 39.
- They might experience more difficulty interpreting expressions like $10 - m$ (how much less m is than 10) than $m - 10$ (take away 10 from the value of m).
- They might be weak in number sense and have difficulty with the arithmetic involved in solving equations.
- They might solve the equation $30 + 4 = m + 10$ incorrectly as $34 = m$, by adding 30 and 4. Rather than thinking about the equation as a balance, they might think of the equals sign as signalling an answer to the calculation on the left.
- They might reverse the variables in a relationship (e.g., instead of writing $h = 24d$ to show that the number of hours is 24 times the number of days, they might write $d = 24h$, thinking a day has 24 hours).
- They might have more difficulty modelling equations like $2m - 1 = 21$ than $2m + 1 = 21$.

Professional Learning Connections

PRIME: Pattern & Algebra, Background and Strategies (Nelson Education Ltd., 2005), pages 61–66, 68–69, 72–76

Making Math Meaningful to Canadian Students K–8 (Nelson Education Ltd., 2008), pages 582–591

Big Ideas from Dr. Small Grades 4–8 (Nelson Education Ltd., 2009), pages 7–14

Good Questions (dist. by Nelson Education Ltd., 2009), pages 131, 134–135, 146–148

More Good Questions (dist. by Nelson Education Ltd., 2010), pages 20, 22–23, 25, 28, 47, 49

 Copyright © 2012 by Nelson Education Ltd.

- When using guess and test, they might not understand how the result of an initial guess can be used to improve the next guess.
- They might not recognize the algebra required to represent a real-life problem (i.e., how to write an equation to represent a problem situation), perhaps not understanding what the coefficient of the variable represents in the problem or the role of the constant value, if there is one.

Diagnostic Tool: Algebra

Use the diagnostic tool to determine the most suitable intervention pathway for algebra. Provide Diagnostic Tool: Algebra, Teacher's Resource pages 160 to 162, and have students complete it in writing or orally. Ontario teachers might choose to omit Question 11, since graphing is optional at this level in the Ontario curriculum.

See solutions on Teacher's Resource pages 163 to 165.

Intervention Pathways

The purpose of the intervention pathways is to help students recognize how variables are used in linear situations and how to model problems involving linear relationships and solve simple linear equations. Eventually, these concepts are extended to linear equations that are somewhat more complex and to non-linear situations.

There are 3 pathways:
- Pathway 1: Solving Problems Using Equations
- Pathway 2: Solving Simple Equations
- Pathway 3: Using Variables

Use the chart below (or the Key to Pathways on Teacher's Resource pages 163 to 165) to determine which pathway is most suitable for each student or group of students.

Diagnostic Tool Results	Intervention Pathway
If students struggle with Questions 9 to 11	use Pathway 1: Solving Problems Using Equations *Teacher's Resource pages 166–167* *Student Resource pages 210–215*
If students struggle with Questions 5 to 8	use Pathway 2: Solving Simple Equations *Teacher's Resource pages 168–169* *Student Resource pages 216–221*
If students struggle with Questions 1 to 4	use Pathway 3: Using Variables *Teacher's Resource pages 170–171* *Student Resource pages 222–227*

Name:________________________ Date:________________

Algebra

Diagnostic Tool

1. What is the value of each expression when $m = 4$ and when $m = -4$?

   a) When $m = 4$, $3m - 3 =$ ________.

   When $m = -4$, $3m - 3 =$ ________.

   b) When $m = 4$, $m + 8 =$ ________.

   When $m = -4$, $m + 8 =$ ________.

   c) When $m = 4$, $10 - 2m =$ ________.

   When $m = -4$, $10 - 2m =$ ________.

   d) When $m = 4$, $5 + 3m =$ ________.

   When $m = -4$, $5 + 3m =$ ________.

2. What does each expression mean?

   a) $k + 5$ ______________________________

   b) $3k$ ______________________________

   c) $40 - 5k$ ______________________________

3. Write an algebraic expression to match each description.

   a) Add 4 to a number. ________________

   b) Multiply a number by 3. Then add 5.

   ____

   c) Triple a number. Then subtract it from 100.

   ____

4. Write an algebraic equation to match each description.

   a) If you double a number and add 3, you get 25.

   ____

   b) If you subtract a number from 20, you end up with 13.

   ____

   c) Adding 4 to a number is the same as adding 3 and then adding 1 to a number.

   ____

Copyright © 2012 by Nelson Education Ltd.

Name:____________________ Date:____________

5. Each balance model below represents an equation with a variable. Write an equation for each model. Keep in mind the following:

- There is an unknown number of cubes in each bag.
- Bags on the same balance with the same variable have the same number of cubes in them.

a)

____

c)

____

b)

____

d)

____

6. Sketch bags and cubes on the balance scale to the right to model the equation $5m + 6 = 11$.

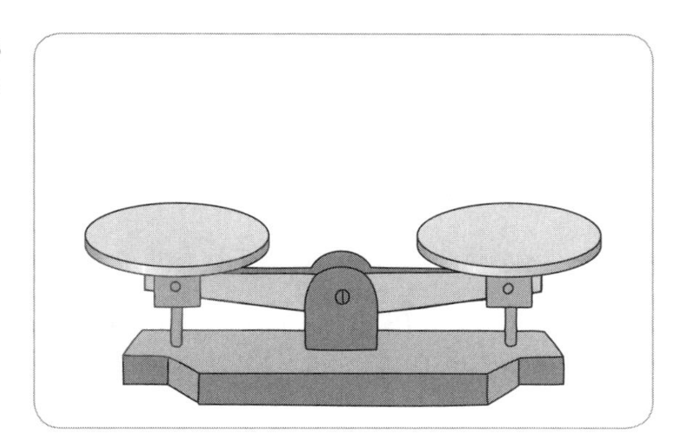

7. a) Why can you add 3 to both sides of the equation $2n - 3 = 51$ without changing the meaning of the equation or its solution?

____

____

b) How might adding 3 to both sides help you solve $2n - 3 = 51$?

____

____

Copyright © 2012 by Nelson Education Ltd.

Name:______________________________ Date:______________________

8. Solve each equation.

a) $h + 8 = 22$

c) $2h + 6 = 38$

b) $3h = 60$

d) $3h - 2 = 28$

9. Match each problem with an equation. One equation has no match.

You had \$30 in the bank and then you deposited \$10 each week. How long will it take to have \$300?	$10 + 30w = 300$
You had \$10 in the bank and then you deposited \$30 more each week. How long will it take to have \$300?	$30 + 10w = 300$
You had \$300 in the bank and then you deposited \$30 each week. How much will you have in 10 weeks?	$300 + 10w = 30$
	$m = 300 + (30 \times 10)$

10. Write an equation to represent this problem. Do not solve it.

A restaurant charges \$150 to rent the party room and \$25 per guest for food. How many guests were there if the total cost was \$1025?

____

11. This graph relates the perimeter and width of a rectangle that has a length of 20 cm for any width greater than 0. How would you use this graph to estimate the width of a rectangle with a perimeter of 76 cm?

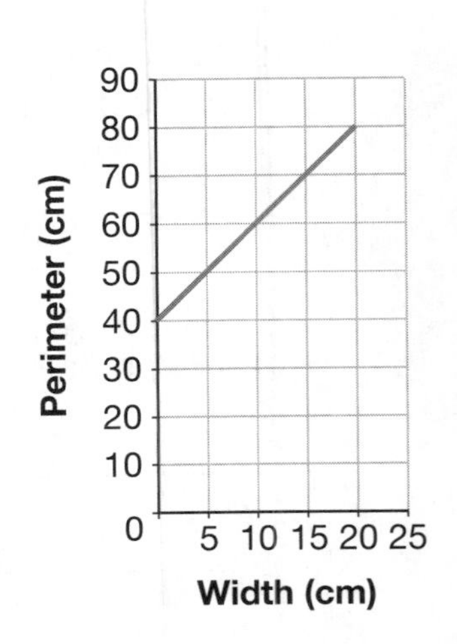

____

____

____

 Copyright © 2012 by Nelson Education Ltd.

Solutions and Key to Pathways

Name:________________________ Date:________________

Algebra

Diagnostic Tool

Pathway 3

1. What is the value of each expression when $m = 4$ and when $m = -4$?

   a) When $m = 4$, $3m - 3 =$ 9.

   When $m = -4$, $3m - 3 =$ −15.

   b) When $m = 4$, $m + 8 =$ 12.

   When $m = -4$, $m + 8 =$ 4.

   c) When $m = 4$, $10 - 2m =$ 2.

   When $m = -4$, $10 - 2m =$ 18.

   d) When $m = 4$, $5 + 3m =$ 17.

   When $m = -4$, $5 + 3m =$ −7.

2. What does each expression mean?

   a) $k + 5$ e.g., Add 5 to a number.

   b) $3k$ e.g., Multiply a number by 3.

   c) $40 - 5k$ e.g., Multiply a number by 5 and take it away from 40.

3. Write an algebraic expression to match each description.

   a) Add 4 to a number. e.g., $n + 4$

   b) Multiply a number by 3. Then add 5.

   e.g., $3n + 5$

   c) Triple a number. Then subtract it from 100.

   e.g., $100 - 3n$

4. Write an algebraic equation to match each description.

   a) If you double a number and add 3, you get 25.

   e.g., $2n + 3 = 25$

   b) If you subtract a number from 20, you end up with 13.

   e.g., $20 - m = 13$

   c) Adding 4 to a number is the same as adding 3 and then adding 1 to a number.

   e.g., $n + 4 = n + 3 + 1$

Copyright © 2012 by Nelson Education Ltd.

Copyright © 2012 by Nelson Education Ltd.

Name:_______________________ Date:_______________

5. Each balance model below represents an equation with a variable. Write an equation for each model. Keep in mind the following:
   - There is an unknown number of cubes in each bag.
   - Bags on the same balance with the same variable have the same number of cubes in them.

a)

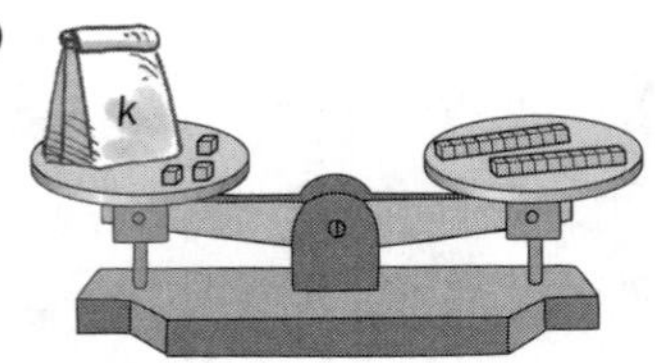

$k + 3 = 20$

c)

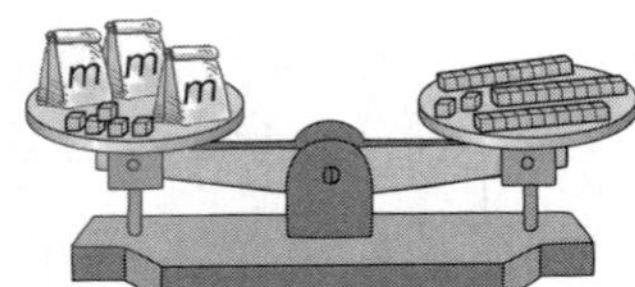

$3m + 5 = 32$

b)

$2s = 20$

d)

$4b + 1 = 21$

Pathway 2

6. Sketch bags and cubes on the balance scale to the right to model the equation $5m + 6 = 11$.

e.g.,

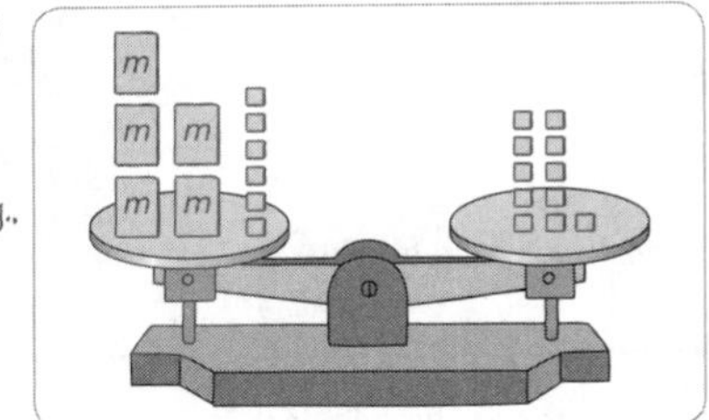

7. **a)** Why can you add 3 to both sides of the equation $2n - 3 = 51$ without changing the meaning of the equation or its solution?

e.g., If you do the same thing to each side, you don't change the balance.

b) How might adding 3 to both sides help you solve $2n - 3 = 51$?

e.g., It's easier to figure out the solution to $2n = 54$ than to $2n - 3 = 51$.

Copyright © 2012 by Nelson Education Ltd.

Copyright © 2012 by Nelson Education Ltd.

Name:______________________ Date:______________

Pathway 2

8. Solve each equation.

a) $h + 8 = 22$

e.g., $h = 14$

b) $3h = 60$

e.g., $h = 60 \div 3$
$h = 20$

c) $2h + 6 = 38$

e.g.,
$2h + 6 - 6 = 38 - 6$
$2h = 32$
$2h \div 2 = 32 \div 2$
$h = 16$

d) $3h - 2 = 28$

e.g.,
$3h - 2 + 2 = 28 + 2$
$3h = 30$
$h = 10$

Pathway 1

9. Match each problem with an equation. One equation has no match.

Problem	Equation
You had \$30 in the bank and then you deposited \$10 each week. How long will it take to have \$300?	$30 + 10w = 300$
You had \$10 in the bank and then you deposited \$30 more each week. How long will it take to have \$300?	$10 + 30w = 300$
You had \$300 in the bank and then you deposited \$30 each week. How much will you have in 10 weeks?	$m = 300 + (30 \times 10)$

Equations listed: $10 + 30w = 300$; $30 + 10w = 300$; $300 + 10w = 30$; $m = 300 + (30 \times 10)$

10. Write an equation to represent this problem. Do not solve it.

A restaurant charges \$150 to rent the party room and \$25 per guest for food. How many guests were there if the total cost was \$1025?

$150 + 25g = 1025$

11. This graph relates the perimeter and width of a rectangle that has a length of 20 cm for any width greater than 0. How would you use this graph to estimate the width of a rectangle with a perimeter of 76 cm?

e.g., Go to $P = 76$ cm on the vertical axis. Then go right to the graph and then down to the horizontal axis and read the number. It's about 18 cm.

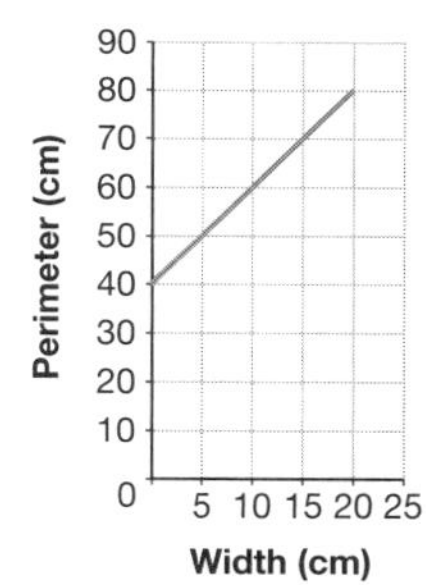

Copyright © 2012 by Nelson Education Ltd.

Copyright © 2012 by Nelson Education Ltd.

Pathway 1 OPEN-ENDED

Solving Problems Using Equations

You will need

- pan balance, paper bags, and base ten blocks or linking cubes (optional)
- 1 cm Grid Paper (BLM 12) or 0.5 cm Grid Paper (BLM 13)
- algebra tiles: *x*-tiles and 1-tiles (optional)
- calculators (optional)
- Student Resource pages 210–211

$2l + 8 = 60$

Open-Ended Intervention

Before Using the Open-Ended Intervention

Tell students that you are imagining a room with 60 people. There is a large group at the back of the room and a group of 8 in front. Ask:

- What equation would you write to figure out the number of people in the large group? Why?
  (e.g., *$l + 8 = 60$; There are 60 altogether, made up of a large group (l) and a group of 8.*)
- How could you use a balance model to represent and solve the equation?
  (e.g., *I'd put 1 empty bag and 8 cubes on one side and 60 cubes on the other side, and then add the same number of cubes to each bag until it balances. The number of cubes in each bag is the solution.*)
- How else could you figure out the number of people in the large group?
  (e.g., *Subtract 8 from both sides of the equation.*)
- What equation and balance model would represent the situation if there were 2 groups in the back and a group of 8 in the front?
  (e.g., *$2l + 8 = 60$; 2 empty bags and 8 cubes on one side, 60 cubes on the other*)

Using the Open-Ended Intervention (Student Resource pages 210–211)

Read through the tasks and the Remember Box on the student pages together. Ensure students have materials for modelling equations. Students might choose to read through the instructional sections of Pathways 1, 2, and 3 guided interventions for descriptions of various models for solving equations (graphing, length model, balance model, and algebra tiles).

Give students time to work, ideally in pairs.

Observe whether students

- can write a linear equation to represent a problem situation
- can solve a linear equation in order to solve a problem

Consolidating and Reflecting

Ensure understanding by asking questions such as these based on students' work:

- Which problem was easiest to model with an equation? Why?
  (e.g., *$2p + 1 = 1$; It's obvious that, if you had pairs of socks and 1 extra, you'd write $2p + 1$.*)
- Which model did you use to solve each equation? Why?
  (e.g., *Sometimes I used a balance model, but I visualized it instead of drawing it. Sometimes I graphed it, if it wasn't hard to draw.*)
- How can you tell from a problem what the equation should be?
  (e.g., *If something grows by a fixed amount, the fixed amount is multiplied by the variable. And if there's a starting amount, it's included in the equation.*)

Copyright © 2012 by Nelson Education Ltd.

Solving Problems Using Equations

Pathway 1 GUIDED

Guided Intervention

Before Using the Guided Intervention

Tell students that you are imagining a room with 60 people. There is a large group at the back of the room and a group of 8 in front. Ask:

- What equation would you write to figure out the number of people in the large group? Why?
  (e.g., *$l + 8 = 60$; There are 60 altogether, made up of a large group (l) and a group of 8.*)
- How could you use a balance model to represent and solve the equation?
  (e.g., *I'd put 1 empty bag and 8 cubes on one side and 60 cubes on the other side, and then add cubes to the bag until it balances. The number of cubes in the bag is the solution.*)
- How else could you figure out the number of people in the large group?
  (e.g., *Subtract 8 from both sides of the equation.*)

You will need

- pan balance, paper bags, and base ten blocks or linking cubes (optional)
- 1 cm Grid Paper (BLM 12) or 0.5 cm Grid Paper (BLM 13)
- algebra tiles: x-tiles and 1-tiles (optional)
- calculators (optional)
- Student Resource pages 212–215

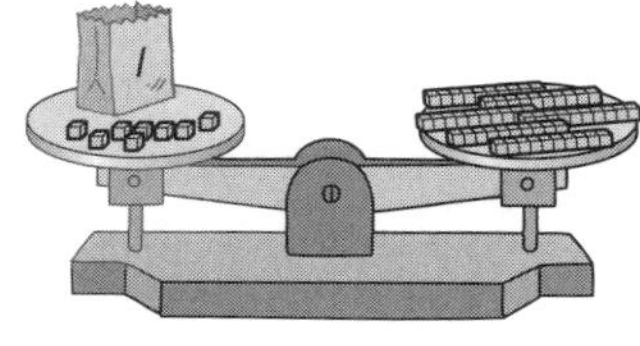

$l + 8 = 60$

Using the Guided Intervention Student Resource pages 212–215

Work through the instructional section of the student pages together. Ensure students have grid paper for graphing and materials for modelling equations. Students might choose to read through the instructional sections of Pathways 2 and 3 guided interventions for descriptions of other models for solving equations (balance model and algebra tiles).

Have students work through the **Try These** questions individually or in pairs.

Observe whether students

- can relate an equation to a linear situation (Questions 1, 2, 3, 4, 5, 7)
- can solve a linear equation (Questions 2, 3, 4, 5, 6, 7)
- can create a problem for an equation (Questions 6, 7)

Consolidating and Reflecting

Ensure understanding by asking questions such as these based on students' work:

- How did you solve your equation in Question 5?
  (e.g., *I imagined a balance model, took away the same amount from both sides, and then divided both sides by the same amount, all in my head.*)
- Which model did you use to solve each equation?
  (e.g., *Sometimes I used a balance model, but I visualized it instead of drawing it. Sometimes I graphed it, if it wasn't hard to draw.*)
- How can you tell from a problem what the equation should be?
  (e.g., *If something grows by a fixed amount, the fixed amount is multiplied by the variable. And if there's a starting amount, it's included in the equation.*)

Pathway 2
OPEN-ENDED

Solving Simple Equations

You will need

- pan balance, paper bags, and base ten blocks or linking cubes
- Student Resource pages 216–217

Open-Ended Intervention

Before Using the Open-Ended Intervention

Ask:

- What do you think an equation is? (e.g., *something with an equals sign*)
- What does the equals sign mean? (e.g., *Both sides are worth the same.*)
- When is each of these equations true?
  $4 + 3 = 7$ ■ $+ 3 = 7$ $n + 3 = 7$ $n + 2 = n + 1 + 1$
  (e.g., *The first one is a true statement. The last one is true no matter what value you substitute for n. The middle ones are only true when* ■ *or* $n = 4$.)
- What do you think $3 + m = 5$ means?
  (e.g., *If you add a number to 3, you get 5.*)

Using the Open-Ended Intervention Student Resource pages 216–217

Read through the tasks and the Remember Box on the student pages together. Ensure students have materials for modelling the equations. Students could read through the instructional section of the Pathway 2 guided intervention for more in-depth descriptions of the length and balance models. Ensure students understand that equations like $3k - 2 = 22$ (with a negative constant) are easier to solve using an equivalent equation, $3k = 24$, or using a length model, instead of a balance model.

Give students time to work, ideally in pairs.

Observe whether students

- can interpret a model of an equation
- can model an equation
- can solve a simple equation
- understand that equations resulting from solving an equation are equivalent (e.g., $3k + 2 = 20$ is equivalent to $3k = 18$, which is equivalent to $k = 6$)

Consolidating and Reflecting

Ensure understanding by asking questions such as these based on students' work:

- Which model did you prefer? Why?
  (e.g., *the balance model, since I could use real objects*)
- Which model did you use for the equation involving subtracting? Why?
  (e.g., *the length model; I wasn't sure how to show* $k - 12$ *on a balance.*)
- How did your model show the number that was multiplied by the variable?
  (e.g., *The 4 bags on one side showed that the variable was multiplied by 4.*)
- Why did you sometimes take cubes away from both sides of the balance?
  (e.g., *It's easier to see what the solution is, and it doesn't change the equation because both sides still balance.*)

 Copyright © 2012 by Nelson Education Ltd.

Solving Simple Equations

Pathway 2 GUIDED

Guided Intervention

You will need
- pan balance, paper bags, and base ten blocks or linking cubes
- Student Resource pages 218–221

Before Using the Guided Intervention

Ask:
- What do you think an equation is?
  (e.g., *something with an equals sign*)
- What does the equals sign mean?
  (e.g., *Both sides are worth the same.*)
- When is each of these equations true?
  $4 + 3 = 7$ $\blacksquare + 3 = 7$ $n + 3 = 7$ $n + 2 = n + 1 + 1$
  (e.g., *The first one is a true statement. The last one is true no matter what value you substitute for n. The middle ones are only true when* $\blacksquare$ *or* $n = 4$.)
- What do you think $3 + m = 5$ means?
  (e.g., *If you add a number to 3, you get 5.*)

Using the Guided Intervention (Student Resource pages 218–221)

Work through the instructional section of the student pages together. Ensure students have materials for modelling the equations. Ensure students understand that equations like $3k - 2 = 22$ (with a negative constant) are easier to solve using an equivalent equation ($3k = 24$) or a length model, instead of a balance model.

Have students work through the **Try These** questions individually or in pairs.

Observe whether students
- can interpret a model of an equation (Questions 1, 2)
- can model an equation (Question 3)
- understand when various forms of equations are equivalent (Questions 4, 5, 6, 7, 10)
- can solve equations (Questions 8, 9, 11)

Consolidating and Reflecting

Ensure understanding by asking questions such as these based on students' work:
- How did the model in Question 1b) show the number that was multiplied by the variable?
  (e.g., *The 2 bags on one side showed that the variable was multiplied by 2.*)
- Why did the length model make sense for Question 5?
  (e.g., *If I used a balance, I'd have to take something out of one of the x bags but not the other, and then they wouldn't be worth the same amount.*)
- Which model did you prefer? Why?
  (e.g., *the length model, since I could use it for any equation*)
- Why did you sometimes take cubes away from both sides of the balance?
  (e.g., *It's easier to see what the solution is, and it doesn't change the equation because both sides still balance.*)

Copyright © 2012 by Nelson Education Ltd.

Using Variables

You will need

- algebra tiles: x-tiles and 1-tiles
- Student Resource pages 222–223

Open-Ended Intervention

Before Using the Open-Ended Intervention

Put out 3 long algebra tiles (x-tiles) and 2 square tiles (1-tiles). Ask:

- The square tile is worth 1. Let's name the long tile k. How do you know this model represents the expression $3k + 2$?
  (e.g., *There are 3 k's and 2 ones.*)
- How would you model $2k + 5$?
  (*2 long tiles and 5 square tiles*)
- Suppose the long tile was worth 10. What is $2k + 5$ worth? (*25*)
- The dark sides of the tiles represent $-k$ and -1. How would you represent $2k - 5$?
  (*2 white long tiles and 5 dark square tiles*)
- What does the equation $P = 2l + 2w$ mean?
  (e.g., *The perimeter of a rectangle is the sum of 2 times the length and 2 times the width.*)

Using the Open-Ended Intervention (Student Resource pages 222–223)

Read through the tasks on the student pages together, including the definitions and Remember box. Provide algebra tiles. Give students time to work, ideally in pairs.

Observe whether students

- recognize when a variable is used in an equation to represent a single unknown value
- recognize when variable(s) are used in an equation that represents a relationship
- can describe algebraic expressions and equations in words
- can substitute for a variable in an expression to evaluate it
- can solve simple equations

Consolidating and Reflecting

Ensure understanding by asking questions such as these based on students' work:

- How do you know what words to use when describing an algebraic expression or equation?
  (e.g., *The number with the variable tells what happens to the variable, so if the number is 2, it's doubled. I also look to see if there is adding or subtracting.*)
- Compare the expressions $2j + 4$ and $4j + 2$. When are they the same? When are they different?
  (e.g., *Both expressions are worth 6 if $j = 1$; otherwise, they are worth different values. If j is a lot, $4j + 2$ is worth more than $2j + 4$.*)
- What things did you change about your expression or equation?
  (e.g., *the operation sign and then one of the numbers*)
- How are the descriptions of an equation and an expression different?
  (e.g., *For an equation, you have to say that 2 things are the same or equal.*)

 Copyright © 2012 by Nelson Education Ltd.

Using Variables

Pathway 3 GUIDED

Guided Intervention

Before Using the Guided Intervention

Put out 3 long algebra tiles (x-tiles) and 2 square tiles (1-tiles). Ask:

- The square tile is worth 1. Let's name the long tile k. How do you know this model represents the expression $3k + 2$? (e.g., *There are 3 k's and 2 ones.*)
- How would you model $2k + 5$? (*2 long tiles and 5 square tiles*)
- Suppose the long tile was worth 10. What is $2k + 5$ worth? (*25*)
- The dark sides of the tiles represent $-k$ and -1. How would you represent $2k - 5$? (*2 white long tiles and 5 dark square tiles*)

Using the Guided Intervention Student Resource pages 224–227

Provide algebra tiles and work through the instructional section of the student pages together.

Have students work through the **Try These** questions individually or in pairs.

Observe whether students

- recognize when a variable is used in an equation to represent a single unknown value (Question 1)
- recognize when variable(s) are used in an equation that represents a relationship (Question 1)
- relate verbal expressions/statements to algebraic expressions/equations (Questions 2, 3, 7, 9)
- can model algebraic expressions (Question 4)
- can substitute for a variable in an expression to evaluate it (Question 5)
- can write an equivalent expression by simplifying (Question 6)
- can solve simple equations (Question 8)
- recognize how expressions and equations differ (Question 9)

Consolidating and Reflecting

Ensure understanding by asking questions such as these based on students' work:

- What did you think about when you matched the expressions in Question 2? (e.g., *The number with the letter tells what happens to the variable, so if the number next to the letter is 2, the variable is doubled. I also look to see if there is adding or subtracting.*)
- Why were your models for Questions 4b) and c) different? (e.g., *One had more variable tiles and the other had more 1-tiles.*)
- How could you have used a model for Question 6a)? (e.g., *I'd show 3 squares and 1 rectangle, and 4 more squares and 1 rectangle for $3 + m + 4 + m$. Then I'd combine like tiles to get 7 squares and 2 rectangles, which is $2m + 7$.*)
- How are descriptions of an equation and an expression different? (e.g., *For an equation, you have to say that 2 things are the same or equal.*)

You will need

- algebra tiles: x-tiles and 1-tiles
- Student Resource pages 224–227

Copyright © 2012 by Nelson Education Ltd.

Strand: Geometry

Geometry Strand Overview

How were the geometry topics chosen?

This resource provides materials for assisting students with 5 geometry topics. These topics were drawn from the curriculum outcomes from across the country for Grades 5 to 7. Topics are divided into distinct levels, called pathways, that address gaps in students' prerequisite skills and knowledge.

How were the pathways determined?

Geometry topics in Grades 5 to 7 include continued exploration of 2-D and 3-D shapes, location, and transformations, as well as the newer topic of drawings.

The 3-D geometry pathways focus on representations. There are 3 pathways to cover nets, different points of view, and isometric drawings.

The 2-D geometry pathways cover sorting and classifying polygons, determining congruence, and recognizing and applying properties of similarity.

Four pathways for geometric drawings include triangles, circles, lines and polygons, and bisecting angles and line segments. Traditionally, constructions were limited to using straightedges and compasses, but current curriculums acknowledge the value of using additional tools such as transparent mirrors, paper folding, protractors, and dynamic geometry software. The topic reflects that flexibility.

The location pathways focus on plotting points in Quadrant 1 and in all 4 quadrants.

The transformations pathways examine either single transformations that do not change sizes of shapes, combined transformations, dilatations (which preserve shape but not size), and the use of all of these transformations in designs.

What geometry topics have been omitted?

Some topics related to geometry appear in the measurement pathways or are imbedded in other topics: angle measurement; angle sums in triangles and quadrilaterals; the relationships between radius, diameter, and circumference of circles; shapes that tile planes; and the angles between the faces of a prism.

 Copyright © 2012 by Nelson Education Ltd.

Materials

Materials for assisting students with geometry topics are likely in the classroom or easily accessible. Blackline masters are provided at the back of this resource. Many of these materials are optional and are listed only in the Teacher's Resource.

- linking cubes, pattern blocks
- models of 3-D shapes
- sticky notes or coloured dots
- a digital or document camera
- a box (e.g., an empty cereal box)
- connecting faces (e.g., Polydrons)
- scissors, tape, string
- protractors, rulers, compasses
- calculators
- transparent mirrors (e.g., Miras)
- transparent centimetre grids
- dynamic geometry software (optional)

BLMs 11, 12, 13: Grid Paper
BLMs 14 and 15: Square Dot Paper
BLM 16: Triangle Dot Paper
BLM 17: Circular Geoboard Paper
BLM 18: Fraction Circles/Spinners
BLM 26: Regular Polygons
BLM 27: Polygons
BLM 28: Star

Measurement Topics and Pathways

Topics and pathways in this strand are shown below.
Each pathway has an open-ended intervention and a guided intervention.

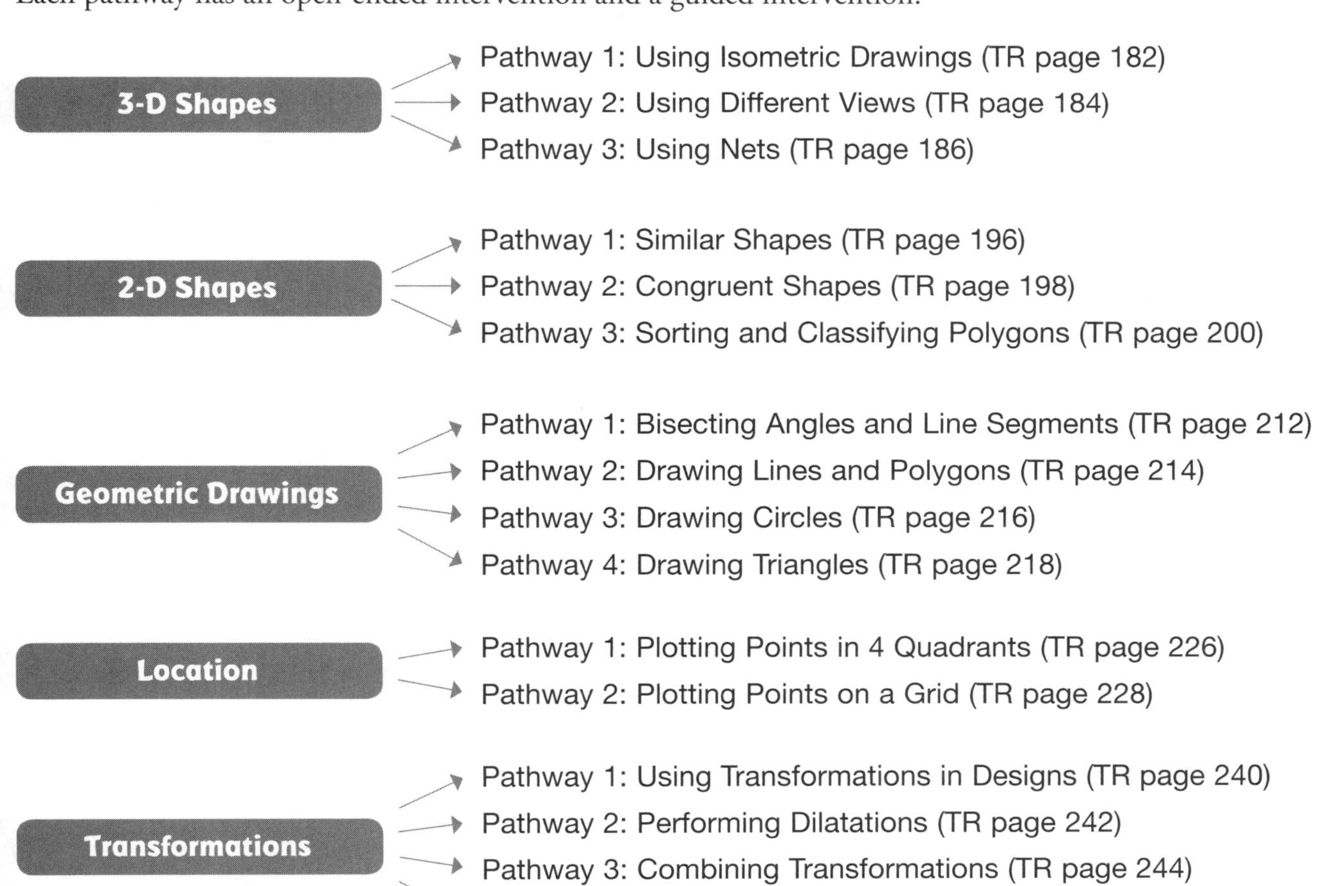

3-D Shapes

Planning For This Topic

Materials for assisting students with 3-D shapes consist of a diagnostic tool and 3 intervention pathways. Pathway 1 focuses on isometric drawings of rectangular structures. Pathway 2 focuses on different views of rectangular structures (i.e., top, front, side). Pathway 3 focuses on nets of 3-D shapes.

Each pathway has an open-ended intervention and a guided intervention. Choose the type of intervention more suitable for your students' needs and your particular circumstances.

Professional Learning Connections

PRIME: Geometry, Background and Strategies (Nelson Education Ltd., 2007), pages 67–71
Making Math Meaningful to Canadian Students K–8 (Nelson Education Ltd., 2008), pages 305–309
Big Ideas from Dr. Small Grades 4–8 (Nelson Education Ltd., 2009), pages 107–110
Good Questions (dist. by Nelson Education Ltd., 2009), pages 69, 89–90
More Good Questions (dist. by Nelson Education Ltd., 2010), pages 96, 98

Curriculum Connections

Grades 5 to 8 curriculum connections for this topic are provided online. See www.nelson.com/leapsandbounds. The content in all 3 pathways is introduced in Grade 8 of the WNCP curriculum, so students at that level can use the material in their core texts, rather than these pathways.

Why might students struggle with 3-D Shapes?

Students might struggle with 3-D shapes for any of the following reasons:

- They might have difficulty visualizing how a net folds into a 3-D shape and how a 3-D shape unfolds into a net.
- They might not realize that a net must not only have the correct number and shapes of faces, but the shapes must be connected and arranged so that when the net is folded, it forms a closed 3-D shape with no overlapping faces.
- They might not be able to recognize the different views of a 3-D shape (e.g., the picture here shows the top, front, and right-side views).

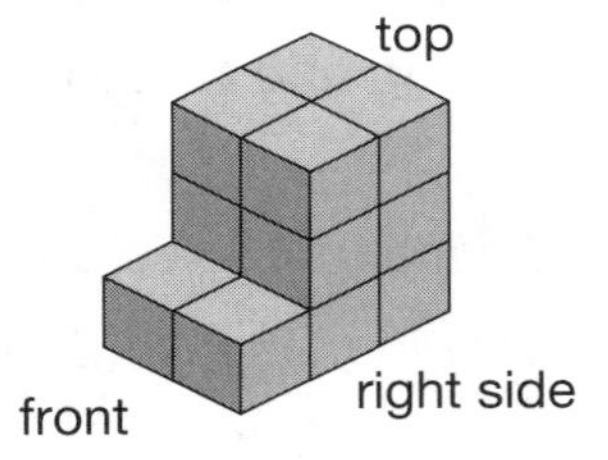

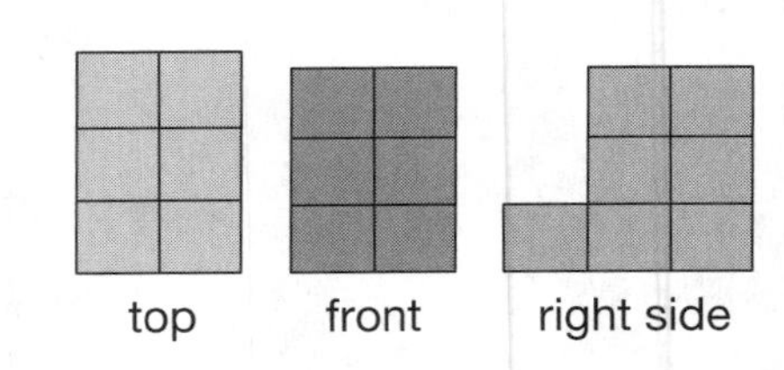

- They might not realize that some views (i.e., top, front, side) do not show all parts of a 3-D shape.
- They might have difficulty drawing an isometric view of a 3-D shape using triangle dot paper, also called isometric paper. (The drawing at the left has the front faces shaded, which may help students see the 3-D shape better.)

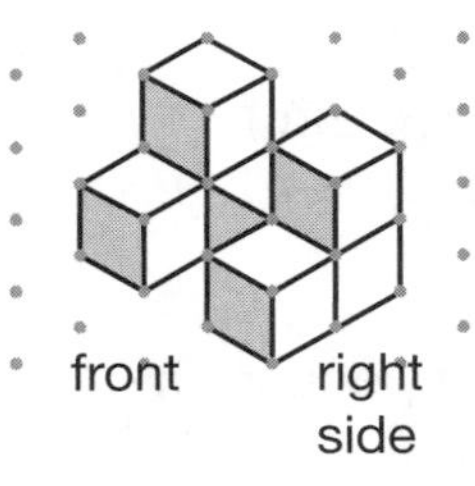

 Copyright © 2012 by Nelson Education Ltd.

Diagnostic Tool: 3-D Shapes

Use the diagnostic tool to determine the most suitable intervention pathway for 3-D Shapes. Provide Diagnostic Tool: 3-D Shapes, Teacher's Resource pages 176 to 178, and have students complete it in writing or orally. Have linking cubes and 3-D models available.

See solutions on Teacher's Resource pages 179 to 181.

Intervention Pathways

The purpose of the intervention pathways is to help students visualize the shape and number of the faces of 3-D shapes, so that they can represent 3-D shapes using isometric drawings, drawings of different views (also called orthographic drawings), and nets. This will be useful, at later grades, in measurement for determining volume and surface area of a wider variety of 3-D shapes (e.g., right prisms, pyramids, cylinders).

There are 3 pathways:
- Pathway 1: Using Isometric Drawings
- Pathway 2: Using Different Views
- Pathway 3: Using Nets

Use the chart below (or the Key to Pathways on Teacher's Resource pages 179 to 181) to determine which pathway is most suitable for each student or group of students. Pathways 1, 2, and 3 are somewhat independent. Use the diagnostic tool to decide which are required. Begin with Pathway 3 if more than one pathway is required; it is generally viewed as the simplest of the three.

Diagnostic Tool Results	Intervention Pathway
If students struggle with Questions 7 to 9	use Pathway 1: Using Isometric Drawings *Teacher's Resource pages 182–183* *Student Resource pages 228–233*
If students struggle with Questions 4 to 6	use Pathway 2: Using Different Views *Teacher's Resource pages 184–185* *Student Resource pages 234–239*
If students struggle with Questions 1 to 3	use Pathway 3: Using Nets *Teacher's Resource pages 186–187* *Student Resource pages 240–244*

Copyright © 2012 by Nelson Education Ltd.

Name:______________________________ Date:______________________

3-D Shapes

Diagnostic Tool

1. Will this net fold into a pyramid or a prism? Tell how you know.

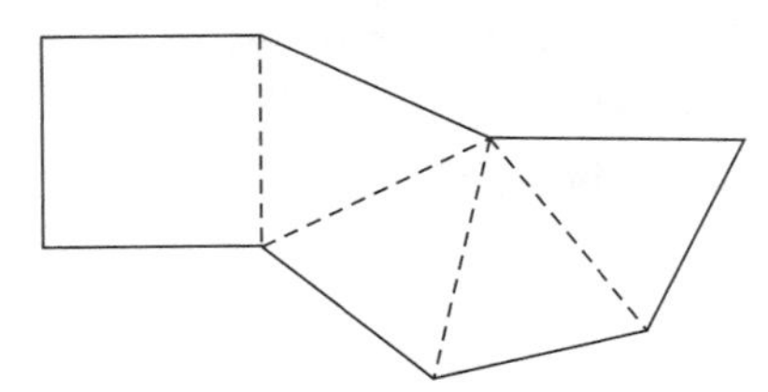

____

____

2. How would the nets of these 2 prisms be the same and different?

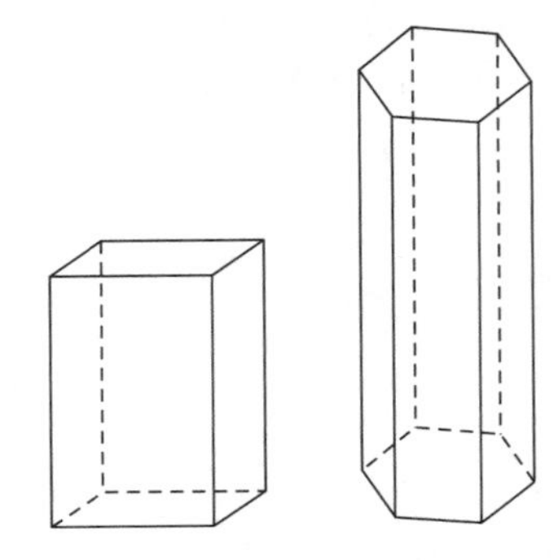

same: ______________________________

____

different: ______________________________

____

____

3. Sketch 2 different nets for a box like this triangle-based prism.

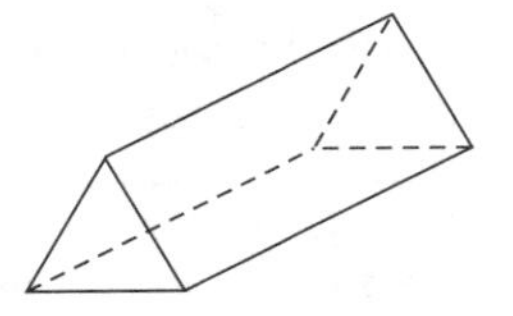

You will need

- linking cubes
- models of 3-D shapes
- a ruler

Remember

- A net is a 2-D model of a 3-D shape that shows all of its faces and can be folded to make the 3-D shape.
- A prism has 2 identical polygons at opposite ends, and other faces that are rectangles.

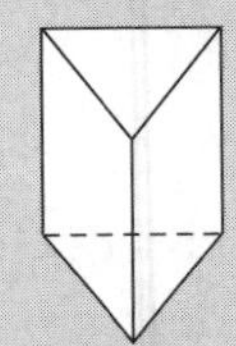

triangle-based prism

- A pyramid has 1 polygon base, and other faces that are triangles.

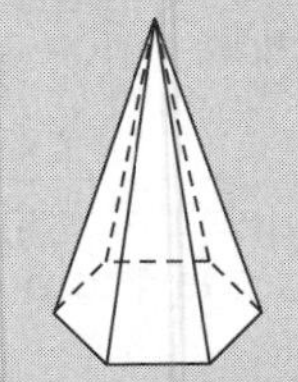

hexagon-based pyramid

 Copyright © 2012 by Nelson Education Ltd.

Name:______________________________ Date:______________

4. Which top view below matches the cube structure at the right? Circle the correct top view.

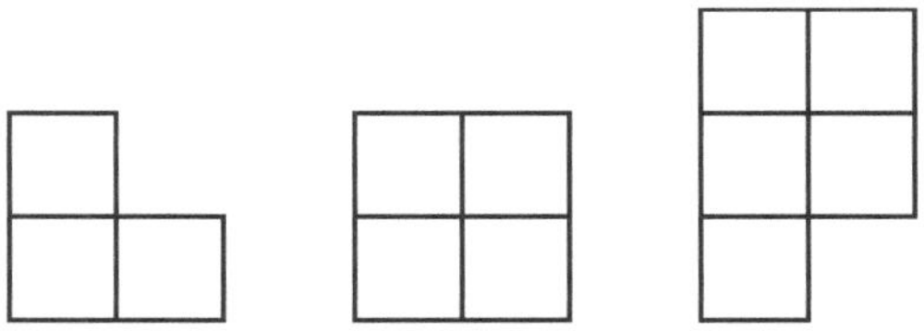

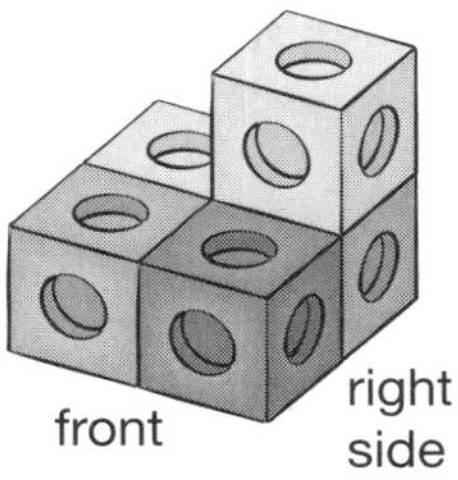

front

right side

5. Which cube structure below matches the 3 views at the right? Circle the correct cube structure.

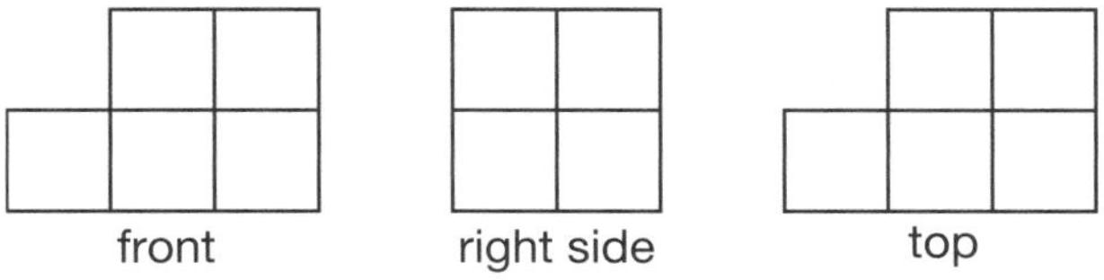

front

right side

top

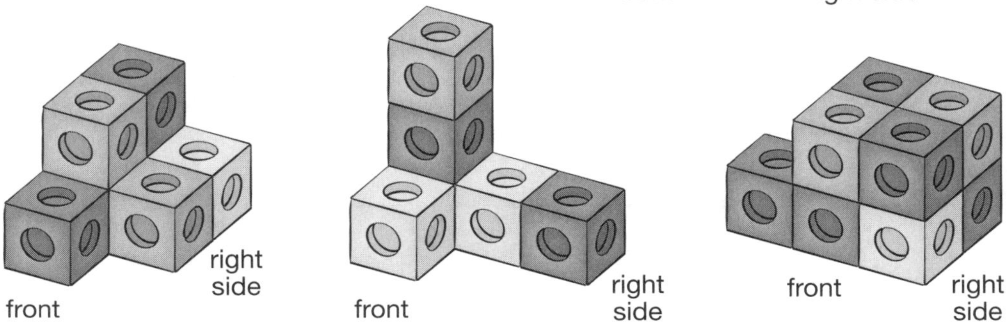

front

right side

front

right side

front

right side

6. Sketch the 3 different views—left side, front, and top—for these stairs.

front

Name: ______________________________ Date: ____________________

7. Use the 2 drawings below to build 2 cube structures using linking cubes. How are your cube structures the same and different?

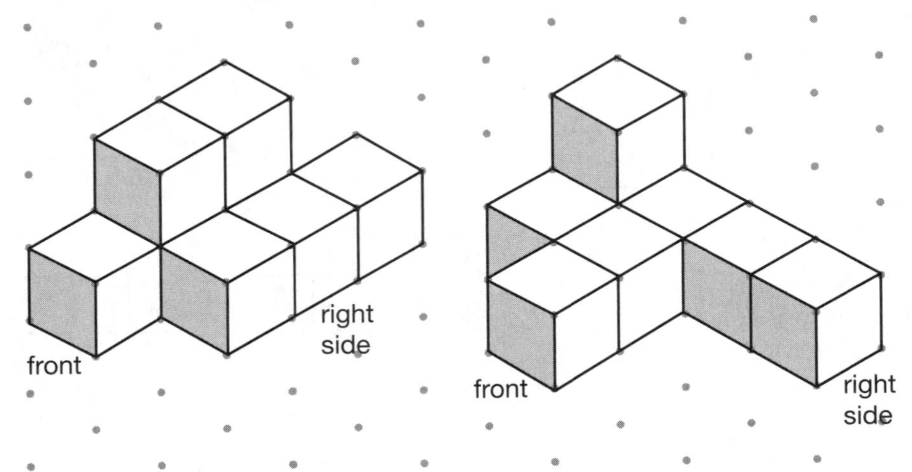

same: ______________________________

____

different: ______________________________

____

____

8. Examine the cube structure drawing at the right. How many linking cubes might this structure have? List the possibilities.

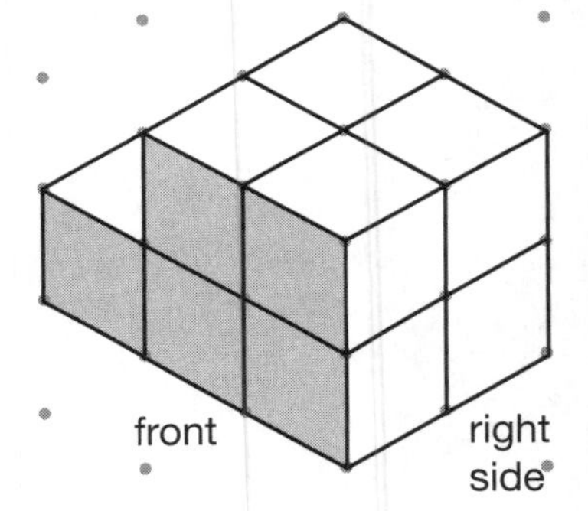

____

9. a) Build a cube structure using linking cubes where
   - the bottom layer shows 3 rows of 3 cubes, and
   - the 2nd layer shows 1 cube in the middle on top of the 1st layer.

   How many linking cubes did you use?

   ____

   b) Sketch an isometric drawing (like the ones in Questions 7 and 8) of your cube structure on the triangle dot paper.

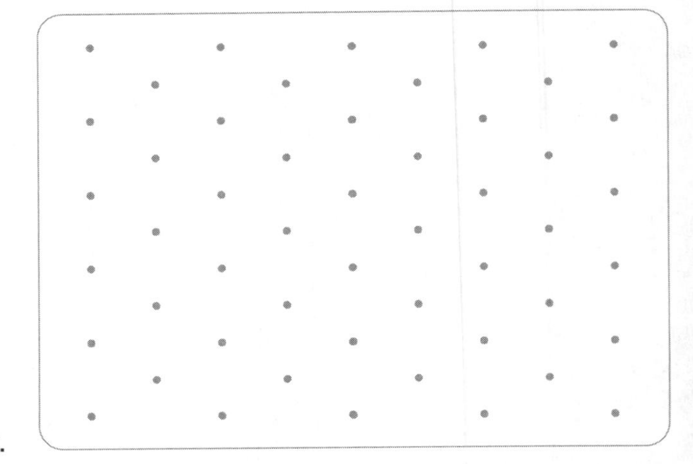

 Copyright © 2012 by Nelson Education Ltd.

Solutions and Key to Pathways

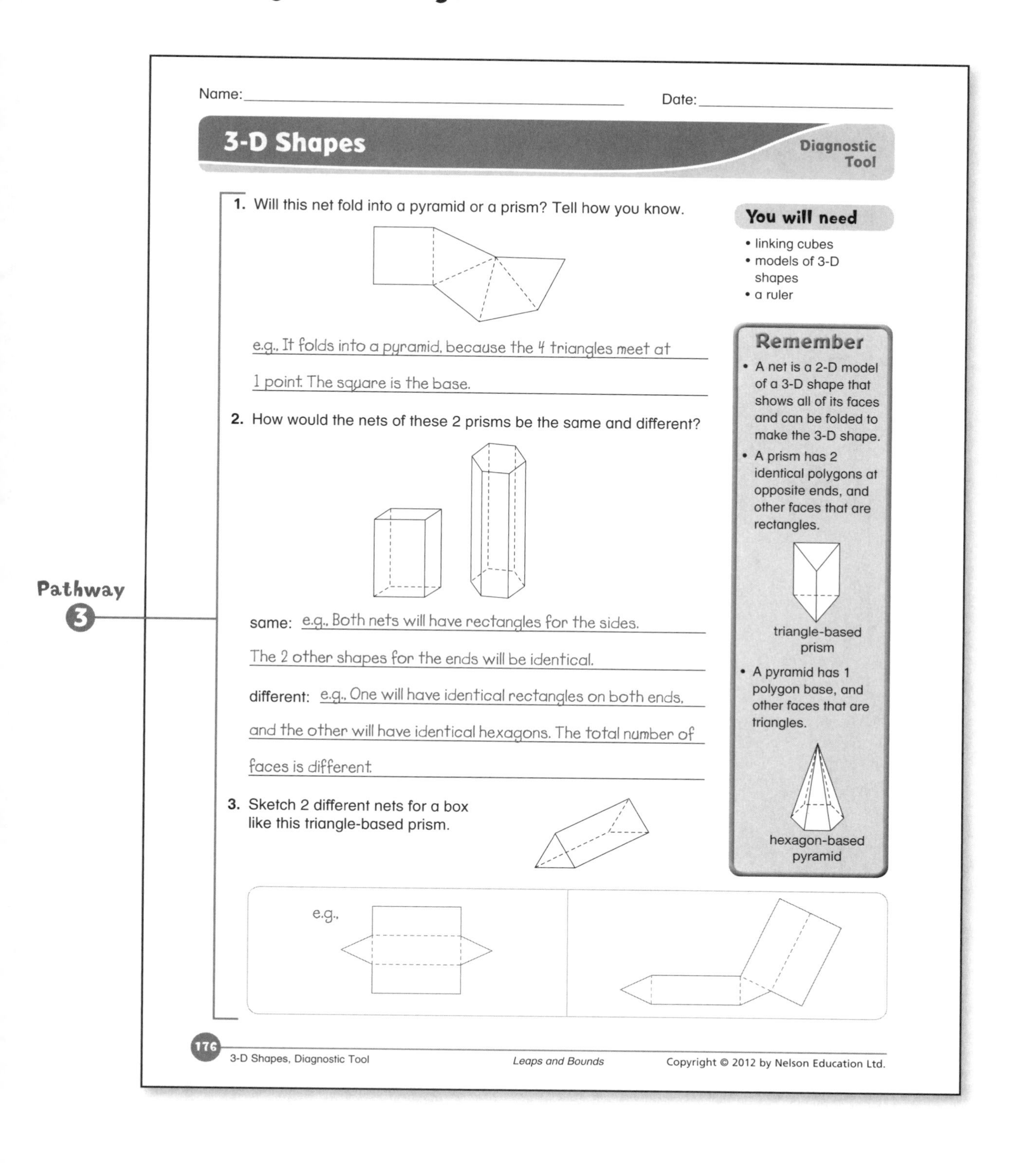

Name: ______________________ Date: ______________

3-D Shapes

Diagnostic Tool

Pathway 3

1. Will this net fold into a pyramid or a prism? Tell how you know.

   e.g., It folds into a pyramid, because the 4 triangles meet at 1 point. The square is the base.

2. How would the nets of these 2 prisms be the same and different?

   same: e.g., Both nets will have rectangles for the sides. The 2 other shapes for the ends will be identical.

   different: e.g., One will have identical rectangles on both ends, and the other will have identical hexagons. The total number of faces is different.

3. Sketch 2 different nets for a box like this triangle-based prism.

   e.g.,

You will need

- linking cubes
- models of 3-D shapes
- a ruler

Remember

- A net is a 2-D model of a 3-D shape that shows all of its faces and can be folded to make the 3-D shape.
- A prism has 2 identical polygons at opposite ends, and other faces that are rectangles.

triangle-based prism

- A pyramid has 1 polygon base, and other faces that are triangles.

hexagon-based pyramid

176 3-D Shapes, Diagnostic Tool *Leaps and Bounds* Copyright © 2012 by Nelson Education Ltd.

Name: ______________________ Date: ______________

4. Which top view below matches the cube structure at the right? Circle the correct top view.

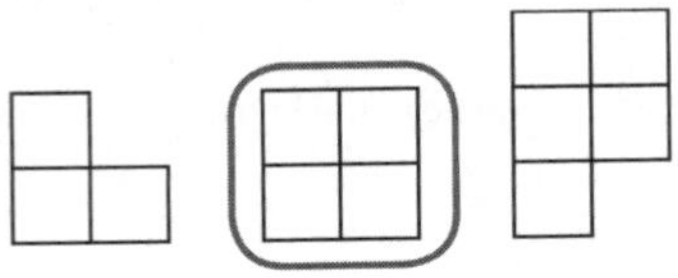

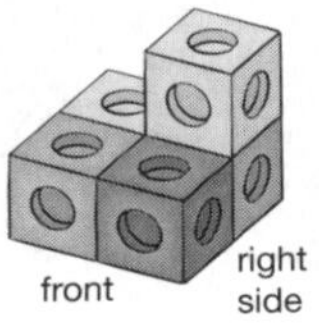

front right side

5. Which cube structure below matches the 3 views at the right? Circle the correct cube structure.

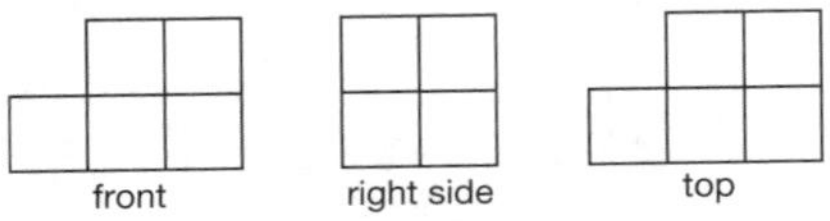

front right side top

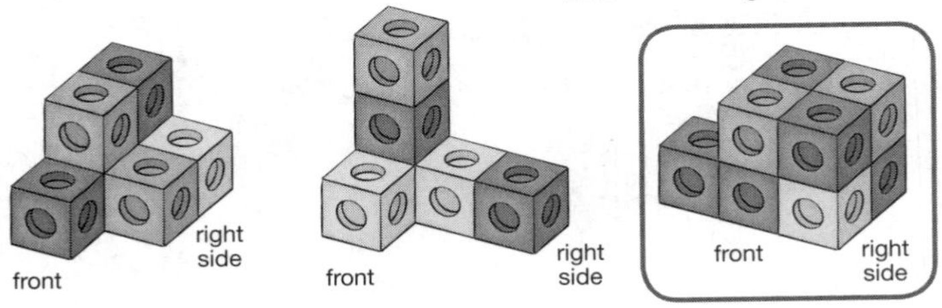

front right side front right side front right side

Pathway 2

6. Sketch the 3 different views—left side, front, and top—for these stairs.

front

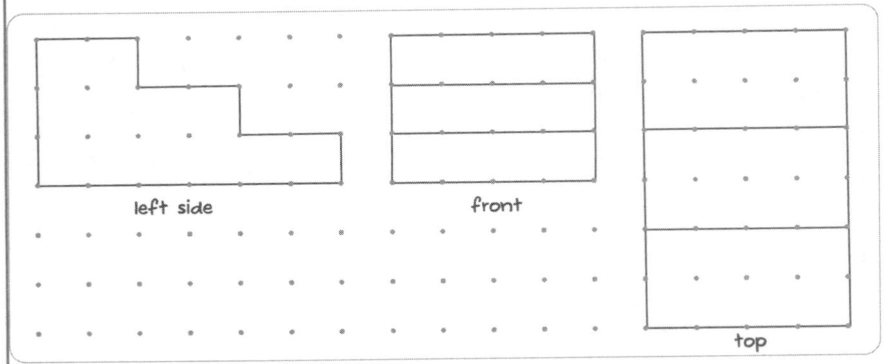

left side front top

Copyright © 2012 by Nelson Education Ltd.

 Copyright © 2012 by Nelson Education Ltd.

Name:______________________________ Date:__________________

7. Use the 2 drawings below to build 2 cube structures using linking cubes. How are your cube structures the same and different?

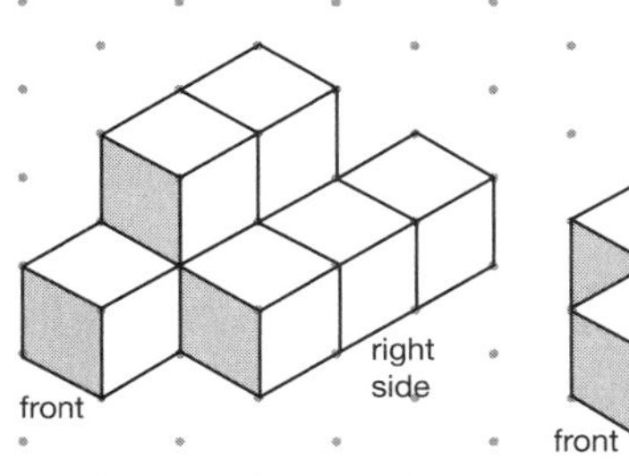

same: e.g., There are 2 layers of cubes. Both structures are no more than 4 cubes long or wide.

different: e.g., I used 7 cubes for one and 8 cubes for the other. The front view of one structure is 2 cubes wide and the other is 4 cubes wide.

Pathway 1

8. Examine the cube structure drawing at the right. How many linking cubes might this structure have? List the possibilities.

8, 9, or 10

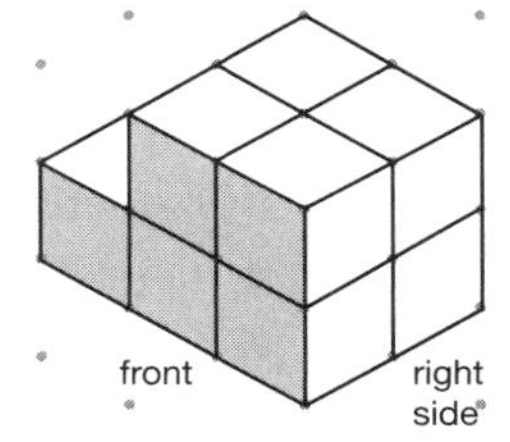

9. a) Build a cube structure using linking cubes where
- the bottom layer shows 3 rows of 3 cubes, and
- the 2nd layer shows 1 cube in the middle on top of the 1st layer.

How many linking cubes did you use?

10

b) Sketch an isometric drawing (like the ones in Questions 7 and 8) of your cube structure on the triangle dot paper.

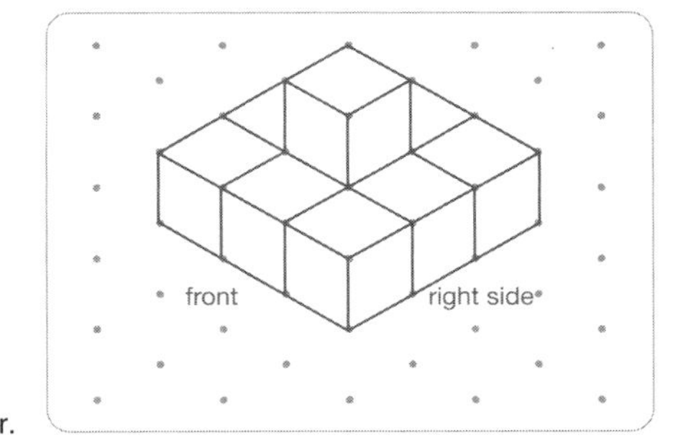

 Copyright © 2012 by Nelson Education Ltd.

Copyright © 2012 by Nelson Education Ltd.

Pathway 1 OPEN-ENDED

Using Isometric Drawings

You will need

- linking cubes
- Triangle Dot Paper (BLM 16)
- 1 cm Square Dot Paper (BLM 15)
- rulers
- sticky notes or coloured dots (optional)
- a digital camera (optional)*
- a document camera (optional)*
- Student Resource pages 228–229

**Note: If a digital camera is available, students can take photos of their cube structures to keep with their work. They might share their drawings using a document camera and let other students guess what the cube structures looked like.*

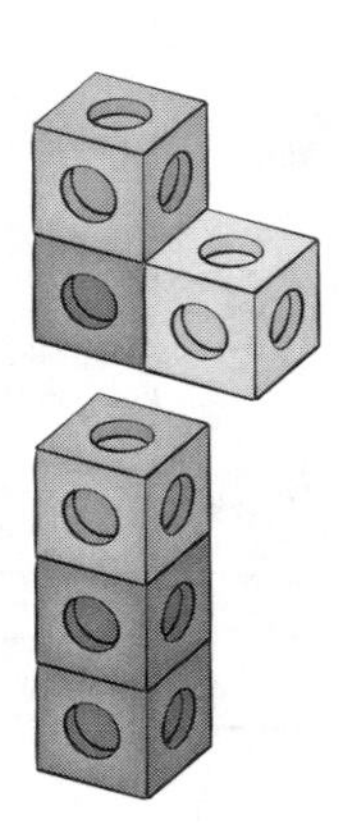

Open-Ended Intervention

Before Using the Open-Ended Intervention

Have students use linking cubes to make the 2 models shown in the margin. Ask:

- ▸ How do your 2 cube structures look different when you look at them from the front? (e.g., *One looks like an L, and the other looks like an I.*)
- ▸ Do they look different if you look from the side? (e.g., *Yes, one is 2 squares high and the other is 3 squares high.*)

Show students triangle dot paper (or isometric paper) and square dot paper. Ask:

- ▸ How does this dot paper look different from regular grid paper and square dot paper? (e.g., *It has dots instead of lines, and the dots form triangles.*)

Using the Open-Ended Intervention (Student Resource pages 228–229)

Read through the tasks on the student pages together. Have students model the 2-cube structure at the top of the page and sketch it on triangle dot paper. Students might use sticky notes or coloured dots to label the front, right side, and left side on their cube structures to help them check their drawings.

Give students time to work, ideally in pairs.

Observe whether students

- can create and label an isometric drawing that accurately represents the cube structure they built
- identify measurements of the cube structure from the drawing (e.g., number of cube units, height, length)
- recognize that some parts of the cube structure are not shown in the isometric drawings when sketched from one corner perspective
- can build a cube structure, given criteria

Consolidating and Reflecting

Ensure understanding by asking questions such as these based on students' work:

- ▸ What does an isometric drawing help you see better than other drawings might? What is still hard to see? (e.g., *An isometric drawing helps me see different sides of the cube structure at the same time. I usually can't see 1 side, and sometimes if the structure has a part sticking out, I can't see that part.*)
- ▸ What strategies did you use to sketch an isometric drawing of your cube structure? (e.g., *I drew vertical edges with vertical lines, and horizontal edges with diagonal lines. If I have many cubes in a row to draw, I just extend the diagonal line. If there's more than 1 layer, I extend the vertical line.*)
- ▸ Is it certain that an isometric drawing shows all parts of a cube structure? Why? (e.g., *No, if the cube structure has many parts sticking out, I might not see all the parts from 1 corner.*)

Copyright © 2012 by Nelson Education Ltd.

Using Isometric Drawings

Pathway 1 GUIDED

Guided Intervention

Before Using the Guided Intervention

Have students use linking cubes to make the models shown in the margin. Ask:

- ▸ How do your 2 cube structures look different when you look at them from the front? (e.g., *One looks like an L, and the other one looks like an I.*)
- ▸ Do they look different if you look from the side? (e.g., *Yes, one is 2 squares high and the other is 3 squares high.*)
- ▸ Which side did you look from: the right side or the left side? (e.g., *right side*)

Show students triangle dot paper (or isometric dot paper) and square dot paper. Ask:

- ▸ How does this dot paper look different from regular grid paper and square dot paper? (e.g., *It has dots instead of lines. Regular grid paper shows squares. In square dot paper, 4 dots make a square, like a geoboard. In triangle dot paper, 3 dots make an equilateral triangle, and 4 dots make a rhombus.*)

Using the Guided Intervention Student Resource pages 230–233

Work through the instructional section of the student pages together. Students might use sticky notes or coloured dots to label the front, right side, and left side on their cube structures to help them check their drawings.

Have students work through the **Try These** questions individually or in pairs.

Observe whether students

- can build a cube structure given criteria or models (Questions 1, 2, 3, 4, 5, 7)
- recognize that some parts of the cube structure are not shown in the isometric drawings when sketched from one corner perspective (Questions 2, 4, 5, 6, 8)
- can create and label isometric drawings that accurately represent the cube structure they built (Questions 3, 5, 7)

Consolidating and Reflecting

Ensure understanding by asking questions such as these based on students' work:

- ▸ In Question 3b), what strategies did you use to sketch isometric drawings of your cube structures? (e.g., *I drew vertical edges with vertical lines, and horizontal edges with diagonal lines. If I have many cubes in a row to draw, I just extend the diagonal line. If there's more than 1 layer, I extend the vertical line.*)
- ▸ Is it certain that an isometric drawing shows all parts of a cube structure? Why? (e.g., *No, if the cube structure has many parts sticking out, I might not see all the parts from 1 corner.*)
- ▸ What does an isometric drawing help you see better than other drawings might? What is still hard to see? (e.g., *I can see different sides of the cube structure at the same time. I usually can't see 1 side, and sometimes if the structure has a part sticking out, I can't see that part.*)

You will need

- linking cubes
- Triangle Dot Paper (BLM 16)
- 1 cm Square Dot Paper (BLM 15)
- rulers
- sticky notes or coloured dots (optional)
- a digital camera (optional)*
- a document camera (optional)*
- Student Resource pages 230–233

**Note: If a digital camera is available, students can take photos of their cube structures to keep with their work. They might share their drawings using a document camera and let other students guess what the cube structures looked like.*

Using Different Views

You will need

- linking cubes
- 2 cm Square Dot Paper (BLM 14) or 1 cm Square Dot Paper (BLM 15)
- rulers
- a digital camera (optional)*
- a document camera (optional)*
- Student Resource pages 234–235

**Note: If a digital camera is available, students can take photos of their cube structures to keep with their work. They might share their drawings using a document camera and let other students guess what the cube structures looked like.*

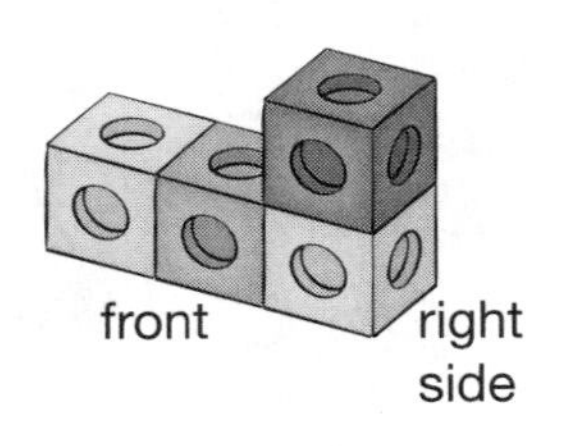

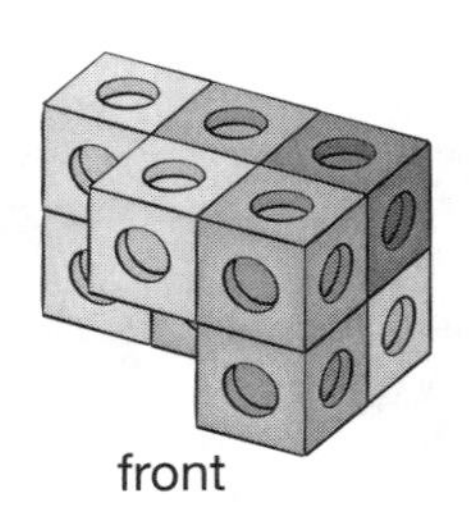

Open-Ended Intervention

Before Using the Open-Ended Intervention

Have students use 4 linking cubes to make the model shown in the margin. Ask:

- What do you see when you look at the top, front, right, and left sides of this cube structure, in terms of shape and number of cubes?
  (e.g., *From the top, I see 3 squares in a row. From the front, I see 3 squares in the bottom row and 1 square at the right end of the 2nd row. From the right side, I see a stack of 2 squares. From the left side, I see a stack of 2 squares.*)
- Can you be sure of what the cube structure looks like if I just show you the top view? (e.g., *no, because you can't tell how tall it is*)

Using the Open-Ended Intervention (Student Resource pages 234–235)

Read through the tasks on the student pages together. Provide the materials. Give students time to work, ideally in pairs.

Observe whether students

- can identify the number of cubes in a cube structure from a top view
- can build a cube structure and then sketch and label the various views (top, front, left-side and right-side views)
- can use different views to build a cube structure
- recognize that some parts of a cube structure are not visible in drawings of a single view
- understand the purpose of 3 different views (e.g., top, front, left or right side)

Consolidating and Reflecting

Ensure understanding by asking questions such as these based on students' work:

- What strategies did you use to sketch different views of your cube structure?
  (e.g., *I looked at the front of the cube structure and drew the number of square faces that I saw in its shape. Then I looked down on the top and drew the number of square faces. Then I turned the structure and looked at it from the side and drew the number of square faces I saw.*)
- Why do you think a drawing could match cube structures that are similar, except that one structure may have 1 or more cubes than the other?
  (e.g., *Some cubes are hidden behind other cubes.*)
- Why does it help to write the number of cubes on a top view?
  (e.g., *It tells how many cubes are in a column. If you include the numbers, you are more sure of the structure.*)
- Why are 3 views usually needed to determine a 3-D shape?
  (e.g., *3 views usually show the cubes that would be hidden if you had only 1 view.*)
- When might 3 views not be enough? If necessary, show a cube structure like the one in the margin. (e.g., *when there are a lot of cubes and lots are hidden*)

 Copyright © 2012 by Nelson Education Ltd.

Using Different Views

Pathway 2
GUIDED

Guided Intervention

Before Using the Guided Intervention

Have students use linking cubes to make the model shown in the margin. Ask:

- What do you see when you look at the top, front, right, and left sides of this cube structure, in terms of shape and number of cubes?
  (e.g., *From the top, I see 3 squares in a row. From the front, I see 3 squares in the bottom row and 1 square at the right end of the 2nd row. From the right side, I see a stack of 2 squares. From the left side, I see a stack of 2 squares.*)
- Can you be sure of what the cube structure looks like if I just show you the top view?
  (e.g., *no, because you can't tell how tall it is*)

You will need

- linking cubes
- 2 cm Square Dot Paper (BLM 14) or 1 cm Square Dot Paper (BLM 15)
- rulers
- a digital camera (optional)*
- a document camera (optional)*
- Student Resource pages 236–239

**Note: If a digital camera is available, students can take photos of their cube structures to keep with their work. They might share their drawings using a document camera and let other students guess what the cube structures looked like.*

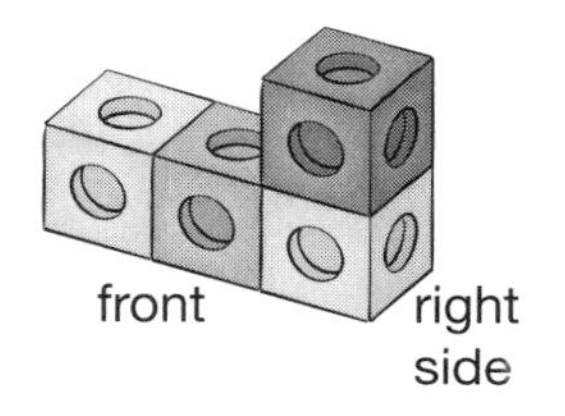

Using the Guided Intervention (Student Resource pages 236–239)

Work through the instructional section of the student pages together. Provide the materials for building the structures and sketching the views.

Have students work through the **Try These** questions individually or in pairs.

Observe whether students

- can identify the different views of a cube structure (Questions 1, 2, 5)
- recognize that some parts of a cube structure are hidden in the 2-D views (Questions 2, 3, 7)
- can use different views to build a cube structure (Questions 3, 4, 7)
- can sketch and label top, front, and side views (Questions 5, 6)
- understand the purpose of using 3 different views (Questions 7, 8)

Consolidating and Reflecting

Ensure understanding by asking questions such as these based on students' work:

- In Question 3, why might more than 1 cube structure be possible?
  (e.g., *Some cubes are hidden behind other cubes; views can't show that.*)
- In Question 4, were you unsure about any cubes that were hidden?
  (e.g., *No, there weren't any hidden cubes in part a). In part b), the back right cube is hidden, but you know it has to be there since there's a cube on top of it.*)
- Why does it help to write the number of faces on a top view?
  (e.g., *It makes you more sure of the structure.*)
- Why do you think that 3 views are usually needed to determine a 3-D shape?
  (e.g., *They usually provide enough information about what a 3-D shape looks like. I think that if there aren't too many cubes, there's only 1 possible structure you can build using 3 views.*)
- When might 3 views not be enough?
  (e.g., *when there are a lot of cubes and lots are hidden, as in Question 7c)*)

Pathway 3 OPEN-ENDED

Using Nets

You will need

- a box that can be folded down to a net (e.g., an empty cereal box)
- scissors and tape
- connecting faces (e.g., Polydrons or cardboard polygons*)
- linking cubes
- geometric models
- blank paper or 2 cm Square Dot Paper (BLM 14)
- rulers
- Regular Polygons (BLM 26, optional)*
- Student Resource page 240

**Note: You can use these regular polygons as templates for making your own set of cardboard polygons. Also include related rectangles and isosceles triangles so that prisms and pyramids can be made.*

Open-Ended Intervention

Before Using the Open-Ended Intervention

Show an empty box, folded down, or a net of a rectangular prism made of Polydrons or cardboard polygons joined with tape. Ask:

- ▸ If this net was folded up, what would it look like? (e.g., *a rectangular prism*)
- ▸ How do you know? (e.g., *It has 6 rectangular faces. Two opposite faces are the same size, so they must be the bases, and the other 4 faces connect to them.*)

Demonstrate how a net can be built in different ways by rearranging its faces. (If you are using a box, you will need to cut apart the rectangles and join them with tape.) Emphasize that just because you use all the faces, you have not made a net if the edges do not meet correctly.

Using the Open-Ended Intervention Student Resource page 240

Provide the materials and read through the tasks on the student page together. Offer students the choice of drawing the nets on blank paper or dot paper, or making them using connecting faces. Give students time to work, ideally in pairs.

Observe whether students

- can create nets to show that the number, shape, size, and orientation of the faces are directly related to the 3-D shape
- show an understanding of how faces need to be connected to make a net of a particular 3-D shape
- explain the strategies used to identify or create a net for 3-D shapes
- realize that there are different nets possible for the same 3-D shape

Consolidating and Reflecting

Ensure understanding by asking questions such as these based on students' work:

- ▸ How did you make sure that you included all the faces of the 3-D shape? (e.g., *I counted the number and type of faces on the 3-D shape and the net to make sure they were the same.*)
- ▸ How did you create different nets for your prisms? (e.g., *I looked at the 1st net I created and then changed the position of the end shape.*)
- ▸ What do you know about opposite ends in a prism? How does that affect the net? (*The ends are the same size and shape.* e.g., *There have to be 2 identical polygons as part of the net to be the bases.*)
- ▸ How can you tell whether a net can fold into the 3-D shape? (e.g., *I check if it has the right number, shape, and size of faces, and then in my mind I fold up each side.*)
- ▸ Do you agree or disagree with the following statement? Why? "To make a net, you need to include only the right number and shape of faces." (*Disagree.* e.g., *You have to make sure the faces are arranged so that they fold up with no overlapping or gaps, and that opposite faces are the same shape and same size.*)

 Copyright © 2012 by Nelson Education Ltd.

Using Nets

Pathway 3 GUIDED

Guided Intervention

Before Using the Guided Intervention

Show an empty box, folded down, or a net of a rectangular prism made of Polydrons or cardboard polygons joined with tape. Ask:

- If this net was folded up, what would it look like? (e.g., *a rectangular prism*)
- How do you know? (e.g., *It has 6 rectangular faces. Two opposite faces are the same size, so they must be the bases, and the other 4 faces connect to them.*)
- Suppose the 3-D shape is a cube. How would the net look different from this one? (e.g., *The 6 faces would all be the same size squares.*)

Demonstrate how a net can be built in different ways by rearranging its faces. (If you are using a box, cut apart the rectangles and join them with tape.) Emphasize that using all the faces doesn't make a net if the edges do not meet correctly.

Using the Guided Intervention Student Resource pages 241–244

Provide the materials. Work through the instructional section of the student pages together. Have students make and fold up the nets shown using dot paper and scissors. They can use these as models for comparison. They might use Regular Polygons (BLM 26) to help them make the nets for Question 5.

Have students work through the **Try These** questions individually or in pairs.

Observe whether students

- show an understanding of how faces need to be connected to make a net of a particular 3-D shape (Questions 1, 3, 4, 5, 6)
- can create nets to show that the number, shape, size, and orientation of the faces are directly related to a 3-D shape (Questions 2, 4, 5)
- realize that different nets are possible for the same 3-D shape (Questions 3, 4)
- explain strategies used to identify and create a net for 3-D shapes (Questions 4, 7)

Consolidating and Reflecting

Ensure understanding by asking questions such as these based on students' work:

- In Question 1, how did you match up a net with a 3-D shape? (e.g., *First I made sure it had the right shape of faces and then the right number of them.*)
- In Question 4, how did you create the second net for each of the 3-D shapes? (e.g., *I looked at the first net I created and changed the position of the end shape.*)
- How can you tell whether a net can fold into the 3-D shape? (e.g., *I check that it has the right number, shape, and size of faces, and then in my mind I fold each fold line until I have a 3-D shape. Then I check it by actually folding the net.*)
- Do you agree or disagree with the following statement? Why? "To make a net, you need to include only the right number and shape of faces." (*Disagree.* e.g., *You have to make sure the faces are arranged so that they fold up with no overlapping or gaps, and that opposite faces are the same shape and same size.*)

You will need

- a box that can be folded down to a net (e.g., an empty cereal box)
- scissors and tape
- connecting faces (e.g., Polydrons or cardboard polygons*)
- linking cubes
- 2 cm Square Dot Paper (BLM 14)
- rulers
- geometric models
- Regular Polygons (BLM 26, optional)*
- Student Resource pages 241–244

**Note: You can use these regular polygons as templates for making your own set of cardboard polygons. Also include related rectangles and isosceles triangles so that prisms and pyramids can be made.*

Copyright © 2012 by Nelson Education Ltd.

2-D Shapes

Planning For This Topic

Materials for assisting students with 2-D shapes consist of a diagnostic tool and 3 intervention pathways. Pathway 1 focuses on similarity of 2-D shapes. Pathway 2 involves congruence of 2-D shapes. Pathway 3 focuses on sorting and classifying polygons.

Each pathway has an open-ended intervention and a guided intervention. Choose the type of intervention more suitable for your students' needs and your particular circumstances.

Curriculum Connections

Grades 5 to 8 curriculum connections for this topic are provided online. See www.nelson.com/leapsandbounds. The WNCP curriculum does not cover similarity or congruence, so you may omit Pathways 1 and 2 with students in WNCP jurisdictions.

Why might students struggle with 2-D shapes?

Students might struggle with properties of 2-D shapes for any of the following reasons:

- They might focus on irrelevant attributes such as colour and orientation when sorting or grouping to determine shape classifications.
- They might have difficulty with overlapping classifications for triangles (e.g., a triangle can be both scalene and right-angled).
- They might not realize that some quadrilaterals can have more than one name based on its attributes (e.g., a square is a quadrilateral, a parallelogram, a rectangle, and a rhombus).
- They might have difficulty recognizing congruent shapes when those shapes are rotated or reflected.
- They might believe that if the angles in 2 shapes are equal, the shapes are congruent, even though they might not be (e.g., the triangles below have equal angle measures but not equal side lengths).

4 cm
3 cm
30° 60°
5 cm

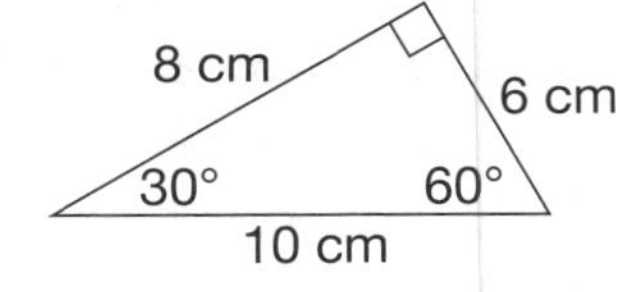

- They might believe that it is not possible to test for congruence without physically matching shapes (e.g., they cannot compare by measuring angle measures and side lengths).
- They might think that all polygons of the same type (e.g., all rectangles or all triangles or all hexagons, etc.) are similar, without realizing that the corresponding side lengths all have to be based on the same ratio.

Professional Learning Connections

PRIME: Geometry, Background and Strategies (Nelson Education Ltd., 2007), pages 43–44, 50–54, 59–62, 80–84, 86

Making Math Meaningful to Canadian Students K–8 (Nelson Education Ltd., 2008), pages 292–296, 298–302, 315–321

Big Ideas from Dr. Small Grades 4–8 (Nelson Education Ltd., 2009), pages 95–105, 112–115

Good Questions (dist. by Nelson Education Ltd., 2009), pages 62, 73, 77

 Copyright © 2012 by Nelson Education Ltd.

Diagnostic Tool: 2-D Shapes

Use the diagnostic tool to determine the most suitable intervention pathway for 2-D shapes. Provide Diagnostic Tool: 2-D Shapes, Teacher's Resource pages 190 to 192, and have students complete it in writing or orally. Provide rulers, protractors, pattern blocks, and tracing paper. If you are using the WNCP curriculum, you might choose to use only the questions related to Pathway 3.

See solutions on Teacher's Resource pages 193 to 195.

Intervention Pathways

The purpose of the intervention pathways is to develop understanding of the geometric properties of triangles, quadrilaterals, and regular polygons (e.g., angle and side properties of polygons, line symmetry). Also, students develop strategies for determining congruence and similarity of polygons (e.g., using sufficient conditions for determining whether triangles are congruent by looking at the corresponding side lengths and angles in the triangles).

There are 3 pathways:

- Pathway 1: Similar Shapes
- Pathway 2: Congruent Shapes
- Pathway 3: Sorting and Classifying Polygons

Use the chart below (or the Key to Pathways on Teacher's Resource pages 193 to 195) to determine which pathway is most suitable for each student or group of students.

Diagnostic Tool Results	Intervention Pathway
If students struggle with Questions 7 to 9	use Pathway 1: Similar Shapes *Teacher's Resource pages 196–197* *Student Resource pages 245–250*
If students struggle with Questions 4 to 6	use Pathway 2: Congruent Shapes *Teacher's Resource pages 198–199* *Student Resource pages 251–256*
If students struggle with Questions 1 to 3	use Pathway 3: Sorting and Classifying Polygons *Teacher's Resource pages 200–201* *Student Resource pages 257–262*

Copyright © 2012 by Nelson Education Ltd.

Name: ______________________ Date: ______________________

2-D Shapes

Diagnostic Tool

You will need

- a protractor
- a ruler
- tracing paper
- pattern blocks (square and trapezoid)

Remember

- Equal lengths are shown with matching marks.

1. Describe how the quadrilaterals were sorted into groups.

a)

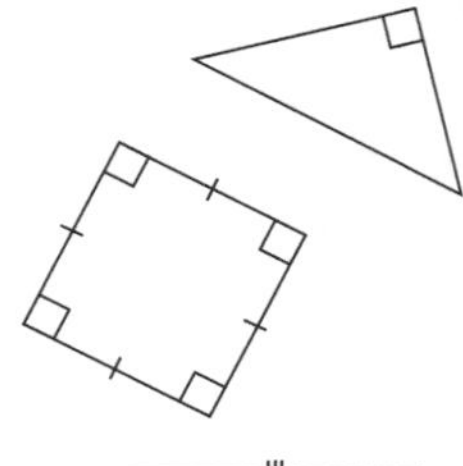

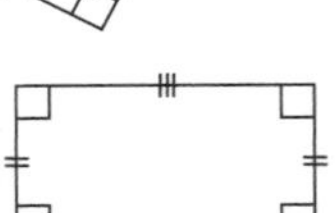

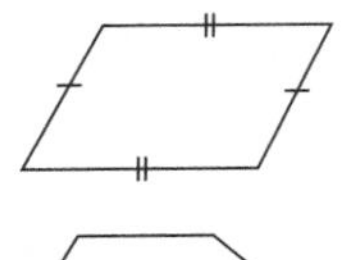

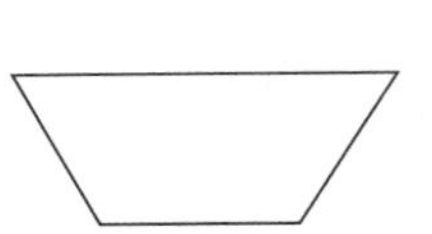

____

____

b)

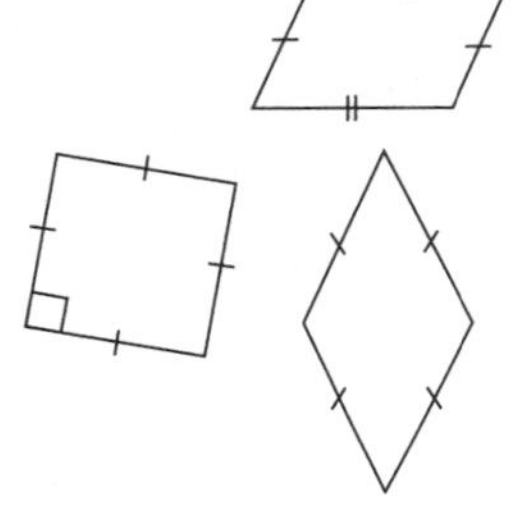

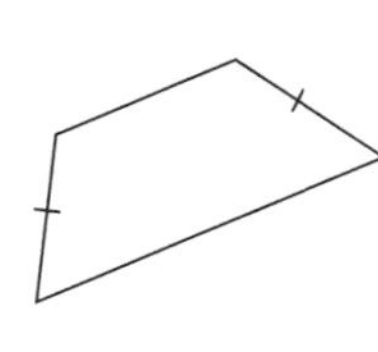

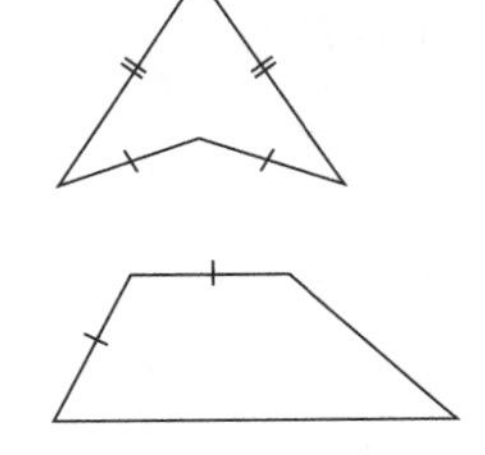

____

____

____

2. Which triangles have each attribute below?

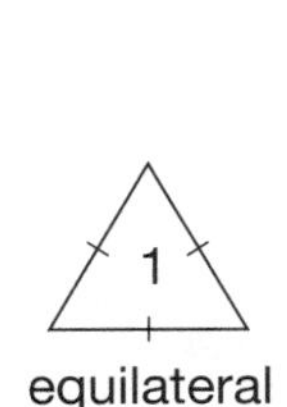

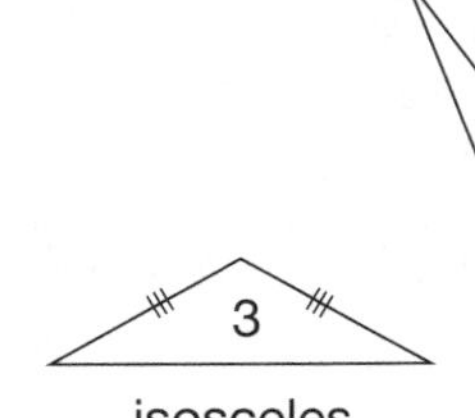

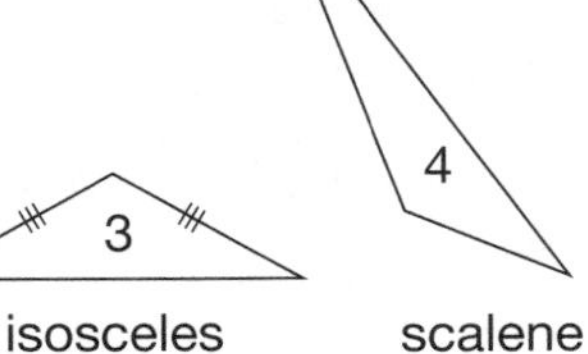

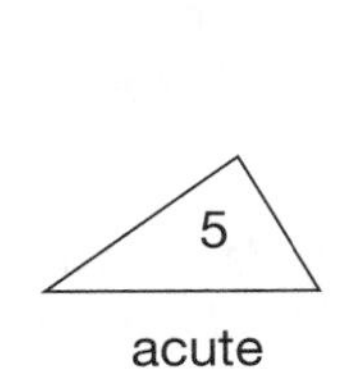

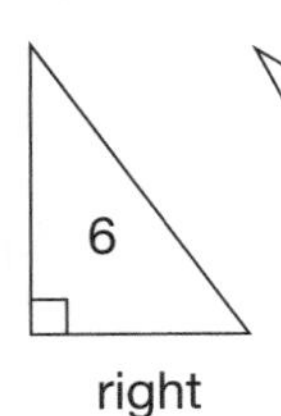

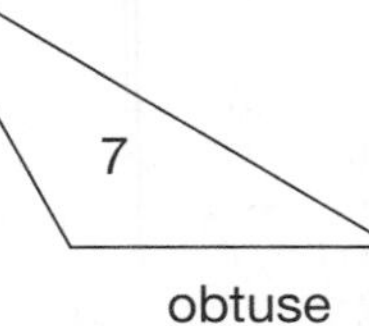

a) Has at least some equal side lengths ______________

b) Has at least 1 obtuse angle ______________

c) Has line (or mirror) symmetry ______________

 Copyright © 2012 by Nelson Education Ltd.

Name:________________________ Date:________________

3. a) Shauna sorted the shapes below into 2 groups: regular polygons (Shapes A and B), and irregular polygons (Shapes C and D). Explain why Shapes C and D are not regular polygons.

____

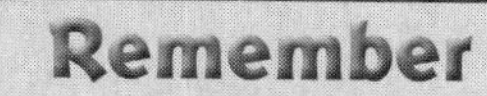

- Regular polygons have all side lengths equal and all angles equal.

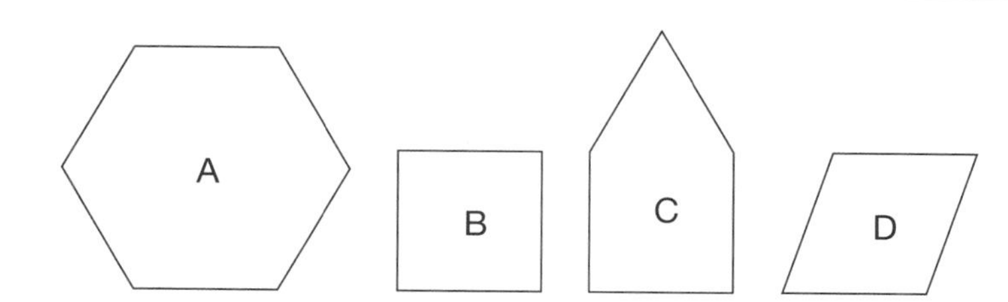

b) How else can you sort shapes A, B, C, and D into 2 groups? Describe at least 2 ways.

____

____

____

4. Are these pictures of pattern blocks congruent to actual pattern blocks? Tell how you know.

____

____

Remember

- Congruent means exactly the same size and shape.

5. This set of triangles includes 2 pairs of congruent triangles. Shade 1 pair and circle the other pair.

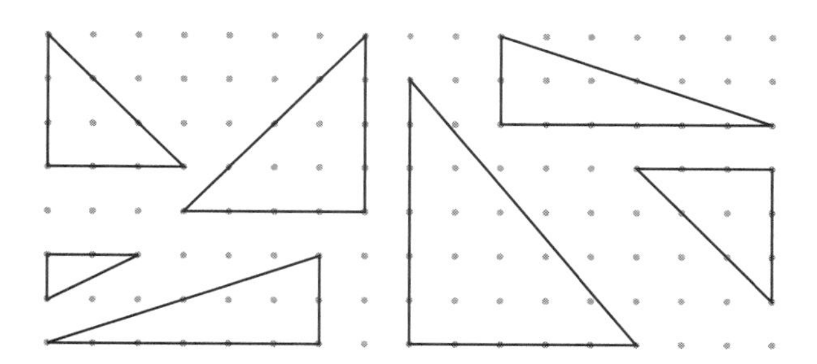

6. Describe 3 ways to decide whether 2 triangles are congruent.

____

____

____

Copyright © 2012 by Nelson Education Ltd.

Name: ______________________ Date: ______________

7. The grey rectangle was reduced in size to make the white rectangle. Are the 2 rectangles similar? Why or why not?

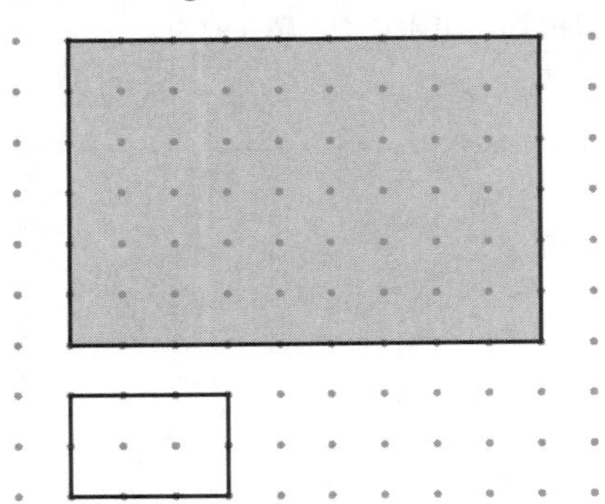

____

____

Remember

- Two shapes are similar if the size is enlarged or reduced by the same factor for all sides of the shape.

8. Are these 2 triangles similar? How can you prove it?

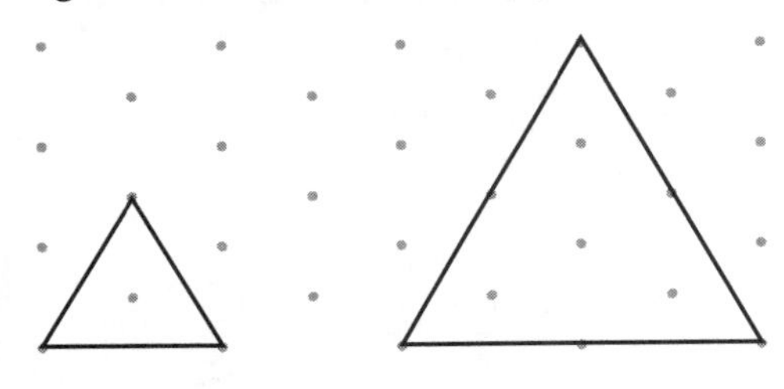

____

____

____

9. Name the triangles below that are congruent and the triangles that are similar. Tell how you know.

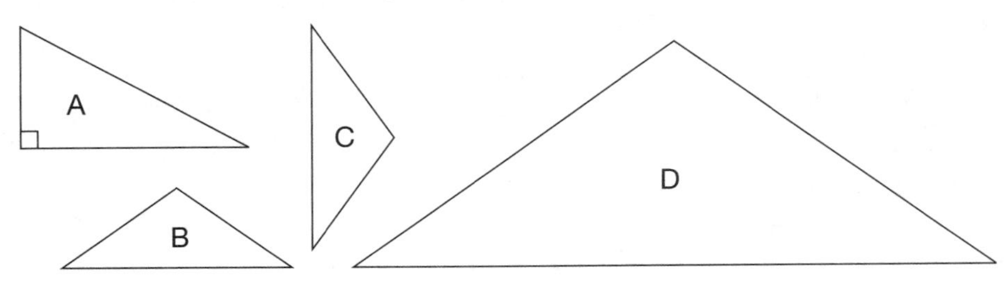

congruent: ______________________________________

____

similar: ____

____

Copyright © 2012 by Nelson Education Ltd.

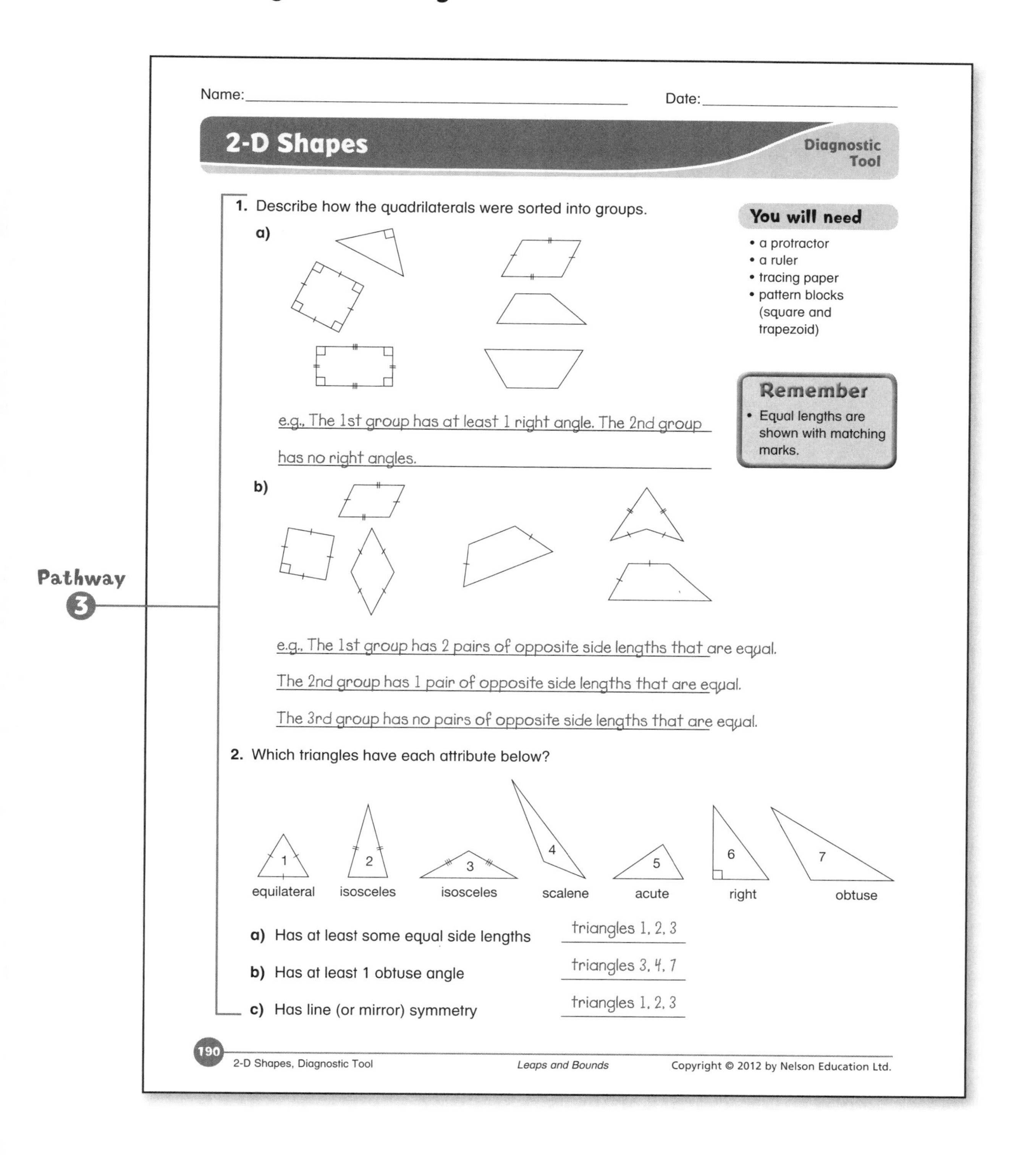

Name:________________________ Date:________________

2-D Shapes

Diagnostic Tool

You will need

- a protractor
- a ruler
- tracing paper
- pattern blocks (square and trapezoid)

Remember

- Equal lengths are shown with matching marks.

Pathway 3

1. Describe how the quadrilaterals were sorted into groups.

a)

e.g., The 1st group has at least 1 right angle. The 2nd group has no right angles.

b)

e.g., The 1st group has 2 pairs of opposite side lengths that are equal.

The 2nd group has 1 pair of opposite side lengths that are equal.

The 3rd group has no pairs of opposite side lengths that are equal.

2. Which triangles have each attribute below?

a) Has at least some equal side lengths triangles 1, 2, 3

b) Has at least 1 obtuse angle triangles 3, 4, 7

c) Has line (or mirror) symmetry triangles 1, 2, 3

190 2-D Shapes, Diagnostic Tool *Leaps and Bounds* Copyright © 2012 by Nelson Education Ltd.

Pathway 3

Name:______________________ Date:______________

3. a) Shauna sorted the shapes below into 2 groups: regular polygons (Shapes A and B), and irregular polygons (Shapes C and D). Explain why Shapes C and D are not regular polygons.

Not all of the angles within each shape have the same measure.

Remember
- Regular polygons have all side lengths equal and all angles equal.

A B C D

b) How else can you sort shapes A, B, C, and D into 2 groups? Describe at least 2 ways.

e.g., has at least 1 obtuse angle (A, C, D), has no obtuse angles (B);

has a right angle (B, C), has no right angles (A, D);

has 4 sides (B and D), has more than 4 sides (A and C)

Pathway 2

4. Are these pictures of pattern blocks congruent to actual pattern blocks? Tell how you know.

No, e.g., When I place the pattern block on top, it doesn't fit exactly. The angles are the same but the side lengths are not.

Remember
- Congruent means exactly the same size and shape.

5. This set of triangles includes 2 pairs of congruent triangles. Shade 1 pair and circle the other pair.

6. Describe 3 ways to decide whether 2 triangles are congruent.

e.g., A tracing of one fits exactly over the other.

The sides of one have the same lengths as the other.

You can reflect one onto the other.

Copyright © 2012 by Nelson Education Ltd. *Leaps and Bounds* 2-D Shapes, Diagnostic Tool 191

Copyright © 2012 by Nelson Education Ltd.

Name: ______________________ Date: ______________

Remember

- Two shapes are similar if the size is enlarged or reduced by the same factor for all sides of the shape.

Pathway 1

7. The grey rectangle was reduced in size to make the white rectangle. Are the 2 rectangles similar? Why or why not?

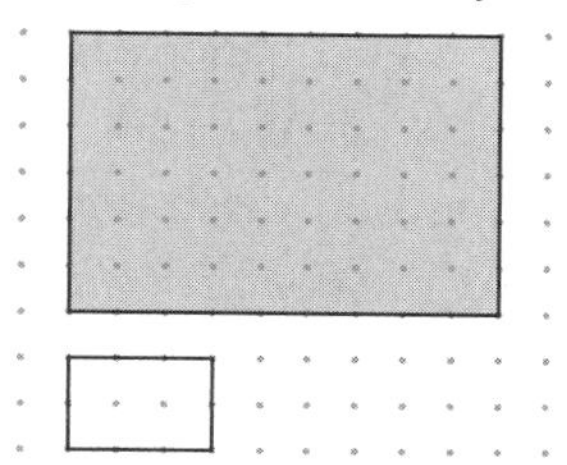

Yes. e.g., The length and width of the little one are both $\frac{1}{3}$ of the length and width of the big one.

8. Are these 2 triangles similar? How can you prove it?

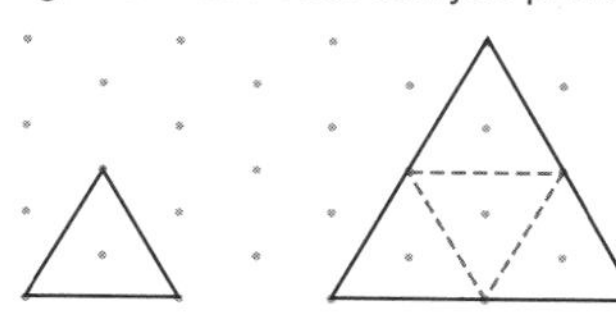

Yes. e.g., Both shapes have 3 angles that measure 60°. The side lengths of the larger triangle are double the side lengths of the smaller triangle.

9. Name the triangles below that are congruent and the triangles that are similar. Tell how you know.

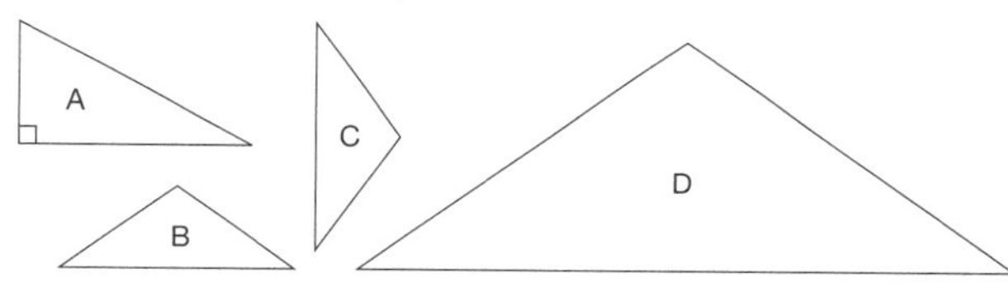

congruent: B and C, e.g., I traced B and it fit exactly onto C.

similar: B, C, and D, e.g., All the angles match, and they all have 2 sides the same length on either side of the largest angle.

 Copyright © 2012 by Nelson Education Ltd.

Copyright © 2012 by Nelson Education Ltd.

Pathway 1 OPEN-ENDED

Similar Shapes

You will need

- 2 cm Grid Paper (BLM 11), 1 cm Grid Paper (BLM 12), 0.5 cm Grid Paper (BLM 13)
- tracing paper
- 2 cm Square Dot Paper (BLM 14) or 1 cm Square Dot Paper (BLM 15)
- rulers
- protractors
- calculators
- Student Resource pages 245–246

Open-Ended Intervention

Before Using the Open-Ended Intervention

Sketch any polygon on 1 cm grid paper (BLM 12) and have students sketch it on grid paper of different sizes (such as BLM 11 and BLM 13). Ask:

- When are the shapes congruent? (e.g., *when the grid used is the same size*)
- How can you be sure? (e.g., *When it's placed on top of the original, it matches.*)
- When is your shape an enlargement of the original? (*when the new grid is bigger*)
- When is your shape a reduction of the original? (*when the new grid is smaller*)

Using the Open-Ended Intervention (Student Resource pages 245–246)

Read through the tasks on the student pages together. Provide the materials. Students may check angle measures by tracing one angle and matching it to another, or by using a protractor. Discuss how the ratio of corresponding side lengths is the same for 2 similar shapes, as described in the **Remember** box.

Give students time to work, ideally in pairs.

Observe whether students

- recognize that for similar triangles one triangle is an enlargement or a reduction of the other, or the 2 triangles are congruent
- determine congruence by either matching or measuring all corresponding parts
- can identify corresponding angles and sides in a pair of similar figures
- determine similarity by comparing ratios of all corresponding sides

Consolidating and Reflecting

Ensure understanding by asking questions such as these based on students' work:

- What's the difference between congruent shapes and similar shapes? (e.g., *Congruent shapes have corresponding angles the same and corresponding side lengths the same. Similar shapes look the same and have corresponding side lengths that are multiplied by the same factor for all pairs.*)
- Do you need to measure lengths or angles or both to test for congruence? (e.g., *You could, but it's easier to just fit one on top of the other.*)
- Did you need to measure lengths or angles or both to test for similarity? (e.g., *You can just test to see if the matching side lengths are in the same ratio, but I measured the angles to check and they were always equal in both shapes if the shapes were similar.*)
- Choose 2 of your triangles that were similar but not congruent. Use a calculator to determine the ratio of the longest side to the shortest side for each of the 2 triangles. What do you notice? (*The ratios are the same.*)
- Test to see if it works when you compare 2 different side lengths in each triangle. Does the same thing happen? (*yes*)

 Copyright © 2012 by Nelson Education Ltd.

Similar Shapes

Pathway 1
GUIDED

Guided Intervention

Before Using the Guided Intervention

Sketch any polygon on 1 cm grid paper (BLM 12) and have students sketch it on grid paper of different sizes (such as BLM 11 and BLM 13). Ask:

- When are the shapes congruent? (e.g., *when the grid used is the same size*)
- How can you be sure? (e.g., *When it's placed on top of the original, it matches.*)
- When is your shape an enlargement of the original? (*when the new grid is bigger*)
- When is your shape a reduction of the original? (*when the new grid is smaller*)

Using the Guided Intervention Student Resource pages 247–250

Provide the materials and work through the instructional section of the student pages together. Students may check angles by tracing one angle and matching it to another, or by using a protractor.

Have students work through the **Try These** questions individually or in pairs. Encourage students to try Question 3 using a quadrilateral of their own. BLM 26 can be used in Question 6 for sketching nested shapes that are similar.

Observe whether students

- recognize that shapes are similar if one is an enlargement or a reduction of the other, or if the shapes are congruent (Questions 1, 2, 3, 4, 5, 6, 8, 9)
- realize that congruence is a subset of similarity (Question 2)
- can determine the measures of sides in a pair of similar figures (Question 7)

Consolidating and Reflecting

Ensure understanding by asking questions such as these based on students' work:

- What's the difference between congruent shapes and similar shapes? (e.g., *Congruent shapes have corresponding angles the same and corresponding side lengths the same. Similar shapes look the same and have corresponding side lengths that are multiplied by the same factor for all pairs.*)
- Do you need to measure lengths or angles or both to test for congruence? (e.g., *You could, but it is easier to just fit one on top of the other.*)
- In Question 2, how do you know Triangle C is similar to larger Triangle A? (e.g., *The side lengths in C are half the side lengths of A.*)
- Did you need to measure lengths or angles or both to test for similarity in Question 2? (e.g., *You can just test to see if the matching side lengths are in the same ratio, but I measured the angles to check and they were always equal.*)
- For the rectangles in Question 4, calculate the ratio of the length to the width in both shapes. What do you notice? (*The ratios are the same.*)
- Test to see if it works when you compare 2 different side lengths in each triangle you drew in Question 7. Does the same thing happen? (*yes*)

You will need

- 2 cm Grid Paper (BLM 11), 1 cm Grid Paper (BLM 12), 0.5 cm Grid Paper (BLM 13)
- tracing paper
- 1 cm Square Dot Paper (BLM 15)
- protractors
- rulers
- calculators
- coloured pencils
- Regular Polygons (BLM 26)
- Student Resource pages 247–250

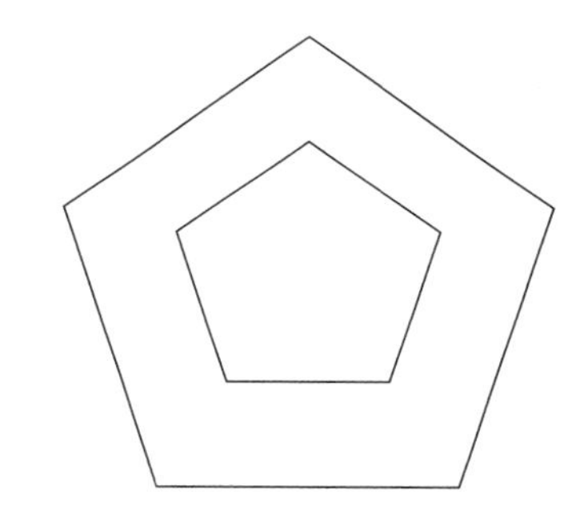

Pathway 2 OPEN-ENDED

Congruent Shapes

You will need

- Fraction Circles (BLM 18)
- scissors
- protractors
- rulers
- thin cardboard that can be cut to make various length strips and angle cutouts
- tracing paper
- compasses (optional)
- Student Resource pages 251–252

Open-Ended Intervention

Before Using the Open-Ended Intervention

Provide cutouts from BLM 18 (use circles divided into 5 or more sections). Ask:

- How do you know that the parts of this circle are equal? (e.g., *If you cut them out and place them all on top of each other, they match.*)
- How many congruent parts are there? (e.g., *5*)
- What can you say about the angle between the 2 straight sides? (e.g., *The angles all have the same measure. I could use a protractor to check.*)

Demonstrate how to cut off the curved part to form a triangle. Ask:

- What type of triangle is this? Why? (*isosceles; 2 sides are the same length*)
- What can you say about the other 2 angles in the triangle? (e.g., *They look the same. I could use a protractor to check.*)

Demonstrate how to fold the triangle in half to fit one angle onto the other.

Using the Open-Ended Intervention **Student Resource pages 251–252**

Read through the tasks on the student pages together. Provide the materials. Students may check angles by tracing one angle and matching it to another, or by using a protractor. Rather than using cardboard strips, students might use a compass and a ruler to create a given length on paper.

Give students time to work, ideally in pairs.

Observe whether students

- can explain ways to prove congruency of 2 or more polygons
- recognize that all of the corresponding measures of congruent shapes are the same, no matter how the shapes are oriented
- can determine the minimum information needed to decide if 2 triangles are congruent and explain why conditions are or are not sufficient

Consolidating and Reflecting

Ensure understanding by asking questions such as these based on students' work:

- How do you know that 2 triangles with the same side lengths are congruent? (e.g., *I used the same 3 strip lengths to make 2 triangles. When I tried to put the strips in a different order, I still came up with the same triangle.*)
- How do you know that 2 triangles with the same angles might not be congruent? (e.g., *I drew a short line and put an angle on each end and made it into a triangle. Then I used the same angles with a longer line between them and made that into a triangle. One triangle was bigger than the other.*)
- What other sets of measurements guaranteed that you get congruent triangles? (e.g., *when I knew 2 of the sides and the angle between them*)
- What other sets of measurements did not guarantee congruent triangles? (e.g., *when I knew 2 of the sides, but the angle was not between them*)

 Copyright © 2012 by Nelson Education Ltd.

Congruent Shapes

Pathway 2
GUIDED

Guided Intervention

Before Using the Guided Intervention

Provide cutouts from BLM 18 (use circles divided into 5 or more sections). Ask:

- How do you know that the parts of this circle are equal?
 (e.g., *If you cut them out and place them all on top of each other, they match.*)
- How many congruent parts are there? (e.g., *5*)
- What can you say about the angle between the 2 straight sides?
 (e.g., *The angles all have the same measure. I could use a protractor to check.*)

Demonstrate how to cut off the curved edge to form a triangle. Ask:

- What type of triangle is this? Why? (*isosceles; 2 sides are the same length*)
- What can you say about the other 2 angles in the triangle?
 (e.g., *They look the same. I could use a protractor to check.*)

Demonstrate how to fold the triangle in half to fit one angle onto the other.

Using the Guided Intervention (Student Resource pages 253–256)

Provide the materials and work through the instructional section of the student pages together. Students may check angles by tracing one angle and matching it to another, or by using a protractor.

Have students work through the **Try These** questions individually or in pairs.

Observe whether students

- can prove congruency of 2 or more polygons (Questions 1, 2, 4, 5)
- can use measurements of one shape to label length and angle measures of a congruent shape (Question 3)
- can determine the minimum information needed to decide if 2 triangles are congruent (Questions 6, 7)

Consolidating and Reflecting

Ensure understanding by asking questions such as these based on students' work:

- How did you show that your triangles in Question 6a) were not congruent?
 (e.g., *I put the 90° angle between the 8 cm and 4 cm sides for one triangle but not for the other, and the triangles were different shapes.*)
- How did you show that your triangles in Question 6c) might not be congruent?
 (e.g., *I drew a short line and put an angle on each end and made it into a triangle. Then I used the same angles with a longer line between them and made that into a triangle. One triangle was bigger than the other, but the angles were equal.*)
- How do you know that your triangles in Question 6d) are congruent?
 (e.g., *I used the 3 strip lengths to make 2 triangles. When I tried to put the strips in a different order, I still came up with the same triangle.*)

You will need

- Fraction Circles (BLM 18)
- scissors
- protractors
- rulers
- transparent mirrors (e.g., Miras)
- tracing paper
- Student Resource pages 253–256

Copyright © 2012 by Nelson Education Ltd.

Sorting and Classifying Polygons

You will need

- Triangle Dot Paper (BLM 16)
- Circular Geoboard Paper (BLM 17)
- rulers
- transparent mirrors (e.g., Miras)
- protractors
- tracing paper (optional)
- Student Resource pages 257–258

Open-Ended Intervention

Before Using the Open-Ended Intervention

Display a variety of polygons, as shown in the margin. Focus students' attention on the equilateral triangle, the right-angled triangle, and the obtuse triangle. Ask:

- ▸ How are these shapes the same? How are they different? (e.g., *They are triangles, but one has an obtuse angle, one has a right angle, and one has neither.*)
- ▸ Which of the other shapes do you think is more like the right-angled triangle than the other triangles? (e.g., *the rectangle, since it has square corners, too*)
- ▸ Which of the other non-triangle shapes do you think is most like the obtuse triangle? Why? (e.g., *the hexagon or the rhombus; both have some obtuse angles*)

Using the Open-Ended Intervention Student Resource pages 257–258

Read through the tasks on the student pages together. Provide the materials. Give each student 2 copies of BLM 17 so they can sketch their 12 shapes. Allow students to classify their shapes using less formal language. You might direct them to the definitions of quadrilaterals in the guided intervention. Give students time to work, ideally in pairs.

Observe whether students

- can identify geometric attributes and properties of polygons (e.g., number of sides or vertices, type of angles, side lengths, line of symmetry)
- can sort and classify regular and irregular polygons according to geometric attributes and properties (e.g., number of sides or vertices, type of angles, side lengths, line of symmetry)
- can explain their sorting and classification rules using geometric attributes and properties

Consolidating and Reflecting

Ensure understanding by asking questions such as these based on students' work:

- ▸ What geometric attributes did you use to sort your polygons? Why did you choose those? (e.g., *number of sides, type of angles; those are easy to compare*)
- ▸ Were there other ways you could have sorted? (e.g., *Yes, I could have figured out whether they had line symmetry or not, or any equal sides or not.*)
- ▸ What made the quadrilaterals in your set different? (e.g., *Some have parallel sides and some don't; some have right angles and some don't.*)
- ▸ How might the attributes of quadrilaterals help you name them? (e.g., *If there are 2 pairs of parallel sides, you call it a parallelogram. If there is only 1 pair of parallel sides, you call it a trapezoid.*)
- ▸ Why might your triangles have been in different categories? (e.g., *It could depend on how many sides were equal or on what kind of angles they had.*)
- ▸ Why is there usually more than one way to sort the same shapes? (e.g., *Shapes have lots of different attributes.*)

 Copyright © 2012 by Nelson Education Ltd.

Sorting and Classifying Polygons

Guided Intervention

Before Using the Guided Intervention

Display a variety of polygons, as shown in the margin. Focus students' attention on the equilateral triangle, the right-angled triangle, and the obtuse triangle. Ask:

- How are these shapes the same? How are they different? (e.g., *They are triangles, but one has an obtuse angle, one has a right angle, and one has neither.*)
- Which of the other non-triangle shapes do you think is most like the obtuse triangle? Why? (e.g., *the hexagon or the rhombus; both have some obtuse angles*)

Using the Guided Intervention (Student Resource pages 259–262)

Provide the materials and work through the instructional section of the student pages together. Students can cut out Shapes 1 to 9 on BLM 27 to help them. They can also use Shapes 10 and 11 for Question 1; Shapes 1 to 6, 10, and 12 to 16 for Question 2; and Shapes 2, 7 to 9, 11 to 13, and 17 to 21 for Question 6. String can be used in Question 2 to model the Venn diagrams.

Have students work through the **Try These** questions individually or in pairs.

Observe whether students

- can identify geometric attributes or properties of polygons (Questions 1, 4)
- can sort and classify shapes according to geometric attributes and properties (Questions 2, 4, 6)
- can describe or sketch shapes that possess certain geometric attributes (Questions 2, 5)
- can compare and contrast shapes (Question 3)
- can identify geometric attributes and properties of polygons that would be good to sort by (Question 7)

Consolidating and Reflecting

Ensure understanding by asking questions such as these based on students' work:

- How are the quadrilaterals in Question 2 different from each other? (e.g., *some have parallel sides and some don't; some have right angles and some don't*)
- How might the attributes of the quadrilaterals help you name them? (e.g., *If there are 2 pairs of parallel sides, you call it a parallelogram, but if there is only 1 pair of parallel sides, you call it a trapezoid.*)
- What geometric attributes did you use to sort the triangles in Question 4? Why? (e.g., *the kind of angles and the number of equal sides, because that's how you name different triangles*)
- What ways did you use to sort the polygons in Question 7c)? (e.g., *number of sides; whether sides were equal; whether they had symmetry*)
- Why is there usually more than one way to sort the same shapes? (e.g., *Shapes have lots of different attributes.*)

You will need

- Triangle Dot Paper (BLM 16)
- protractors
- rulers
- transparent mirrors (e.g., Miras)
- tracing paper
- Polygons (BLM 27)
- scissors
- string or yarn (optional)
- Student Resource pages 259–262

Strand: Geometry

Geometric Drawings

Planning For This Topic

Materials for assisting students with geometric drawings consist of a diagnostic tool and 4 intervention pathways. Pathway 1 focuses on drawing angle bisectors and perpendicular bisectors. Pathway 2 involves drawing parallel lines and intersecting lines, including perpendicular lines, in the context of drawing polygons. Pathway 3 focuses on drawing circles for a given radius or diameter. Pathway 4 is about drawing different types of triangles.

Each pathway has an open-ended intervention and a guided intervention. Choose the type of intervention more suitable for your students' needs and your particular circumstances.

Curriculum Connections

Grades 5 to 8 curriculum connections for this topic are provided online. See www.nelson.com/leapsandbounds. The Ontario curriculum does not cover drawing circles with a specific radius or diameter until Grade 8, so Pathway 3 can be omitted for those students. The other pathways are relevant to both the Ontario and WNCP curriculums.

The traditional view of constructions (using only a straightedge and compass) has given way to a broader view, which also includes the use of technology, protractors, rulers, paper folding, and transparent mirrors, as reflected by the curriculum. Hence, the term "geometric drawings" has been used throughout instead of "constructions."

Professional Learning Connections

PRIME: Geometry, Background and Strategies (Nelson Education Ltd., 2007), pages 71–72
Making Math Meaningful to Canadian Students K–8 (Nelson Education Ltd., 2008), page 320
Big Ideas from Dr. Small Grades 4–8 (Nelson Education Ltd., 2009), page 117
Good Questions (dist. by Nelson Education Ltd., 2009), pages 88, 91
More Good Questions (dist. by Nelson Education Ltd., 2010), pages 115, 117

Why might students struggle with geometric drawings?

Students might struggle with geometric drawings for any of the following reasons:

- They might have difficulty with the geometry terminology, such as *angle bisector* and *perpendicular bisector*, and/or confuse terms such as *radius* versus *diameter*, *acute* versus *obtuse*, and *parallel* versus *perpendicular*.
- They might have difficulty interpreting geometry notation (e.g., symbols to show equal side lengths, equal angles, right angles, and parallel sides).
- They might have difficulty drawing a parallel or perpendicular line or a polygon from a given line segment, if it is not horizontal or vertical.
- They might not be familiar with the different types of triangles and quadrilaterals, and the fact that any triangle or quadrilateral can have multiple names (e.g., a right scalene triangle or a parallelogram that is also a rhombus).
- They might have difficulty drawing different triangles to meet one set of conditions, when only 1 side and 2 angles are given.
- They might struggle more with drawing a circle when the diameter, rather than the radius, is given.

 Copyright © 2012 by Nelson Education Ltd.

Diagnostic Tool: Geometric Drawings

Use the diagnostic tool to determine the most suitable intervention pathway for geometric drawings. Provide Diagnostic Tool: Geometric Drawings, Teacher's Resource pages 204 to 207, and have students complete it. Provide rulers, protractors, compasses, straightedges, scissors, and transparent mirrors (e.g., Miras). Review how to use symbols to show equal lengths (matching tick marks), equal angles (matching dots or arcs), a right angle (a small square), and parallel lines (matching arrowheads).

See solutions on Teacher's Resource pages 208 to 211.

Intervention Pathways

The purpose of the intervention pathways is to help students see how the measurements of shapes and the properties of those shapes interrelate, for example, the relationship between the height of an isosceles triangle and a perpendicular bisector of one of its sides.

There are 4 pathways:

- Pathway 1: Bisecting Angles and Line Segments
- Pathway 2: Drawing Lines and Polygons
- Pathway 3: Drawing Circles
- Pathway 4: Drawing Triangles

Use the chart below (or the Key to Pathways on Teacher's Resource pages 208 to 211) to determine which pathway is most suitable for each student or group of students.

Diagnostic Tool Results	Intervention Pathway
If students struggle with Questions 11 and 12	use Pathway 1: Bisecting Angles and Line Segments *Teacher's Resource pages 212–213* *Student Resource pages 263–269*
If students struggle with Questions 6 to 10	use Pathway 2: Drawing Lines and Polygons *Teacher's Resource pages 214–215* *Student Resource pages 270–276*
If students struggle with Questions 4 and 5	use Pathway 3: Drawing Circles *Teacher's Resource pages 216–217* *Student Resource pages 277–282*
If students struggle with Questions 1 to 3	use Pathway 4: Drawing Triangles *Teacher's Resource pages 218–219* *Student Resource pages 283–289*

Copyright © 2012 by Nelson Education Ltd.

Name: ______________________ Date: ______________________

Geometric Drawings

Diagnostic Tool

You will need

- a ruler
- a protractor
- a compass
- a straightedge
- scissors
- a transparent mirror
- dynamic geometry software (optional)

1. Draw a triangle with 2 angles greater than 50° and 2 side lengths longer than 5 cm. Label the angles that are greater than 50° and the side lengths that are longer than 5 cm.

2. Draw a triangle of each type. Label all angles and side lengths. Use symbols to show equal angles, equal side lengths, and right angles.

a) an equilateral triangle (all sides the same length)

b) an acute triangle that is *not* equilateral (all angles less than 90° and not all sides the same length)

c) a right isosceles triangle (one right angle and 2 sides the same length)

d) an obtuse scalene triangle (one angle greater than 90° and all sides a different length)

Copyright © 2012 by Nelson Education Ltd.

Name: ______________________________ Date: ____________________

3. Darren says that, if he completes the grey triangle, it will be congruent to the black one. Do you agree? Explain.

55° 60°
3 cm

55° 60°
3 cm

____

____

____

4. Draw each circle. Label each circle with the given measurement.

a circle with a diameter of 3 cm	a circle with a radius of 25 mm

5. How did you make sure that your drawings in Question 4 were circles and that they were the right size?

____

____

____

6. Draw the following pairs of lines. *Do not use a protractor.* Use symbols to show parallel lines (matching arrowheads) and right angles.

2 parallel lines	2 perpendicular lines

Copyright © 2012 by Nelson Education Ltd.

Name:_______________________________ Date:_______________

7. Draw the following pairs of lines using a protractor. Use symbols to show parallel lines and right angles.

2 parallel lines	2 perpendicular lines

8. How do you know that your lines in Questions 6 and 7 are parallel and perpendicular?

____

____

____

9. Draw the following pairs of lines. Label the given angle in each.

2 lines that intersect at a 30° angle	2 lines that intersect at a 60° angle

10. Draw a quadrilateral that meets all of these conditions:
- one or more pairs of parallel sides
- one or more pairs of perpendicular sides
- one side longer than 6 cm
- one 120° angle

Label your shape to show each condition above.

Copyright © 2012 by Nelson Education Ltd.

Name:____ Date:____________________

11. a) Draw a **perpendicular bisector** of line segment $\overline{AB}$ below, *without using a ruler or protractor*. Use symbols to show equal lengths (matching tick marks) and right angles (a small square).

> **perpendicular bisector**
> a line that divides another line segment into 2 equal parts at a right angle

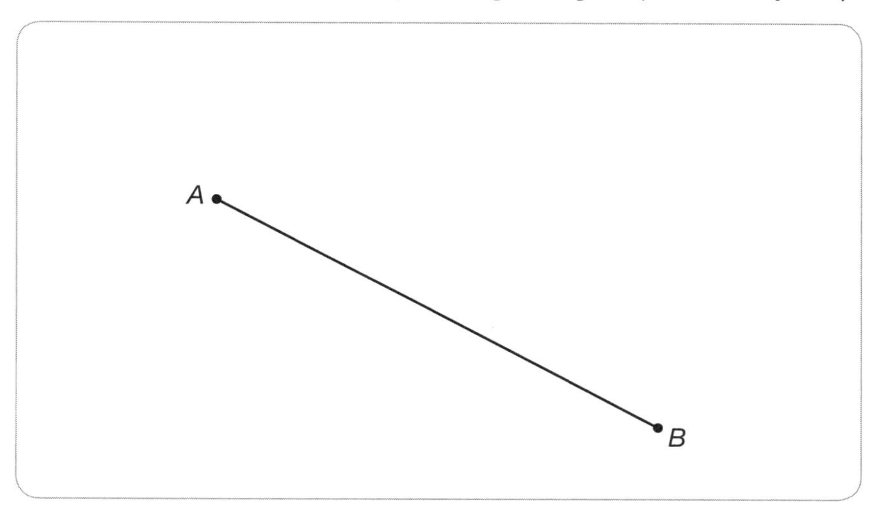

b) How do you know that the line you drew is a perpendicular bisector?

____

____

____

12. a) Draw an **angle bisector** for $\angle EFG$, *without using a protractor*.

> **angle bisector**
> a line that divides an angle into 2 equal parts

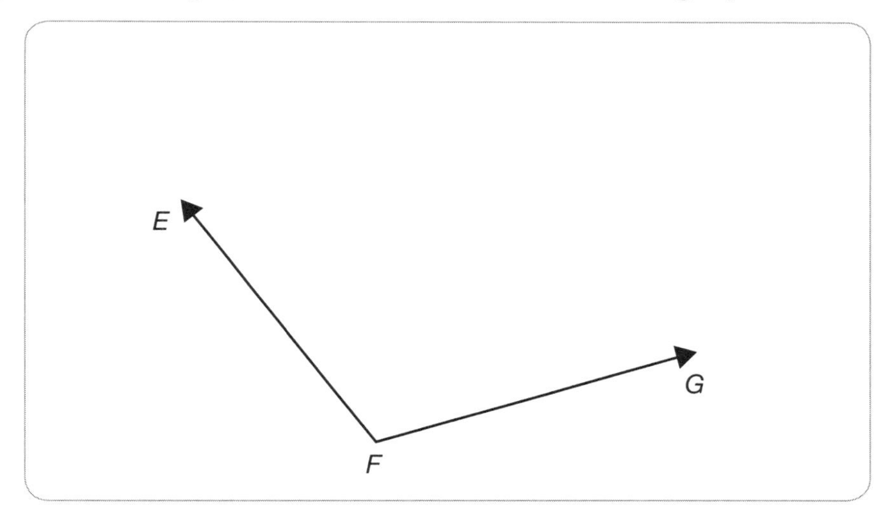

b) How do you know that the line you drew is an angle bisector?

____

____

____

Solutions and Key to Pathways

Name: ____________________ Date: ____________

Geometric Drawings

Diagnostic Tool

You will need

- a ruler
- a protractor
- a compass
- a straightedge
- scissors
- a transparent mirror
- dynamic geometry software (optional)

Pathway 4

1. Draw a triangle with 2 angles greater than 50° and 2 side lengths longer than 5 cm. Label the angles that are greater than 50° and the side lengths that are longer than 5 cm.

   e.g., AB = 6 cm, BC = 6 cm, $m\angle BAC$ = 65°, $m\angle BCA$ = 65°

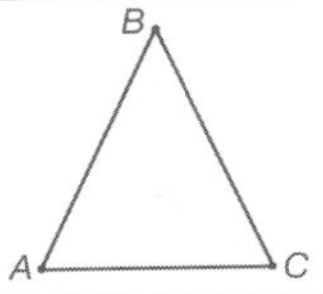

2. Draw a triangle of each type. Label all angles and side lengths. Use symbols to show equal angles, equal side lengths, and right angles.

   a) an equilateral triangle (all sides the same length)

   e.g., 60°, 60°, 60°, 4 cm

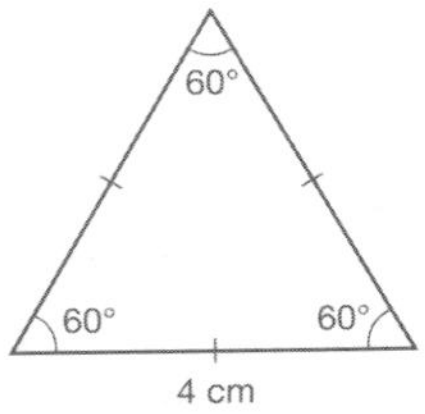

   b) an acute triangle that is *not* equilateral (all angles less than 90° and not all sides the same length)

   e.g., 42°, 56°, 82°, 6 cm, 5 cm, 4 cm

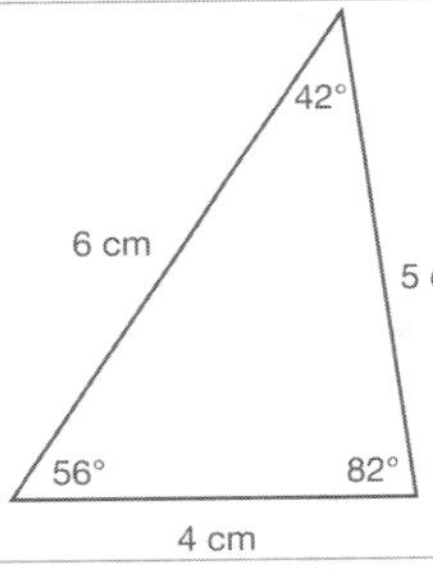

   c) a right isosceles triangle (one right angle and 2 sides the same length)

   e.g., 45°, 45°, 7.1 cm, 5.0 cm

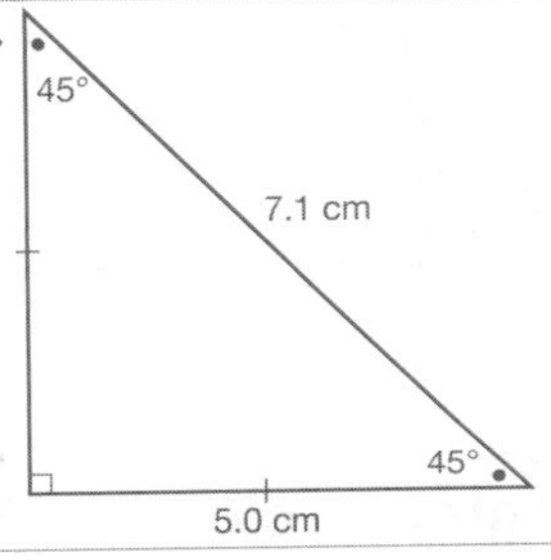

   d) an obtuse scalene triangle (one angle greater than 90° and all sides a different length)

   e.g., 27°, 107°, 46°, 5.0 cm, 6.6 cm, 3.0 cm

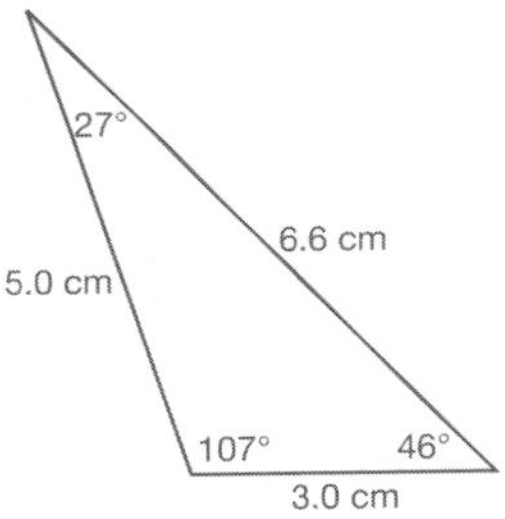

 Copyright © 2012 by Nelson Education Ltd.

 Copyright © 2012 by Nelson Education Ltd.

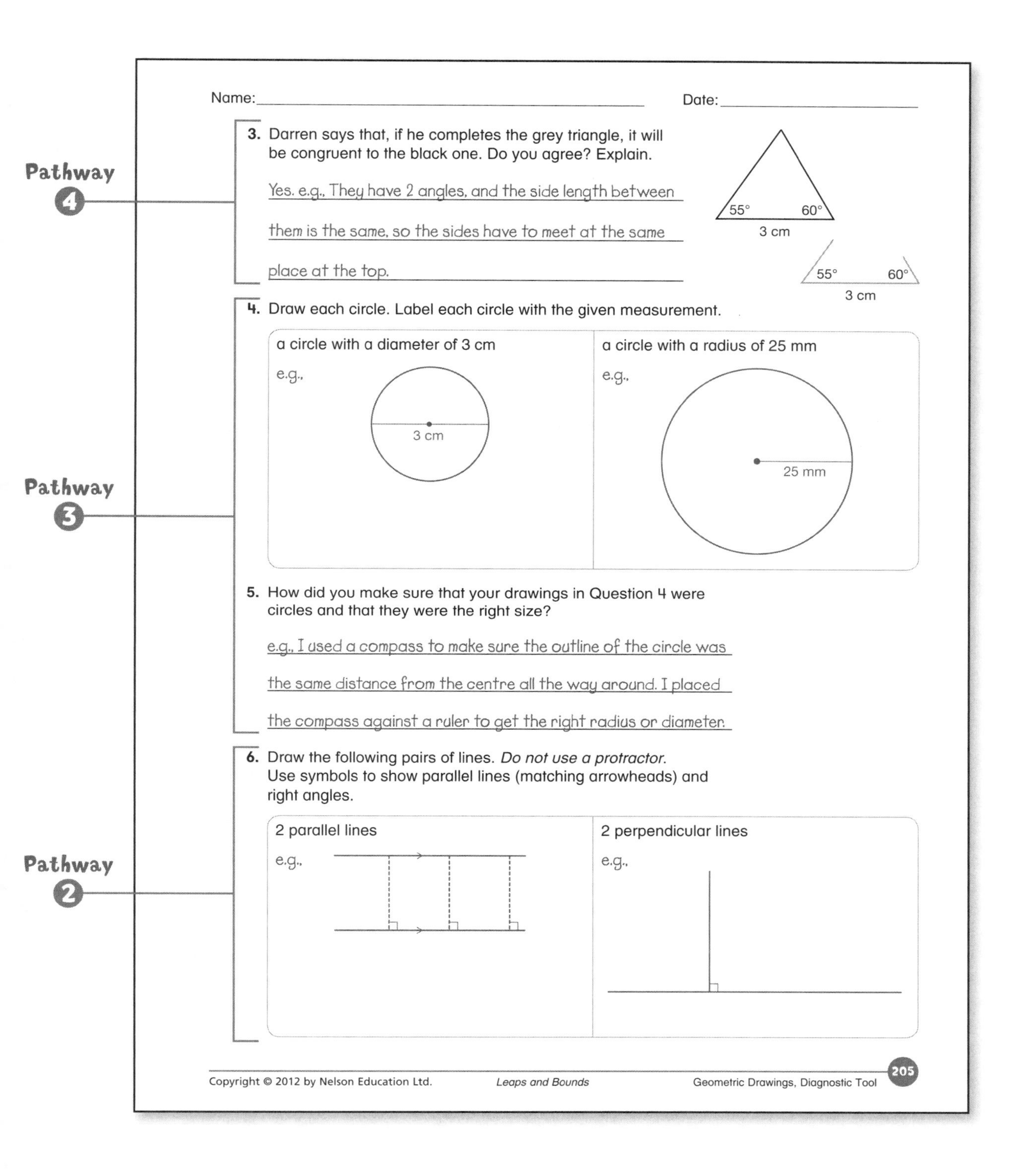

Name:______________________ Date:______________

Pathway 4

3. Darren says that, if he completes the grey triangle, it will be congruent to the black one. Do you agree? Explain.

Yes. e.g., They have 2 angles, and the side length between them is the same, so the sides have to meet at the same place at the top.

Pathway 3

4. Draw each circle. Label each circle with the given measurement.

a circle with a diameter of 3 cm	a circle with a radius of 25 mm
e.g., 3 cm	e.g., 25 mm

5. How did you make sure that your drawings in Question 4 were circles and that they were the right size?

e.g., I used a compass to make sure the outline of the circle was the same distance from the centre all the way around. I placed the compass against a ruler to get the right radius or diameter.

Pathway 2

6. Draw the following pairs of lines. *Do not use a protractor.* Use symbols to show parallel lines (matching arrowheads) and right angles.

2 parallel lines	2 perpendicular lines
e.g.,	e.g.,

Name:______________________ Date:______________

7. Draw the following pairs of lines using a protractor. Use symbols to show parallel lines and right angles.

2 parallel lines	2 perpendicular lines
e.g., 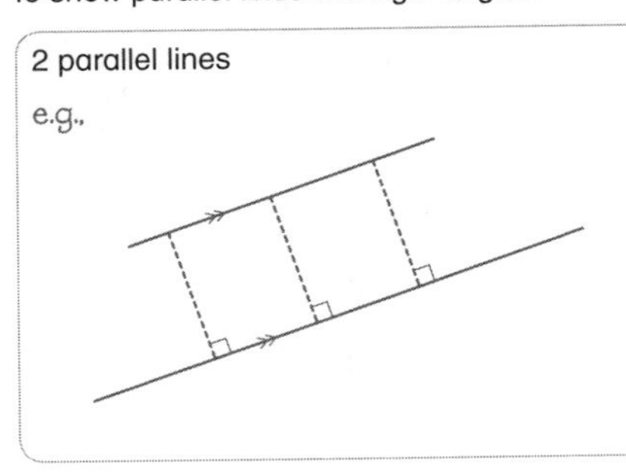	e.g.,

8. How do you know that your lines in Questions 6 and 7 are parallel and perpendicular?

e.g., I can check that the parallel lines are the same distance apart by measuring with a ruler. I can check that the perpendicular lines are at right angles by using a protractor.

Pathway 2

9. Draw the following pairs of lines. Label the given angle in each.

2 lines that intersect at a 30° angle	2 lines that intersect at a 60° angle
e.g., 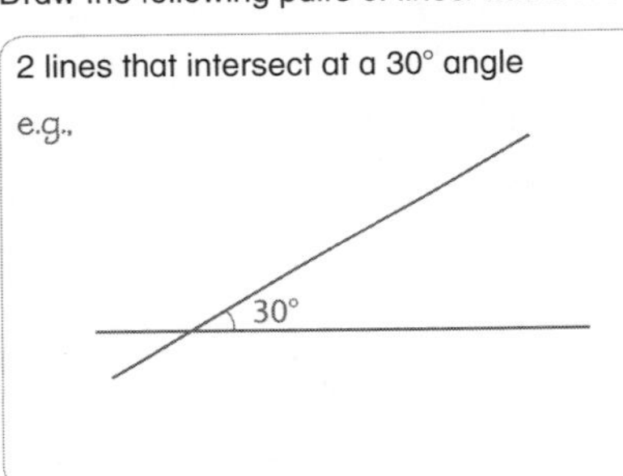	e.g., 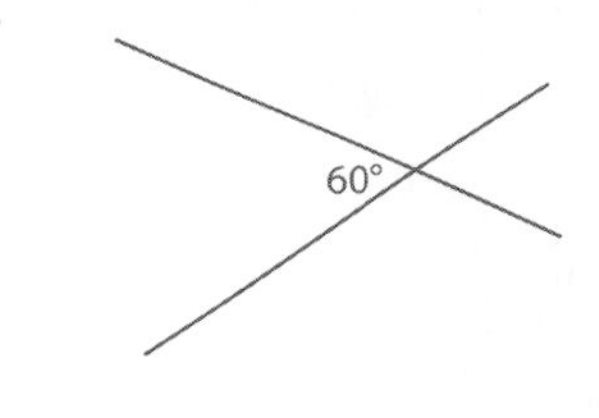

10. Draw a quadrilateral that meets all of these conditions:
- one or more pairs of parallel sides
- one or more pairs of perpendicular sides
- one side longer than 6 cm
- one 120° angle

Label your shape to show each condition above.

e.g.,

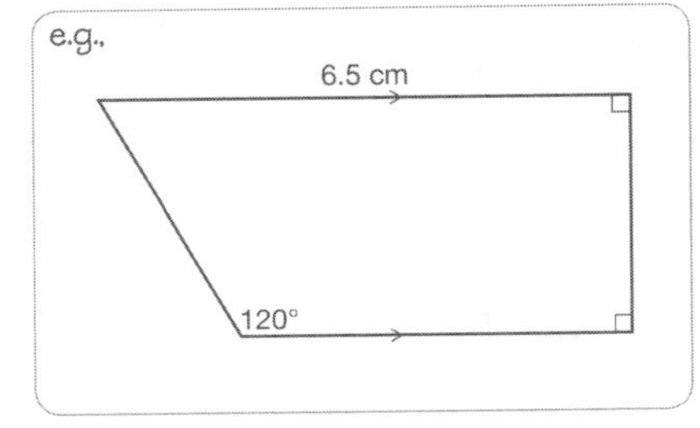

 Copyright © 2012 by Nelson Education Ltd.

 Copyright © 2012 by Nelson Education Ltd.

Name:______________________ Date:______________

11. a) Draw a **perpendicular bisector** of line segment $\overline{AB}$ below, *without using a ruler or protractor*. Use symbols to show equal lengths (matching tick marks) and right angles (a small square).

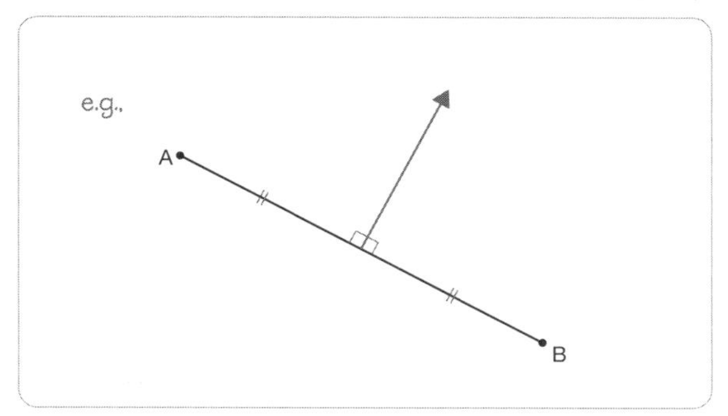

> **perpendicular bisector**
> a line that divides another line segment into 2 equal parts at a right angle

b) How do you know that the line you drew is a perpendicular bisector?

e.g., I made sure it divided the line segment in half and crossed it at a right angle by using a Mira. I found the reflection line that reflected point A exactly onto Point B.

Pathway 1

12. a) Draw an **angle bisector** for ∠*EFG*, *without using a protractor*.

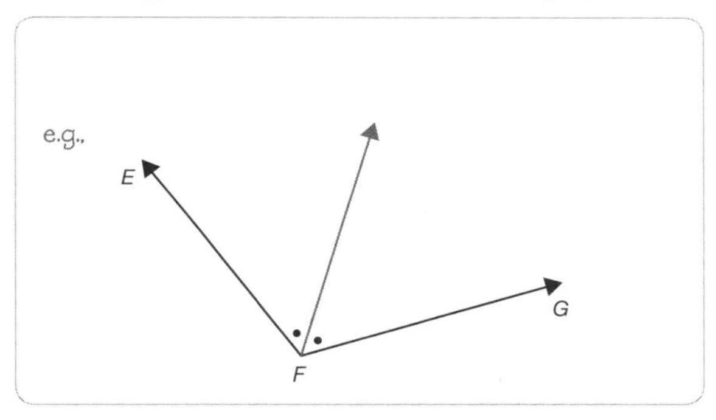

> **angle bisector**
> a line that divides an angle into 2 equal parts

b) How do you know that the line you drew is an angle bisector?

e.g., I made sure it divided the angle in half by tracing the angle and bisector, and then folding along the bisector so that *EF* lined up exactly with *FG*.

Pathway 1 OPEN-ENDED

Bisecting Angles and Line Segments

You will need

- scissors
- rulers
- protractors
- straightedges
- compasses
- transparent mirrors (e.g., Miras)
- dynamic geometry software (optional)
- Student Resource pages 263–264

Open-Ended Intervention

Before Using the Open-Ended Intervention

Draw a square. Mark the vertices with right-angle symbols and the 4 equal sides with matching tick marks. Provide scissors, a ruler, a protractor, a straightedge, a compass, and a transparent mirror. Ask:

- ▸ How could you use any of these tools to draw a line inside the square that would make a 45° angle at any of the vertices? (e.g., *I'd measure a 45° angle at a vertex with the protractor.*)
- ▸ What if you couldn't use a protractor? (e.g., *I'd cut out the square and fold it in half diagonally at one of the vertices—each angle is 90°, so half would be 45°.*)
- ▸ How could you divide one of the side lengths in half? (e.g., *I'd use a ruler and measure the whole length, divide it in half, and then measure that length.*)
- ▸ What if you could not use a ruler? (e.g., *I'd fold one end of the side length onto the other end, and the fold line would be halfway.*)

Using the Open-Ended Intervention (Student Resource pages 263–264)

Some students might benefit from reading the instructional section of the guided intervention before they begin.

Read through the tasks on the student pages together. Provide the materials.

Give students time to work, ideally in pairs.

Observe whether students

- can draw perpendicular bisectors
- can draw angle bisectors
- can explain the drawing process
- realize that geometric drawings are based on the properties of shapes (e.g., congruent triangles have equal angles and side lengths)

Consolidating and Reflecting

Ensure understanding by asking questions such as these based on students' work:

- ▸ Why did your strategy for drawing an angle bisector make sense? (e.g., *When you fold one arm of the angle on top of the other, it's like a reflection. The 2 shapes on either side of the fold are identical, so the angles have to be equal.*)
- ▸ Why did your strategy for drawing a perpendicular bisector make sense? (e.g., *Since you end up making congruent triangles, the angles on either side have to be equal and so do the side lengths. If you cut a straight angle into 2 equal parts, each is 90°.*)
- ▸ How could you draw a 30° angle using a 120° angle and without using a protractor? (e.g., *I could bisect the angle twice.*)
- ▸ What tools and strategies did you prefer for drawing angle bisectors and for perpendicular bisectors? (e.g., *I like using a Mira for both because it's the only tool I need—it's like a protractor and ruler all in one.*)

Copyright © 2012 by Nelson Education Ltd.

Bisecting Angles and Line Segments

Pathway 1 GUIDED

Guided Intervention

Before Using the Guided Intervention

Draw a square. Mark the vertices with right-angle symbols and the 4 equal sides with matching tick marks. Provide scissors, a ruler, a protractor, a straightedge, a compass, and a transparent mirror. Ask:

- How could you use any of these tools to draw a line inside the square that would result in a 45° angle at any of the vertices?
  (e.g., *I'd measure a 45° angle at a vertex with the protractor.*)
- How could you divide one of the side lengths in half? (e.g., *I'd use a ruler and measure the whole length, divide it in half, and then measure that length.*)

Using the Guided Intervention (Student Resource pages 265–269)

Provide the materials and have students do the drawings as you work through the instructional section of the student pages together.

Have students work through the **Try These** questions individually or in pairs.

Observe whether students

- can identify angle and perpendicular bisectors in a diagram (Questions 1, 2)
- can draw angle bisectors and perpendicular bisectors (Questions 3, 4, 5, 7)
- realize that the geometric properties of an isosceles triangle relate to angle bisectors and perpendicular bisectors (Question 6)

Consolidating and Reflecting

Ensure understanding by asking questions such as these based on students' work:

- Use Question 2 to help you explain why a perpendicular bisector can also be an angle bisector. (e.g., *If $\overline{FL}$ is a straight angle with a vertex in the middle at M, then its angle bisector and perpendicular bisector $\overline{MH}$ are the same.*)
- In Question 3a), why can you draw an angle bisector for $\angle ABC$ through B but not a perpendicular bisector of $\overline{AC}$ through B?
  (e.g., *An angle can have arms that are different lengths, so you can bisect $\angle ABC$ straight through B. A perpendicular bisector has to be exactly halfway between the 2 end points of the line segment, and B is not exactly halfway between A and C.*)
- How does thinking about isosceles triangles in Question 6 help make sense of the procedures for drawing angle bisectors and perpendicular bisectors?
  (e.g., *Since an isosceles triangle has 2 equal side lengths and 2 equal angles, its line of symmetry has to divide the straight angle on the non-congruent side and the non-congruent side in half.*)
- How are angle bisectors and perpendicular bisectors alike and different?
  (e.g., *Alike: both divide something in half or into 2 equal parts.*
  Different: angle bisectors divide an angle in half, and perpendicular bisectors divide a line segment in half.)

You will need

- protractors
- straightedges
- compasses
- transparent mirrors (e.g., Miras)
- scissors
- rulers
- coloured pencils
- dynamic geometry software (optional)
- Student Resource pages 265–269

Copyright © 2012 by Nelson Education Ltd.

Drawing Lines and Polygons

You will need

- scissors
- transparent mirrors (e.g., Miras)
- protractors
- rulers
- compasses (optional)
- dynamic geometry software (optional)
- Student Resource pages 270–271

Open-Ended Intervention

Before Using the Open-Ended Intervention

Ask students to tell everything they know about a rectangle.
(e.g., *4 sides, 4 right angles, 2 opposite sides of one length and 2 opposite sides of another length, but for a square all 4 sides are the same; 2 pairs of parallel sides, and each side is perpendicular to another side*)

Provide scissors, a transparent mirror, a protractor, and a ruler. Ask:

- ▸ Which of these tools could you use to help you draw a rectangle? Why?
  (e.g., *a ruler to make sure the pairs of side lengths are equal; a protractor or Mira to make perpendicular sides; scissors to cut paper so I could fold it to make perpendicular sides*)
- ▸ How can you tell that opposite sides are parallel?
  (e.g., *I can measure them to make sure they are the same distance apart all the way along.*)

Using the Open-Ended Intervention (Student Resource pages 270–271)

Some students might benefit from reading the instructional section of the guided intervention before they begin.

Read through the tasks on the student pages together. Provide the materials and give students time to work, ideally in pairs.

Observe whether students

- draw polygons to meet given conditions (i.e., type of polygon, parallel sides, perpendicular sides, side lengths, and angle measures)
- draw parallel line segments and perpendicular line segments
- draw line segments that intersect at 30°, 45°, and 60°
- can describe strategies for drawing parallel and perpendicular line segments

Consolidating and Reflecting

Ensure understanding by asking questions such as these based on students' work:

- ▸ What strategy can you use for drawing parallel lines? Explain why it works.
  (e.g., *Draw a line. Measure and mark 3 points the same distance away from the line on the same side, making sure the ruler is perpendicular to the line each time. Join the points to get a parallel line. It works because parallel lines are always the same distance apart, so they never meet.*)
- ▸ What strategy can you use to draw perpendicular lines? Explain why it works.
  (e.g., *If you draw a line and then fold it somewhere along the line so the parts of the line match, the fold line is perpendicular to the line. It works because a reflection line [the fold line] is perpendicular to points that match on either side of it.*)
- ▸ What polygons can you draw without a protractor? Why?
  (*a square or rectangle;* e.g., *You can create the perpendicular lines using a mirror or by folding.*)

 Copyright © 2012 by Nelson Education Ltd.

Drawing Lines and Polygons

Pathway 2 GUIDED

Guided Intervention

Before Using the Guided Intervention

Provide scissors, a transparent mirror, a protractor, and a ruler. Ask students to tell everything they know about a rectangle.
(e.g., *4 sides, 4 right angles, 2 opposite sides of one length and 2 opposite sides of another length, but for a square all 4 sides are the same; 2 pairs of parallel sides, and each side is perpendicular to another side*)

Ask:

- Which of these tools could you use to help you draw a rectangle? Why? (e.g., *a ruler to make sure the pairs of side lengths are equal; a protractor or Mira to make perpendicular sides; scissors to cut paper so I could fold it to make perpendicular sides.*)

Using the Guided Intervention Student Resource pages 272–276

Provide the materials and have students do the drawings as you work through the instructional section of the student pages together.

Have students work through the **Try These** questions individually or in pairs.

Observe whether students

- can draw parallel and perpendicular line segments (Questions 1, 5)
- can draw line segments that intersect at 30°, 45°, and 60° (Questions 2, 5)
- can describe a strategy for drawing perpendicular and intersecting line segments (Questions 3, 4)
- can draw polygons to meet given conditions (i.e., parallel sides, perpendicular sides, side lengths, and angle measures) (Questions 6 to 10)

Consolidating and Reflecting

Ensure understanding by asking questions such as these based on students' work:

- What strategy did you use to draw perpendicular line segments in Question 5? Explain why it works. (e.g., *I drew a line and then used a Mira to draw a second line perpendicular to the first. It works because a reflection line [the Mira] is always perpendicular to points that match on either side of it.*)
- What strategy did you use to draw the parallel line segment in Question 5? Explain why it works. (e.g., *I measured and marked a point the same distance away from the line in 3 places. I made sure my ruler was perpendicular to the line each time. I joined the points to get a parallel line. It works because parallel lines are always the same distance apart.*)
- Do you think it is easier to draw polygons that have 90° angles than polygons that have 60° angles? Why? (e.g., *Yes; you can just fold or you can just use a square corner, and you don't need a protractor.*)

You will need

- scissors
- transparent mirrors (e.g., Miras)
- protractors
- rulers
- compasses (optional)
- dynamic geometry software (optional)
- Student Resource pages 272–276

Drawing Circles

You will need

- compasses
- rulers
- coloured pencils (optional)
- dynamic geometry software (optional)
- Student Resource pages 277–278

Open-Ended Intervention

Before Using the Open-Ended Intervention

Provide compasses and ask:

- ▸ What sorts of things can you draw with this tool? (e.g., *circles, arcs, curved lines*)
- ▸ What is special about this tool? (e.g., *It makes sure that the mark you make with the pencil end is always the same distance from the centre, or pointed end.*)
- ▸ If a circle had a radius of 3 cm, how could you use the tool to draw it? (e.g., *I'd measure the spread of the pencil and pointed end of the compass against a ruler to make sure it was 3 cm, and then I'd make the circle—the pointed part is the centre, so I'd mark where that was.*)
- ▸ What if the circle had a diameter of 5 cm? (e.g., *I'd divide 5 by 2 to get the radius, 2.5 cm, and then I'd do it the same way but with a 2.5 cm spread.*)

Using the Open-Ended Intervention Student Resource pages 277–278

Some students might benefit from reading the instructional section of the guided intervention before they begin.

Read through the tasks on the student pages together. Provide the materials and give students time to work, ideally in pairs.

Observe whether students

- can draw circles for a given radius or diameter measurement
- can solve problems by drawing circles
- distinguish between circles and polygons

Consolidating and Reflecting

Ensure understanding by asking questions such as these based on students' work:

- ▸ Describe how you drew one of the circles. (e.g., *I measured the spread of the compass against a ruler to get the right radius. Then I placed the pointed end where I wanted it and rotated the pencil end around the circle until I had a closed shape.*)
- ▸ Did your strategy for drawing the circle change if you knew the diameter rather than the radius? (e.g., *No, I made them exactly the same way, but before I started, I divided the diameter by 2 to get the radius.*)
- ▸ How did you make your circle design with the same number of circles but end up with a polygon with a different number of sides? (e.g., *I made 3 of the circles exactly the same but placed the fourth one so that its centre and the centre of 2 others made a straight line. That way I ended up with a polygon with fewer sides.*)
- ▸ Why do you need only one measurement to draw a circle, but more than one for most polygons? (e.g., *Since the circumference of the circle is the same distance from the centre all the way around, you need to know only that one distance.*)

 Copyright © 2012 by Nelson Education Ltd.

Drawing Circles

Pathway 3 GUIDED

Guided Intervention

Before Using the Guided Intervention

Provide compasses and ask:

- ▸ What sorts of things can you draw with this tool?
  (e.g., *circles, arcs, curved lines*)
- ▸ What is special about this tool? (e.g., *It makes sure that the mark you make with the pencil end is always the same distance from the centre, or pointed end.*)
- ▸ If a circle had a radius of 3 cm, how could you use the tool to draw it?
  (e.g., *I'd measure the spread of the pencil and pointed end of the compass against a ruler to make sure it was 3 cm, and then I'd make the circle—the pointed part is the centre, so I'd mark where that was.*)
- ▸ What if the circle had a diameter of 5 cm?
  (e.g., *I'd divide 5 by 2 to get the radius, 2.5 cm, and then I'd do it the same way but with a 2.5 cm spread.*)

You will need

- compasses
- rulers
- string
- coloured pencils (optional)
- dynamic geometry software (optional)
- Student Resource pages 279–282

Using the Guided Intervention (Student Resource pages 279–282)

Provide the materials and have students draw as you work through the instructional section of the student pages together.

Have students work through the **Try These** questions individually or in pairs.

Observe whether students

- can draw circles from a given radius or diameter measurement
  (Questions 1, 5, 6)
- can draw circles from a line segment that represents a given radius or diameter
  (Questions 2, 3)
- recognize how properties of a circle differ from those of polygons
  (Questions 7, 8, 9)

Consolidating and Reflecting

Ensure understanding by asking questions such as these based on students' work:

- ▸ How did you draw the circles in Question 1?
  (e.g., *I measured the spread of the compass against a ruler to get the right radius. Then I placed the pointed end where I wanted it and rotated the pencil end around the circle until I had a closed shape. For 1b), I divided the diameter by 2 first to get the radius.*)
- ▸ How did you know that your flag in Question 5 had to be at least 6 cm high?
  (e.g., *One of the circles had a radius of 3 cm, which means its height is 6 cm.*)
- ▸ Why do you need only one measurement to draw a circle, but more than one for most polygons?
  (e.g., *Since the circumference of the circle is the same distance away from the centre all the way around, you need to know only that one distance.*)

Copyright © 2012 by Nelson Education Ltd.

Pathway 4 OPEN-ENDED

Drawing Triangles

You will need

- rulers
- compasses
- protractors
- transparent mirrors (e.g., Miras)
- coloured pencils (optional)
- dynamic geometry software (optional)
- Student Resource pages 283–284

Open-Ended Intervention

Before Using the Open-Ended Intervention

Draw a line 5 cm long and tell students that it is one side of a triangle. Ask:

- Suppose another side is 6 cm long. What might it look like?
  (e.g., *It would start at one of the end points, but it could go in any direction.*)
- How could you use a compass to figure out all the possible end points for the 6 cm side?
  (e.g., *If I set the compass spread to 6 cm, I could make a circle with its centre at one end of the 5 cm side. The 6 cm side could end anywhere on the circumference of the circle.*)
- Suppose the third side is 5 cm long. What do you know about the triangle? How would you draw that side using a compass?
  (e.g., *It's isosceles, because 2 sides are equal. I'd use a compass again at the other end, but set at 5 cm, and make a circle. Where the 2 circles cross would be the end point of the 6 cm and 5 cm sides.*)

Using the Open-Ended Intervention Student Resource pages 283–284

Some students might benefit from reading the instructional section of the guided intervention before they begin.

Read through the tasks on the student pages together. Provide the materials and give students time to work, ideally in pairs. Alternatively, students could create one of each type of triangle first and then select 4 for their flag.

Observe whether students

- can draw triangles to meet given conditions (i.e., the type of triangle, and given angle measures and side lengths)
- can describe a strategy for drawing specific types of triangles
- recognize that you can create a congruent triangle automatically using only some of the information in the first triangle

Consolidating and Reflecting

Ensure understanding by asking questions such as these based on students' work:

- How does a compass help when drawing an equilateral, isosceles, or scalene triangle? (e.g., *The arc of the circle you make shows all the possible end points of each side length. Without it, you would have to draw all sorts of sides in different directions to see which ones worked.*)
- How is drawing an equilateral, an isosceles, or a scalene triangle different from drawing an acute, a right, or an obtuse triangle? (e.g., *When you know how the side lengths compare, you need only a ruler and compass to make sure you get them right; and, when you know only how the angles compare, you have to use a protractor to make sure the angles are the right size.*)
- Why does it make sense that you don't need all of the information about a triangle to draw a congruent triangle? (e.g., *If, for example, you know 2 side lengths and the angle between them, the third side and other angles are obvious.*)

 Copyright © 2012 by Nelson Education Ltd.

Drawing Triangles

Guided Intervention

Before Using the Guided Intervention

Draw a line 5 cm long and tell students that it is one side of a triangle. Ask:

- Suppose another side is 6 cm long. What might it look like?
  (e.g., *It would start at one of the end points but it could go in any direction.*)
- How could you use a compass to figure out all the possible end points for the 6 cm side?
  (e.g., *If I set the compass spread to 6 cm, I could make a circle with its centre at one end of the 5 cm side. The 6 cm side could end anywhere on the perimeter of the circle.*)

You will need

- rulers
- compasses
- protractors
- transparent mirrors (e.g., Miras)
- dynamic geometry software (optional)
- Student Resource pages 285–289

Using the Guided Intervention Student Resource pages 285–289

Provide the materials and work through the instructional section of the student pages together.

Have students work through the **Try These** questions individually or in pairs.

Observe whether students

- can draw triangles to meet given conditions (i.e., the type of triangle, and given angle measures and side lengths) (Questions 1, 2, 3, 4, 5, 6)
- understand that if only 1 side length and 2 angles are given, more than one triangle is possible (Question 6)
- understand how the properties of the different triangles influence how you draw them (Questions 7, 8, 9)
- recognize that you can create a congruent triangle automatically using only some of the information in the original triangle (Question 10)

Consolidating and Reflecting

Ensure understanding by asking questions such as these based on students' work:

- How did a compass help you draw the triangles in Question 1? (e.g., *The arc of the circle shows all possible end points of each side. Without it, you would have to draw all sorts of sides in different directions to see which ones worked.*)
- In Question 3, why does it make sense that everyone's triangle would look the same if they used the same side lengths and angle?
  (e.g., *After you draw 2 side lengths with that angle in between them, there's no choice about where to draw the third side.*)
- Why does it make sense that more than one triangle is possible in Question 4?
  (e.g., *It doesn't tell you where the angles are in relation to the given side—they could both be at either end of the side or one of them could be opposite to it.*)
- Why is it possible to draw an equilateral triangle without using a ruler?
  (e.g., *You need only a compass to make sure side lengths are equal, as long as you have a straightedge to draw the lines.*)

Copyright © 2012 by Nelson Education Ltd.

Location

Planning For This Topic

Materials for assisting students with location consist of a diagnostic tool and 2 intervention pathways. Pathway 1 focuses on plotting and identifying points in all 4 quadrants of a Cartesian grid. Pathway 2 focuses on plotting and identifying points in the first quadrant of a Cartesian grid.

Each pathway has an open-ended intervention and a guided intervention. Choose the type of intervention more suitable for your students' needs and your particular circumstances.

Curriculum Connections

Grades 5 to 8 curriculum connections for this topic are provided online. See www.nelson.com/leapsandbounds. In both the Ontario and WNCP curriculums, Grade 6 focuses on plotting and identifying points in the first quadrant of a Cartesian grid (addressed in Pathway 2), and Grade 7 focuses on plotting and identifying points in all 4 quadrants of a Cartesian grid (addressed in Pathway 1). The Grade 5 expectations are not addressed in these pathways because those approaches are not used in Grades 7, 8, and higher.

Why might students struggle with location?

Students might struggle with plotting or locating points on a Cartesian grid for any of the following reasons:

- They might count lines rather than spaces when they are creating Cartesian grids and locating points on a Cartesian grid.
- They might mix up the order of the *x*-coordinate and *y*-coordinate when plotting points on a Cartesian grid.
- They might not understand that the first number (or *x*-coordinate) in an ordered pair indicates the number of units left or right of the origin (0, 0) and the second number (or *y*-coordinate) indicates the number of units up or down from the origin when plotting points on a Cartesian grid.
- They might plot points in squares rather than at lattice points where grid lines meet.
- They might have difficulty locating negative integers on a number line and therefore have difficulty locating them on a Cartesian grid.
- They might struggle with which direction (up or down, right or left) to move when working with negative integers.

Professional Learning Connections

PRIME: Geometry, Background and Strategies (Nelson Education Ltd., 2007), pages 97, 98, 111

Making Math Meaningful to Canadian Students K–8 (Nelson Education Ltd., 2008), pages 341–342, 357

Big Ideas from Dr. Small Grades 4–8 (Nelson Education Ltd., 2009), pages 119–120

 Copyright © 2012 by Nelson Education Ltd.

Diagnostic Tool: Location

Use the diagnostic tool to determine the most suitable intervention pathway for location. Provide Diagnostic Tool: Location, Teacher's Resource pages 222 and 223, and have students complete it in writing or orally. Have rulers available. Students could also use dynamic geometry software, if available.

See solutions on Teacher's Resource pages 224 and 225.

Intervention Pathways

The purpose of the intervention pathways is to help students understand the basics of plotting points on a Cartesian grid so they can use these skills later to perform transformations and graph relations, functions, and scatter-plot data.

There are 2 pathways:
- Pathway 1: Plotting Points in 4 Quadrants
- Pathway 2: Plotting Points on a Grid

Use the chart below (or the Key to Pathways on Teacher's Resource pages 224 and 225) to determine which pathway is most suitable for each student or group of students.

Diagnostic Tool Results	Intervention Pathway
If students struggle with Questions 4 to 7	use Pathway 1: Plotting Points in 4 Quadrants *Teacher's Resource pages 226–227* *Student Resource pages 290–295*
If students struggle with Questions 1 to 3	use Pathway 2: Plotting Points on a Grid *Teacher's Resource pages 228–229* *Student Resource pages 296–301*

If students successfully complete Pathway 2, they may or may not need the additional intervention provided by Pathway 1. Either re-administer Pathway 1 questions from the diagnostic tool or encourage students to do a portion of the open-ended intervention for Pathway 1 to decide if work in Pathway 1 would be beneficial.

Name:____________________ Date:____________________

Location

Diagnostic Tool

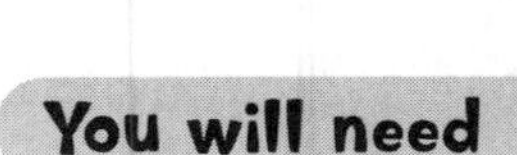

You will need

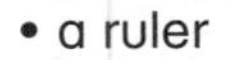

• a ruler

1. a) Label the ordered pair for each point on the Cartesian grid.

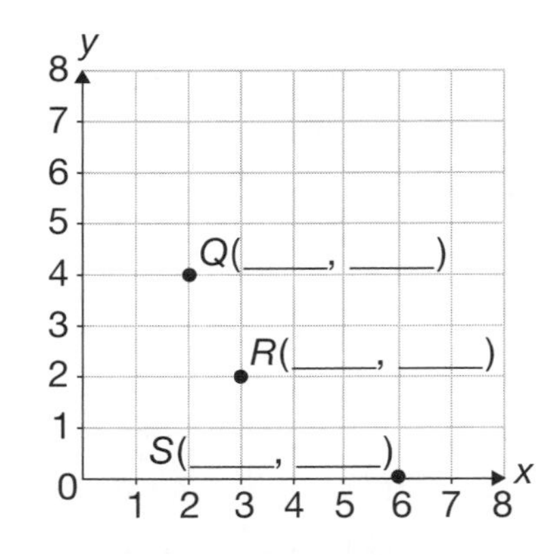

b) Plot these points on the grid above.

T(1, 7) *U*(4, 5) *V*(7, 0)

2. a) Identify 2 points on the grid to the right that have the same *x*-coordinate. Write the letters and the ordered pairs.

____

b) Identify 2 points on the grid with *different y*-coordinates. Write the letters and the ordered pairs.

____

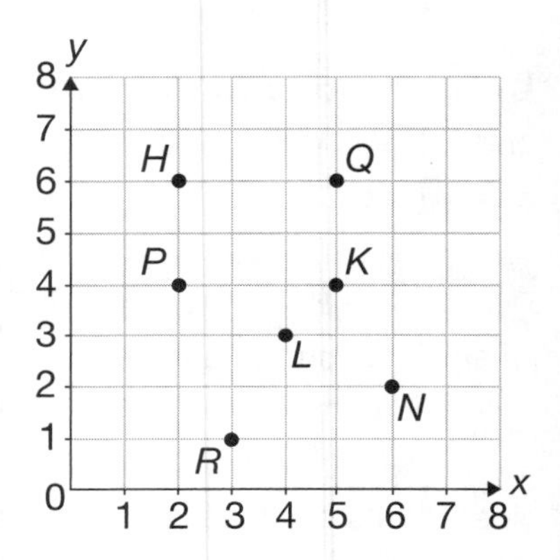

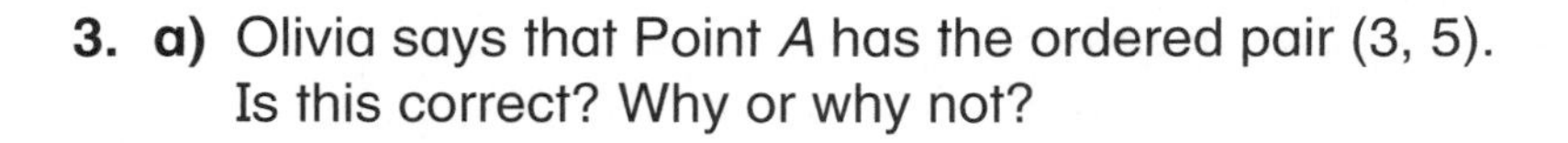

3. a) Olivia says that Point *A* has the ordered pair (3, 5). Is this correct? Why or why not?

____

____

____

b) Jacob says to get from Point *A* to Point *B*, you move 3 spaces right and 2 spaces up. Is this correct? Why or why not?

____

____

____

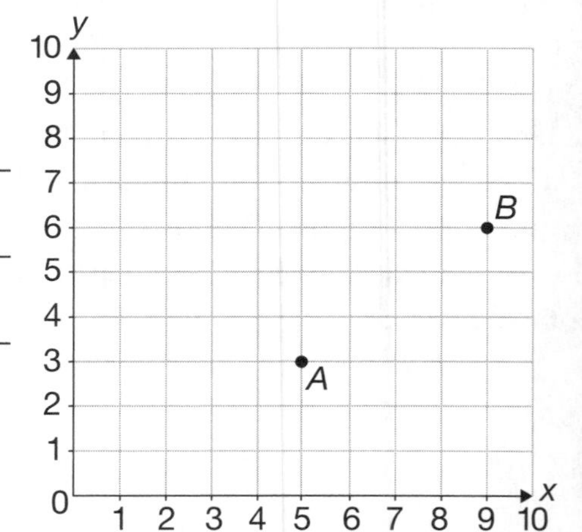

 Copyright © 2012 by Nelson Education Ltd.

Name:______________________________ Date:______________

4. a) Label the points with ordered pairs on the grid.

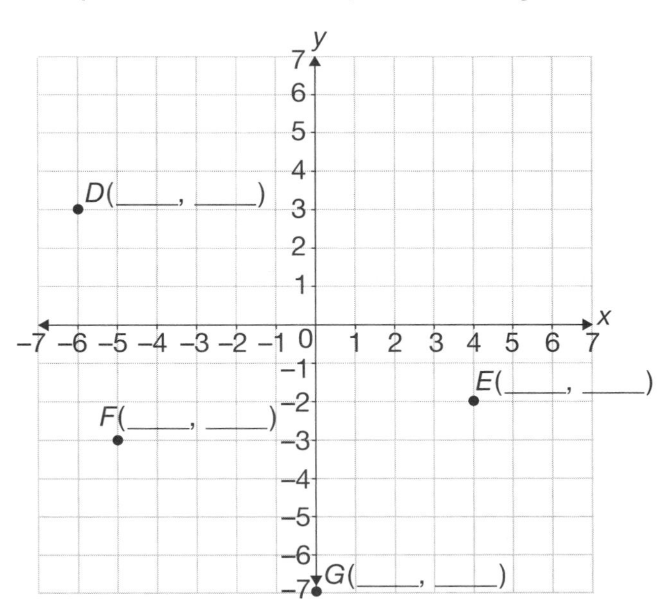

b) Plot these points on the grid above.

$H(1, -3)$ $I(-6, -4)$ $J(-2, 1)$

5. Thomas says that Point C has the ordered pair $(3, -3)$. Is this correct? Why or why not?

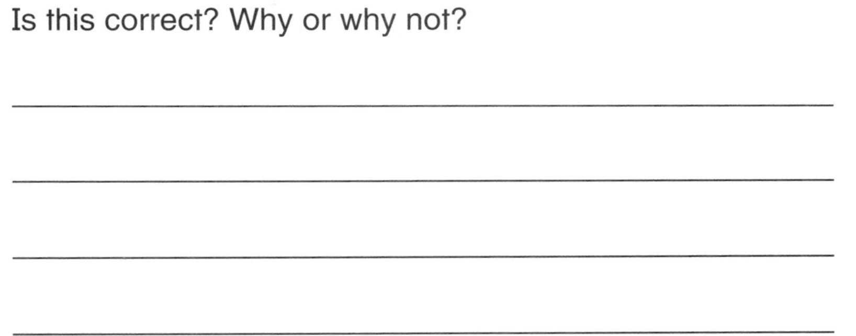

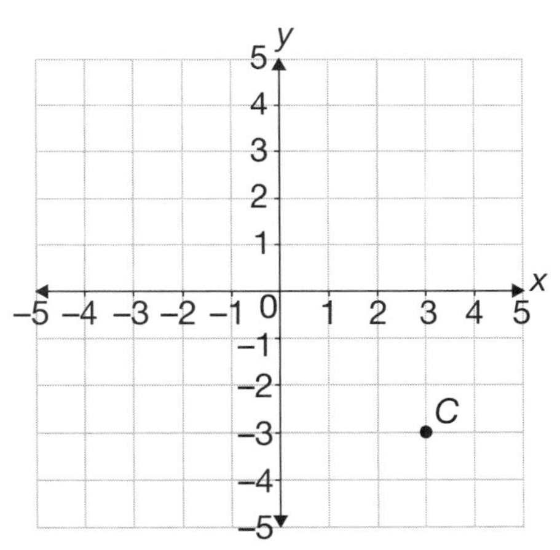

6. How would you move from Point $M(-2, 4)$ to Point $N(3, -2)$ on a Cartesian grid? Plot the points on the grid beside Question 5 and describe the move.

____

____

7. What can you say about any of the numbers in ordered pairs in Quadrant 3?

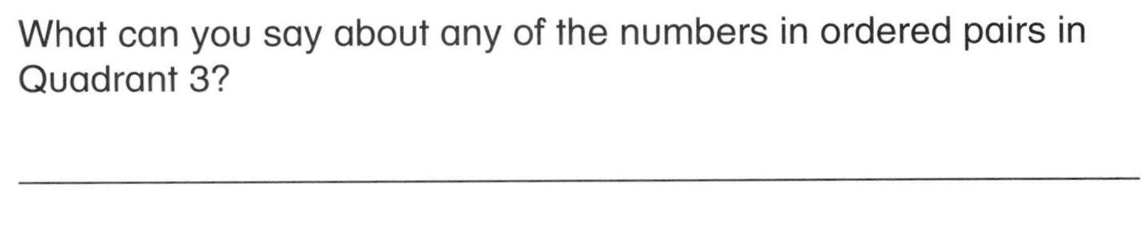

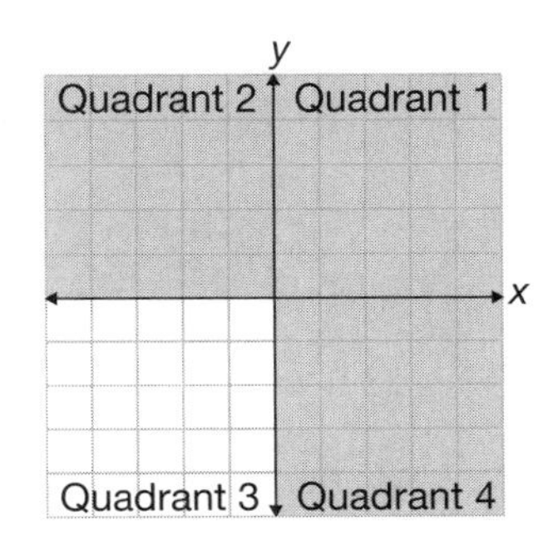

Copyright © 2012 by Nelson Education Ltd.

Solutions and Key to Pathways

Name:______________________ Date:______________

Location

Diagnostic Tool

You will need
- a ruler

Pathway 2

1. a) Label the ordered pair for each point on the Cartesian grid.

T(1, 7)
U(4, 5)
Q(2 , 4)
R(3 , 2)
S(6 , 0)
V(7, 0)

b) Plot these points on the grid above.

T(1, 7) *U*(4, 5) *V*(7, 0)

2. a) Identify 2 points on the grid to the right that have the same *x*-coordinate. Write the letters and the ordered pairs.

e.g., *P*(2, 4) and *H*(2, 6)

b) Identify 2 points on the grid with *different y*-coordinates. Write the letters and the ordered pairs.

e.g., *L*(4, 3) and *N*(6, 2)

3. a) Olivia says that Point *A* has the ordered pair (3, 5). Is this correct? Why or why not?

No, this is incorrect. e.g., The *x*-coordinate and *y*-coordinate are in the wrong order in the ordered pair. It should be (5, 3).

b) Jacob says to get from Point *A* to Point *B*, you move 3 spaces right and 2 spaces up. Is this correct? Why or why not?

No, this is incorrect. e.g., Point *B* is at (9, 6) so you move 4 units to the right and 3 units up. You have to count the spaces, not just the number of points or lines in the middle.

 Copyright © 2012 by Nelson Education Ltd.

 Copyright © 2012 by Nelson Education Ltd.

Name:_______________________________ Date:_______________

4. a) Label the points with ordered pairs on the grid.

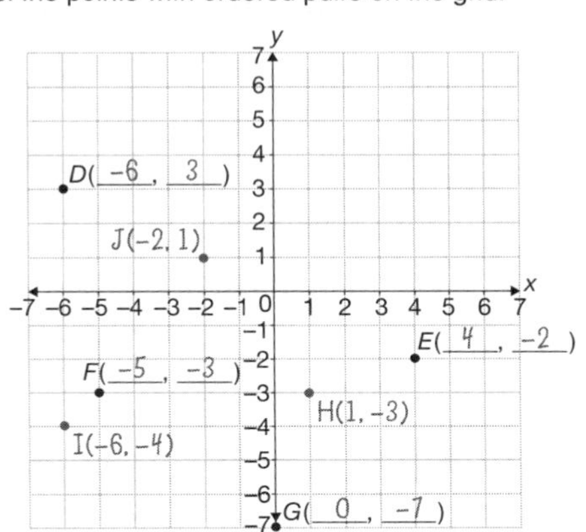

b) Plot these points on the grid above.

$H(1, -3)$ $I(-6, -4)$ $J(-2, 1)$

Pathway 1

5. Thomas says that Point C has the ordered pair $(3, -3)$. Is this correct? Why or why not?

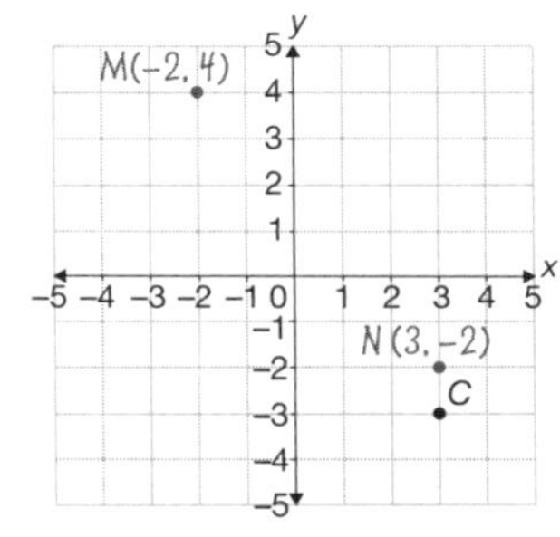

Yes, this is correct. e.g., The y-coordinate should be negative and the x-coordinate should be positive, because it's in Quadrant 4.

6. How would you move from Point $M(-2, 4)$ to Point $N(3, -2)$ on a Cartesian grid? Plot the points on the grid beside Question 5 and describe the move.

You would move 5 spaces right and 6 spaces down.

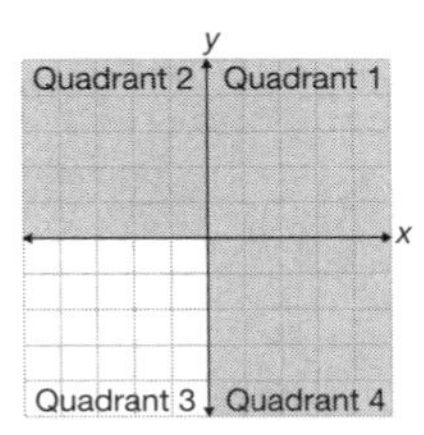

7. What can you say about any of the numbers in ordered pairs in Quadrant 3?

e.g., The numbers in the ordered pairs are all negative, or less than 0.

Pathway 1 OPEN-ENDED

Plotting Points in 4 Quadrants

You will need

- grid chart paper
- a ruler
- dynamic geometry software (optional)
- Student Resource pages 290–291

Open-Ended Intervention

Before Using the Open-Ended Intervention

On chart paper, sketch a Cartesian grid that shows the first quadrant and one that shows all 4 quadrants. Ask students to point out the origin. Ask:

- Where would you plot (5, 2) on each grid?
  (e.g., *5 units to the right of the origin and 2 units up*)
- How is the position of (−5, 2) different from the position of (5, −2)?
  (e.g., *(−5, 2) is 5 units to the left of the origin and 2 units up, and (5, −2) is 5 units to the right of the origin and 2 units down.*)

Using the Open-Ended Intervention Student Resource pages 290–291

Read through the tasks on the student pages together. Give students time to work, ideally in pairs.

Observe whether students

- understand that the first number in an ordered pair tells the number of units right or left of the origin, and the second number tells the number of units up or down from the origin
- can plot points with integer ordered pairs on a 4-quadrant Cartesian grid
- recognize that points (a, b) and $(-a, b)$ are symmetric with respect to the y-axis, and points (a, b) and $(a, -b)$ are symmetric with respect to the x-axis

Consolidating and Reflecting

Ensure understanding by asking questions such as these based on students' work:

- How did you know that your shape would have at least 4 sides?
  (e.g., E, N, *and* K *did not have any x-coordinates or y-coordinates that were opposites, so they were not equally far from the x-axis or y-axis. So I needed to add at least one more point, and 4 vertices usually makes 4 sides.*)
- What do you do to locate the position of an ordered pair, such as (−2, 6)?
  (e.g., *Go to the x-axis to find the point (−2, 0), which is left of the origin, and then go up 6 units.*)
- How do you know which quadrant an ordered pair lies in without plotting it?
  (e.g., *Quadrant 1: both coordinates are positive; Quadrant 2: x-coordinate is negative and y-coordinate is positive; Quadrant 3: both coordinates are negative; Quadrant 4: x-coordinate is positive and y-coordinate is negative*)
- Suppose one of your points was on the y-axis. What is the first number in the ordered pair? How do you know?
  (*0.* e.g., *The first number in an ordered pair tells you where the point is on the x-axis, and the y-axis is at 0 on the x-axis.*)
- Think about 2 points that are equal distances from the x-axis. How are the coordinates related?
  (e.g., *One has a positive y-coordinate and the other has the opposite y-coordinate.*)

 Copyright © 2012 by Nelson Education Ltd.

Plotting Points in 4 Quadrants

Pathway 1 GUIDED

Guided Intervention

Before Using the Guided Intervention

On chart paper, sketch a Cartesian grid that shows the first quadrant and one that shows all 4 quadrants. Ask students to point out the origin. Ask:

- ▸ Where would you plot (5, 2) on each grid?
  (e.g., *5 units to the right of the origin and 2 units up*)
- ▸ How is the position of (−5, 2) different than the position of (5, −2)?
  (e.g., *(−5, 2) is 5 units to the left of the origin and 2 units up, and (5, −2) is 5 units to the right of the origin and 2 units down.*)

Using the Guided Intervention (Student Resource pages 292–295)

Work through the instructional section of the student pages together. Have students work through the **Try These** questions individually or in pairs.

Observe whether students

- use integer ordered pairs to identify the locations of points on a 4-quadrant Cartesian grid (Questions 1, 5, 6)
- can plot points with integer ordered pairs on a 4-quadrant Cartesian grid (Questions 2, 5, 6)
- recognize which numbers in the ordered pairs are always negative or positive in each quadrant (Questions 3, 7)
- understand why points on the y-axis are always in the form $(0, y)$, and points on the x-axis are always in the form $(x, 0)$ (Question 4)
- recognize the symmetry between points with ordered pairs that have the same numerals but different signs (Question 6)

Consolidating and Reflecting

Ensure understanding by asking questions such as these based on students' work:

- ▸ What did you do to locate the position of an ordered pair in Question 2?
  (e.g., *For C (2, −2), go to the x-axis to find the point (2, 0), which is right of the origin, and then go down 2 units.*)
- ▸ Why did it make sense, in Question 4a), that points of the form $(0, a)$ are on the y-axis?
  (e.g., *If the x-coordinate is 0, it means that you don't have to move right or left to get to that point, so the point must be on the y-axis.*)
- ▸ In Question 5, how did you decide which coordinate would change?
  (e.g., *If you went up or down, the y-coordinate changed, but if you went right or left, only the x-coordinate changed.*)
- ▸ How do you know which quadrant an ordered pair lies in without plotting it?
  (e.g., *Quadrant 1: both coordinates are positive; Quadrant 2: x-coordinate is negative and y-coordinate is positive; Quadrant 3: both coordinates are negative; Quadrant 4: x-coordinate is positive and y-coordinate is negative*)

You will need

- grid chart paper
- a ruler
- dynamic geometry software (optional)
- Student Resource pages 292–295

Pathway 2
OPEN-ENDED

Plotting Points on a Grid

You will need

- grid chart paper
- a ruler
- dynamic geometry software (optional)
- Student Resource pages 296–297

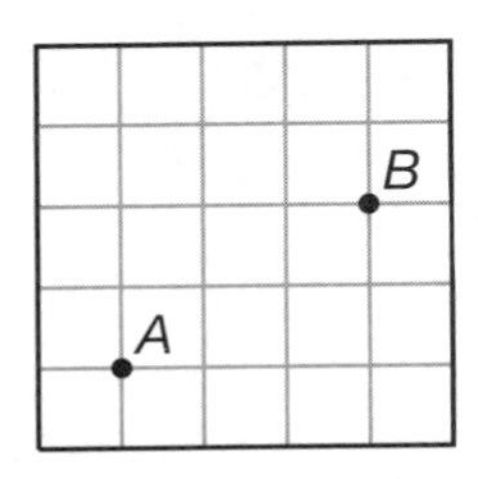

Open-Ended Intervention

Before Using the Open-Ended Intervention

Mark a few points on grid chart paper (not a Cartesian grid). Ask:

- ▸ How can you describe how to get from Point *A* to Point *B*?
  (e.g., *Go right 3 units and up 2 units.*)
- ▸ Put a dot on the point that is 3 units to the left of *B* and label it *C*. How would you get from *C* to *A*? (e.g., *Go down 2 units.*)

Using the Open-Ended Intervention Student Resource pages 296–297

Read through the tasks on the student pages together. Give students time to work, ideally in pairs.

Observe whether students

- understand that the first number in an ordered pair tells the number of units right of the origin, and the second number tells the number of units up from the origin
- can correctly plot points using ordered pairs in the first quadrant of a Cartesian grid
- use whole number ordered pairs to identify the location of points in the first quadrant of a Cartesian grid

Consolidating and Reflecting

Ensure understanding by asking questions such as these based on students' work:

- ▸ How did you plot Point *P*?
  (e.g., *I went to the x-axis to find 5 and then up 2 units parallel to the y-axis.*)
- ▸ Why is Point *C* on the *y*-axis?
  (e.g., *The first number in the ordered pair is 0, so you don't have to move right from the origin.*)
- ▸ Why is Point *B* lower than Point *C*?
  (e.g., *The y-coordinate is less.*)
- ▸ Why are Points *B* and *P* on the same vertical line?
  (e.g., *They have the same x-coordinate.*)
- ▸ How did you decide how far apart 2 points were?
  (e.g., *I counted spaces if the points were on the same horizontal or vertical lines. I used a ruler if they weren't.*)
- ▸ Why can you describe how to get from one point to another using right/left and up/down directions, even if the points are joined with a line that is on a slant?
  (e.g., *Even if they are joined on a slant, you can figure out how far over and how far up to go from one to the other.*)
- ▸ Why do you think ordered pairs can be a useful way to describe location?
  (e.g., *There is only one spot on the grid for each ordered pair, and the numbers tell you exactly where it is.*)

Copyright © 2012 by Nelson Education Ltd.

Plotting Points on a Grid

Pathway 2 GUIDED

Guided Intervention

You will need

- grid chart paper
- a ruler
- dynamic geometry software (optional)
- Student Resource pages 298–301

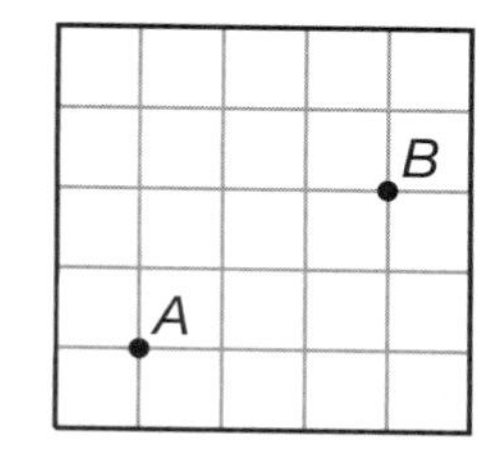

Before Using the Guided Intervention

Mark a few points on grid chart paper (not a Cartesian grid). Ask:

- How can you describe how to get from Point *A* to Point *B*? (e.g., *Go right 3 units and up 2 units.*)
- Put a dot on the point 3 units to the left of *B* and label it *C*. How would you get from *C* to *A*? (e.g., *Go down 2 units.*)

Using the Guided Intervention (Student Resource pages 298–301)

Work through the instructional section of the student pages together. Have students work through the **Try These** questions individually or in pairs.

Observe whether students

- use whole number ordered pairs to identify the location of points in the first quadrant of a Cartesian grid (Questions 1, 5, 6, 8)
- can correctly plot points using ordered pairs in the first quadrant of a Cartesian grid (Questions 2, 7, 8)
- understand why points on the y-axis are always in the form $(0, y)$, and points on the x-axis are always in the form $(x, 0)$ (Question 3)
- can describe how to move from one point to another on a Cartesian grid (Question 4)
- understand why 2 coordinates, not just 1 coordinate, are required (Question 9)

Consolidating and Reflecting

Ensure understanding by asking questions such as these based on students' work:

- How did you plot $C(2, 7)$ in Question 2a)? (e.g., *I went to the x-axis to find 2 and then up 6 units.*)
- How did you figure out how far you would move left or right, or up or down, in Question 4? (e.g., *I subtracted the x-coordinates to figure out how far to move left or right, and the y-coordinates to figure out how far to move up or down. I paid attention to whether it was right or left, or up or down, depending on whether the coordinates of the new point were greater or less than the originals.*)
- Why can you describe how to get from one point to another using right/left and up/down directions, even if the points are joined with a line that is on a slant? (e.g., *Even if they are joined on a slant, you can figure out how far over and how far up to go from one to the other.*)
- In Question 6a), what had to be true about each coordinate? (e.g., *The x-coordinate had to be big, and the y-coordinate had to be small.*)
- Why do you think ordered pairs can be a useful way to describe location? (e.g., *There is only one spot on the grid for each ordered pair, and the numbers tell you exactly where it is.*)

Copyright © 2012 by Nelson Education Ltd.

Transformations

Planning For This Topic

Materials for assisting students with transformations consist of a diagnostic tool and 4 intervention pathways. Pathway 1 encourages students to explore the use of transformations in designs. Pathway 2 focuses on dilatations. Pathway 3 has students explore the result of combining transformations, and Pathway 4 focuses on single transformations—translations, reflections, and rotations. Many of the transformations are performed on a Cartesian grid.

Each pathway has an open-ended intervention and a guided intervention. Choose the type of intervention more suitable for your students' needs and your particular circumstances.

Curriculum Connections

Grades 5 to 8 curriculum connections for this topic are provided online. See www.nelson.com/leapsandbounds. The WNCP curriculum does not mention the use of transformations in designs or the use of dilatations, so you may choose to omit Pathways 1 and 2 with students in those jurisdictions. The Ontario curriculum does not include combinations of transformations specifically, so you could omit Pathway 3 with Ontario students.

Professional Learning Connections

PRIME: Geometry, Background and Strategies (Nelson Education Ltd., 2007), pages 99–111

Making Math Meaningful to Canadian Students K–8 (Nelson Education Ltd., 2008), pages 342–354

Big Ideas from Dr. Small Grades 4–8 (Nelson Education Ltd., 2009), pages 120–128

Good Questions (dist. by Nelson Education Ltd., 2009), pages 72, 78, 87, 90–91

More Good Questions (dist. by Nelson Education Ltd., 2010), pages 111–112, 114–115

Why might students struggle with transformations?

Students might struggle with recognizing or performing transformations for any of the following reasons:

- They might confuse horizontal and vertical reflections.
- They might be comfortable performing horizontal and vertical reflections but not reflections across other lines.
- They might not realize that different motions can have the same effect (e.g., a shape can be moved to its right by 2 reflections or a single translation).
- They might have difficulty visualizing rotation images or holding just one point fixed while performing a rotation.
- They might struggle relating the original position to the final position when a sequence of transformations is performed.
- They might struggle performing dilatations that reduce the size of a shape.
- They might think that all similarity transformations are dilatations, but that is not true: some combine dilatations with other transformations.
- They might have little exposure to transformations in general as this topic may have been given little attention in previous grades.

Copyright © 2012 by Nelson Education Ltd.

Diagnostic Tool: Transformations

Use the diagnostic tool to determine the most suitable intervention pathway for transformations. Provide Diagnostic Tool: Transformations, Teacher's Resource pages 232 to 235, and have students complete it in writing or orally. Students should have rulers, transparent mirrors (e.g., Miras), and tracing paper if possible. You may choose to omit items related to pathways that are not related to your curriculum.

See solutions on Teacher's Resource pages 236 to 239.

Intervention Pathways

The purpose of the intervention pathways is to develop understanding of the similarities and differences among types of transformations, and to see the effects of some transformations as performed on Cartesian grids. Students will see how some motions relate to congruence, whereas others relate to similarity. This work should also help prepare students with later work involving other geometric relationships.

There are 4 pathways:

- Pathway 1: Using Transformations in Designs
- Pathway 2: Performing Dilatations
- Pathway 3: Combining Transformations
- Pathway 4: Performing Single Transformations

Use the chart below (or the Key to Pathways on Teacher's Resource pages 236 to 239) to determine which pathway is most suitable for each student or group of students.

Diagnostic Tool Results	Intervention Pathway
If students struggle with Questions 10 and 11	use Pathway 1: Using Transformations in Designs *Teacher's Resource pages 240–241* *Student Resource pages 302–306*
If students struggle with Questions 7 to 9	use Pathway 2: Performing Dilatations *Teacher's Resource pages 242–243* *Student Resource pages 307–311*
If students struggle with Questions 4 to 6	use Pathway 3: Combining Transformations *Teacher's Resource pages 244–245* *Student Resource pages 312–317*
If students struggle with Questions 1 to 3	use Pathway 4: Performing Single Transformations *Teacher's Resource pages 246–247* *Student Resource pages 318–323*

Name:________________________________ Date:__________________

Transformations

Diagnostic Tool

You will need

- a ruler
- a transparent mirror (optional)
- tracing paper (optional)

1. Suppose the grey shape below was rotated or reflected.

a) Circle the white shapes that show what it *might* look like after a single rotation.

b) Put a rectangle around the white shapes that show what it *might* look like after a single reflection.

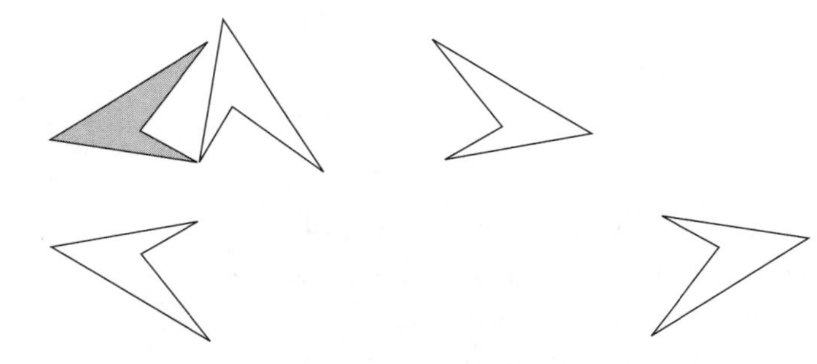

2. Sketch the resulting shape after each transformation.

a) Translate the triangle along the translation arrow.

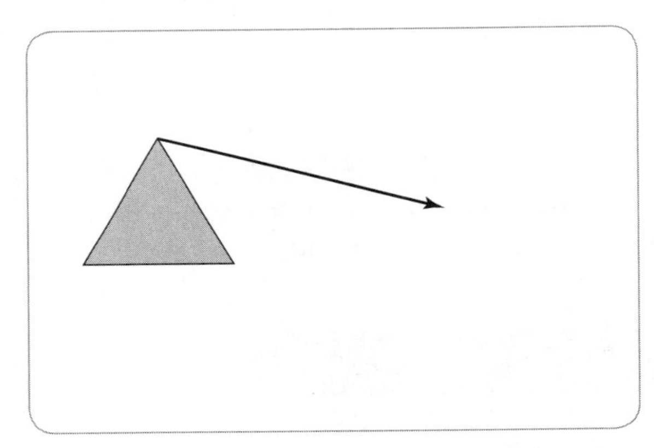

b) Reflect the triangle across the reflection line.

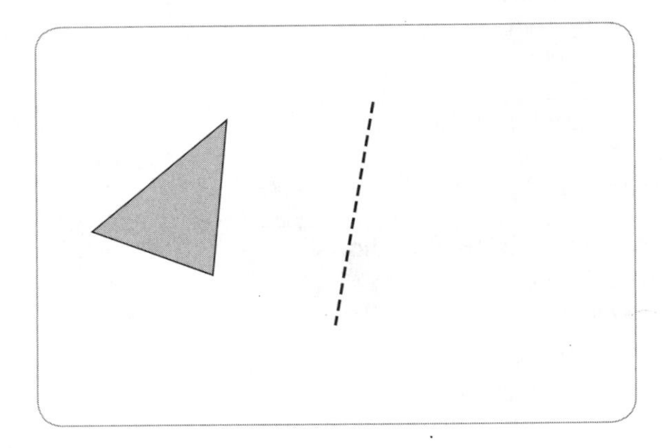

c) Rotate the trapezoid $\frac{1}{4}$ turn clockwise around the bottom-right corner.

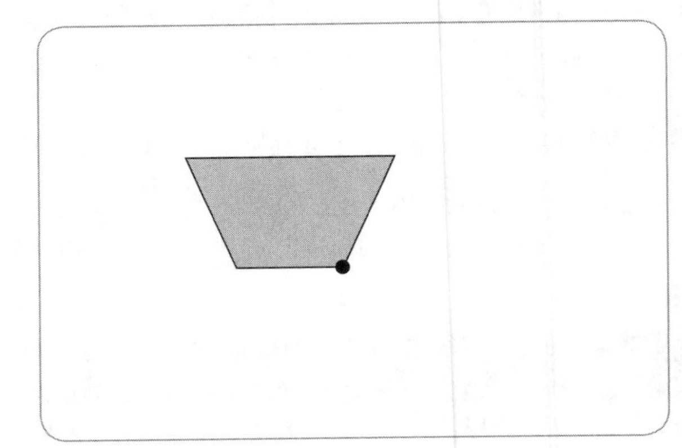

d) Rotate the trapezoid $\frac{1}{4}$ turn counterclockwise around the centre dot.

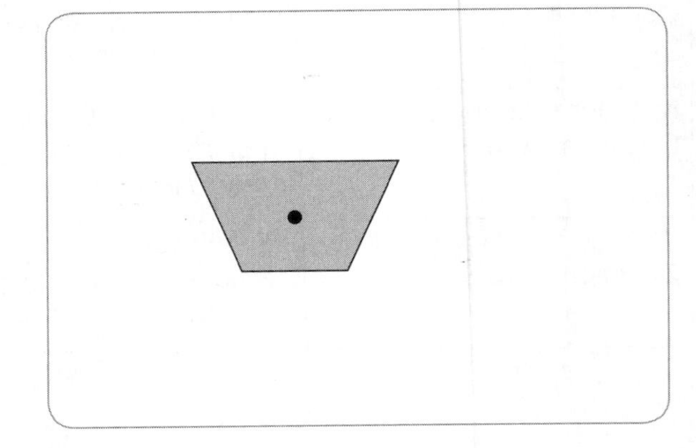

 Copyright © 2012 by Nelson Education Ltd.

Name:______________________________ Date:______________________

3. Draw the result of each transformation. Label the coordinates of the new position of Point *A* after the transformation.

a) translation:

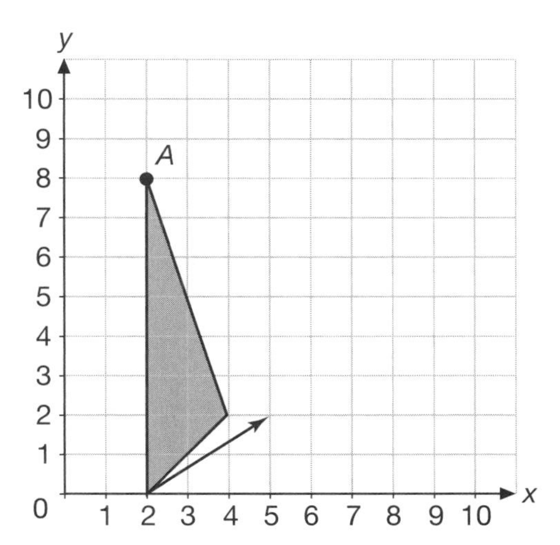

b) translation:

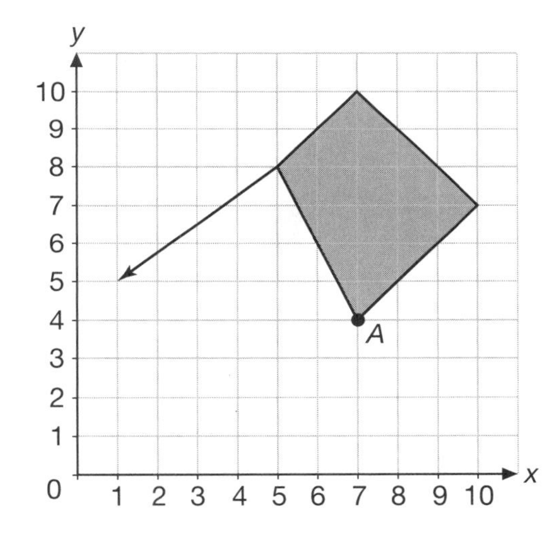

c) reflection:

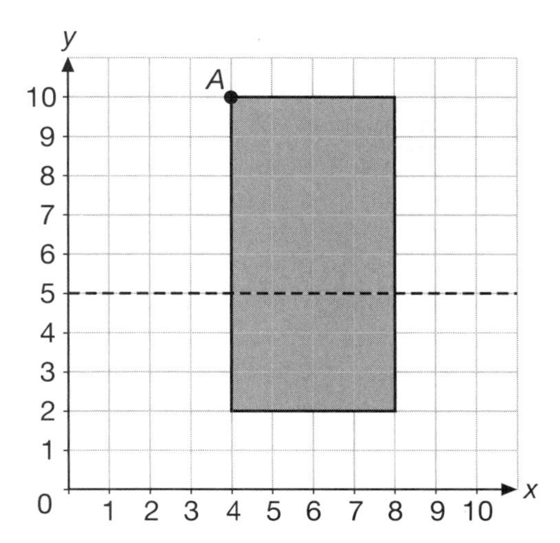

d) rotation of 90° clockwise (cw) around the bottom-right corner:

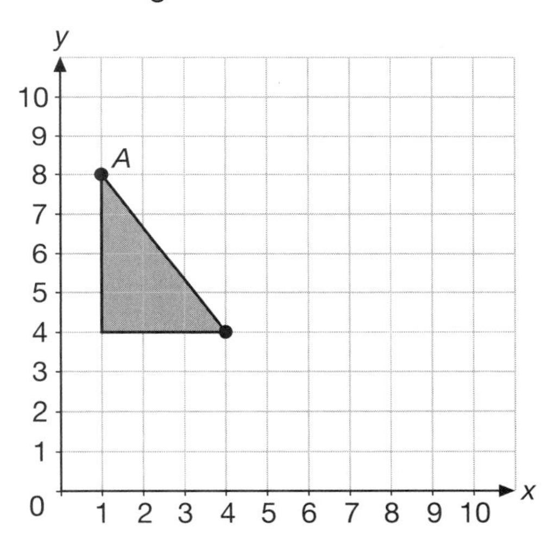

4. Reflect the grey shape across the reflection line and draw the reflection with dashed lines. Then rotate your reflected shape 180° around the top vertex of your shape. Draw the final shape.

a)

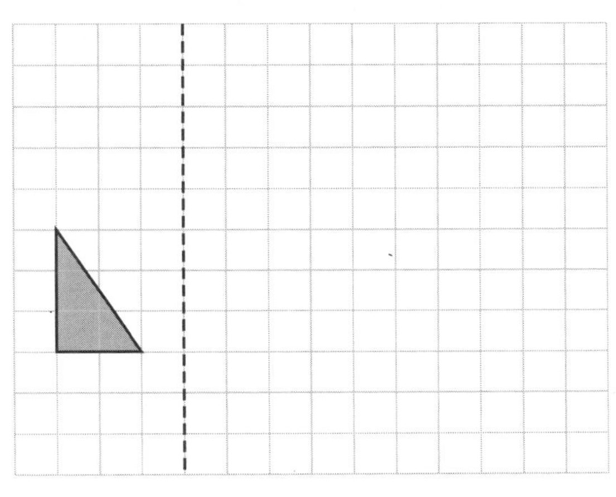

b)

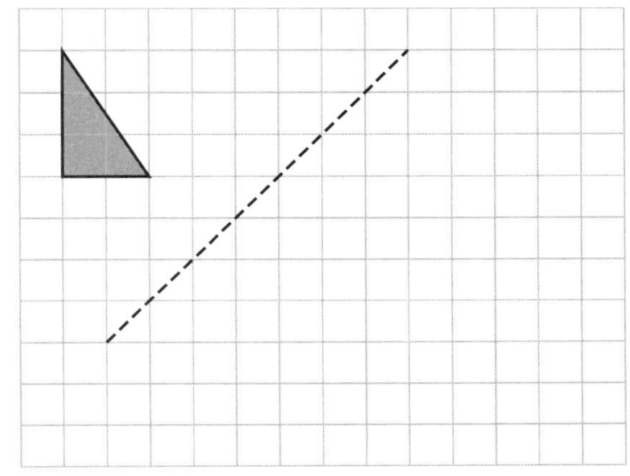

Copyright © 2012 by Nelson Education Ltd.

Name:________________________________ Date:________________

5. Reflect the grey shape across Line 1 and draw the reflection using dashed lines. Then reflect your shape across Line 2. Draw the final shape.

a)

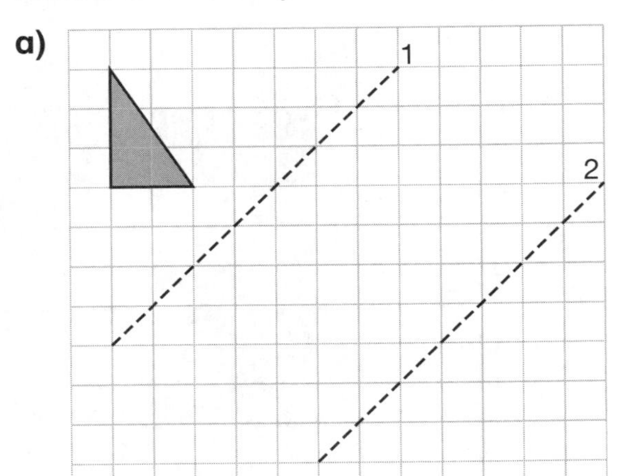

b)

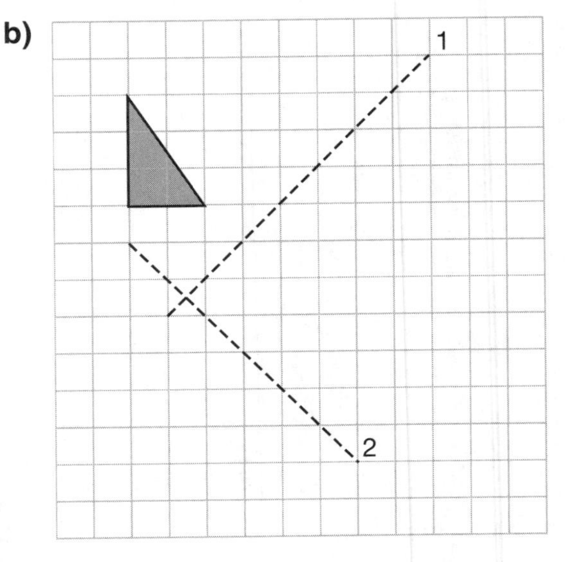

6. Show a sequence of 2 transformations that can be used to move the grey shape (start) to the white shape (final).

a)

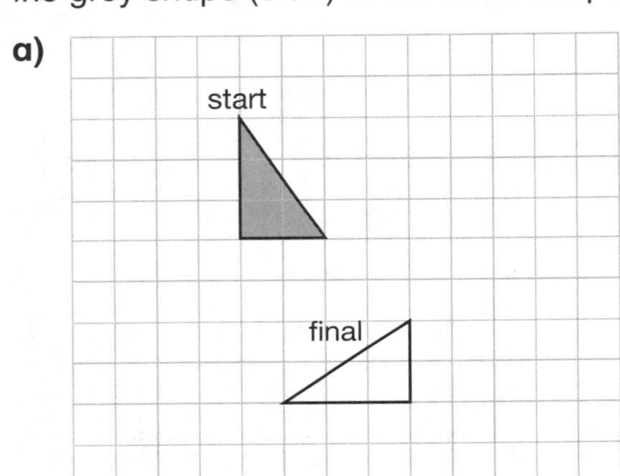

b)

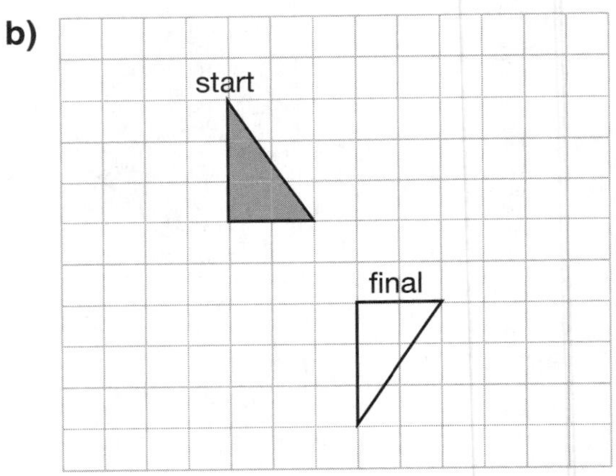

7. a) Use the dot to perform a dilatation (an enlargement) on the grey shape below to make a shape with sides twice as long.

b) Use the same dot to perform a different dilatation (a reduction) to make a shape with sides half as long as the sides of the grey shape.

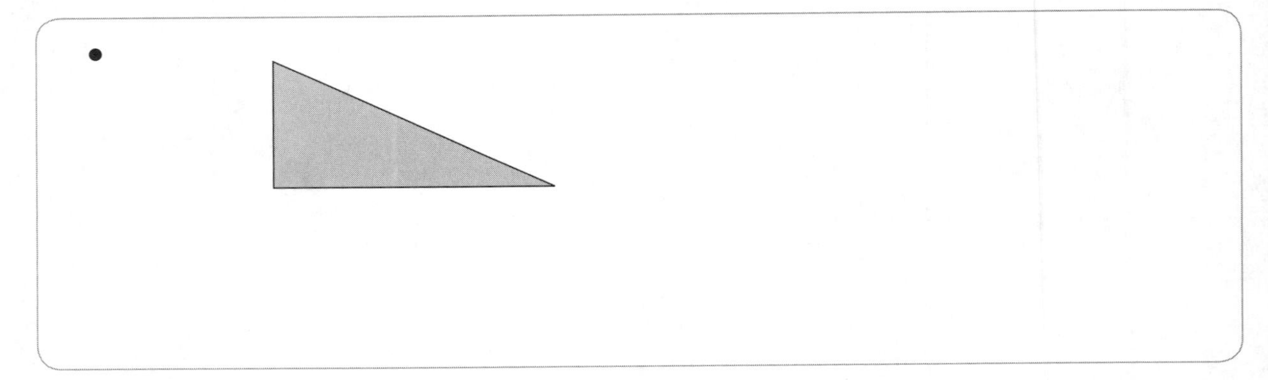

 Copyright © 2012 by Nelson Education Ltd.

Name:______________________ Date:______________

8. Where is the dilatation centre that enlarges the small shape? Show how you know.

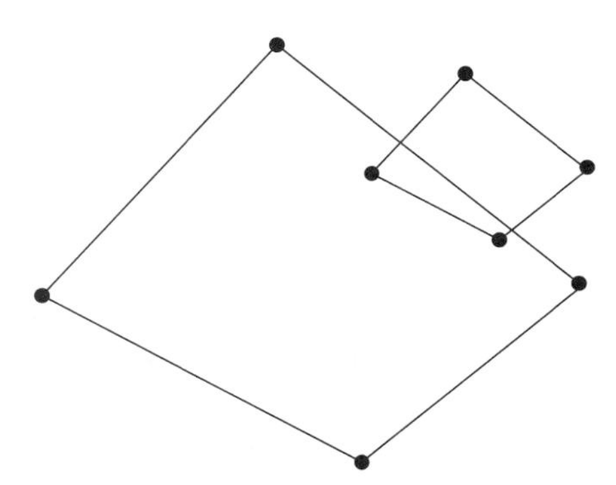

9. Is this a dilatation? Why or why not?

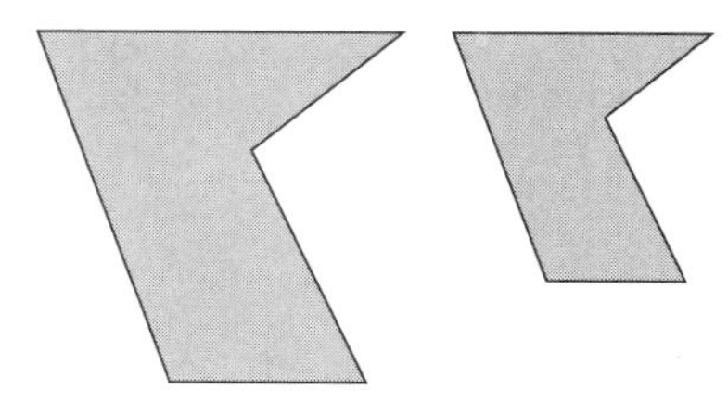

____

____

10. Show that you could use this shape as a tile.

11. Create a design that includes a reflection and a rotation. Mark the reflection lines and the turn angles, including the rotation centre and direction, to show the transformations.

Copyright © 2012 by Nelson Education Ltd.

Solutions and Key to Pathways

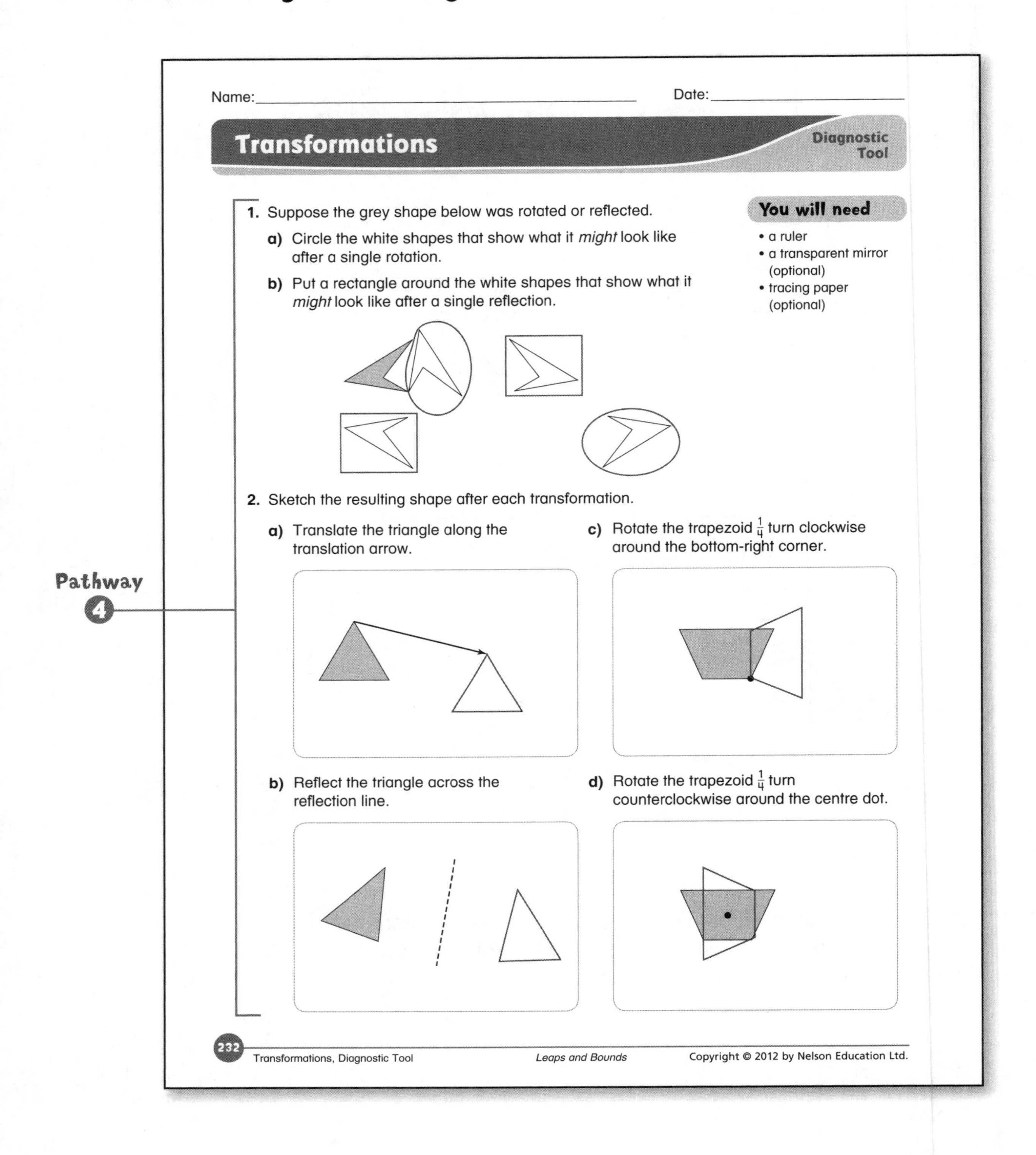

 Copyright © 2012 by Nelson Education Ltd.

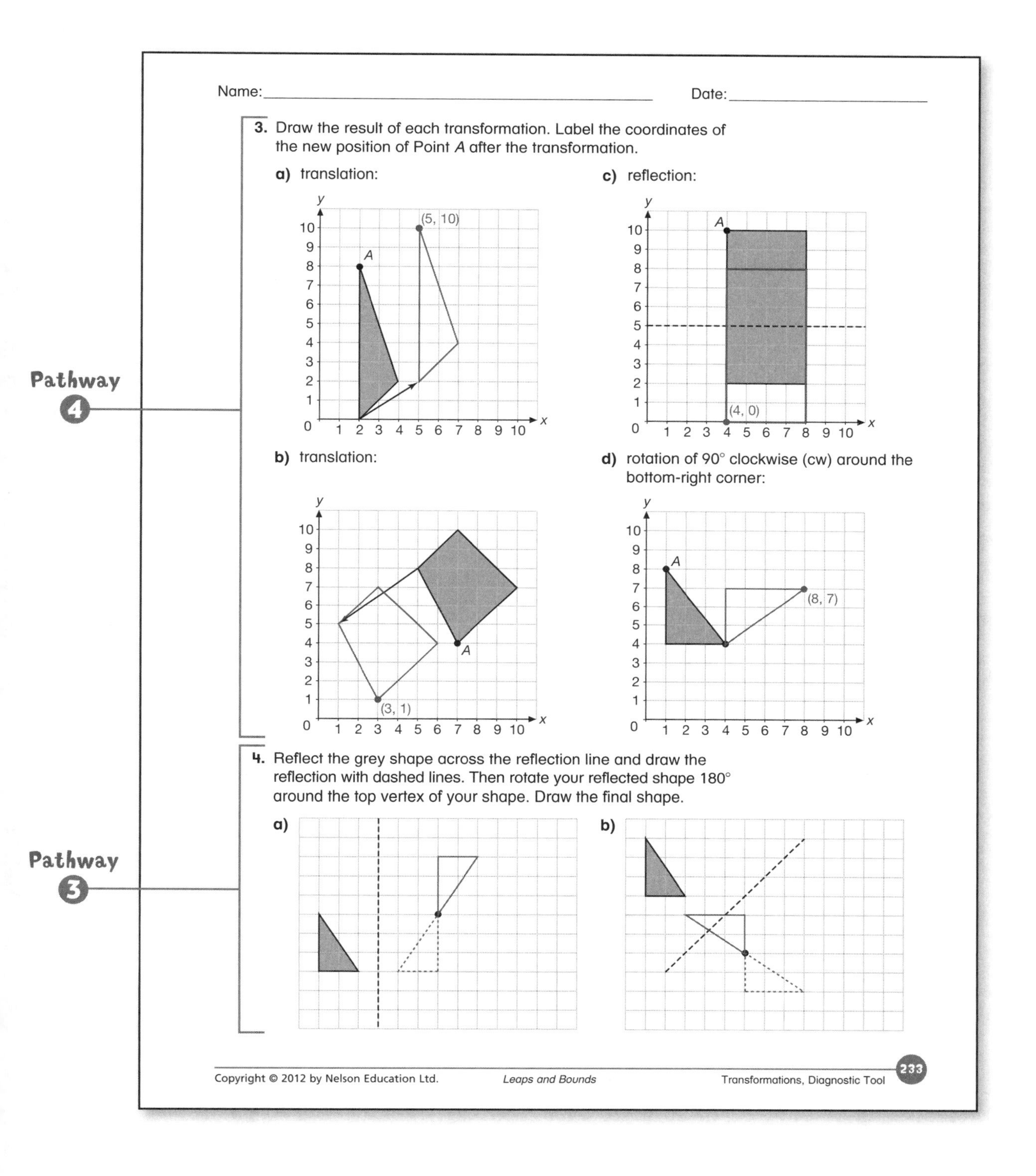

Pathway 4

Pathway 3

Name:______________________ Date:______________

3. Draw the result of each transformation. Label the coordinates of the new position of Point A after the transformation.

a) translation:

b) translation:

c) reflection:

d) rotation of 90° clockwise (cw) around the bottom-right corner:

4. Reflect the grey shape across the reflection line and draw the reflection with dashed lines. Then rotate your reflected shape 180° around the top vertex of your shape. Draw the final shape.

a)

b)

Copyright © 2012 by Nelson Education Ltd. *Leaps and Bounds* Transformations, Diagnostic Tool 233

Copyright © 2012 by Nelson Education Ltd.

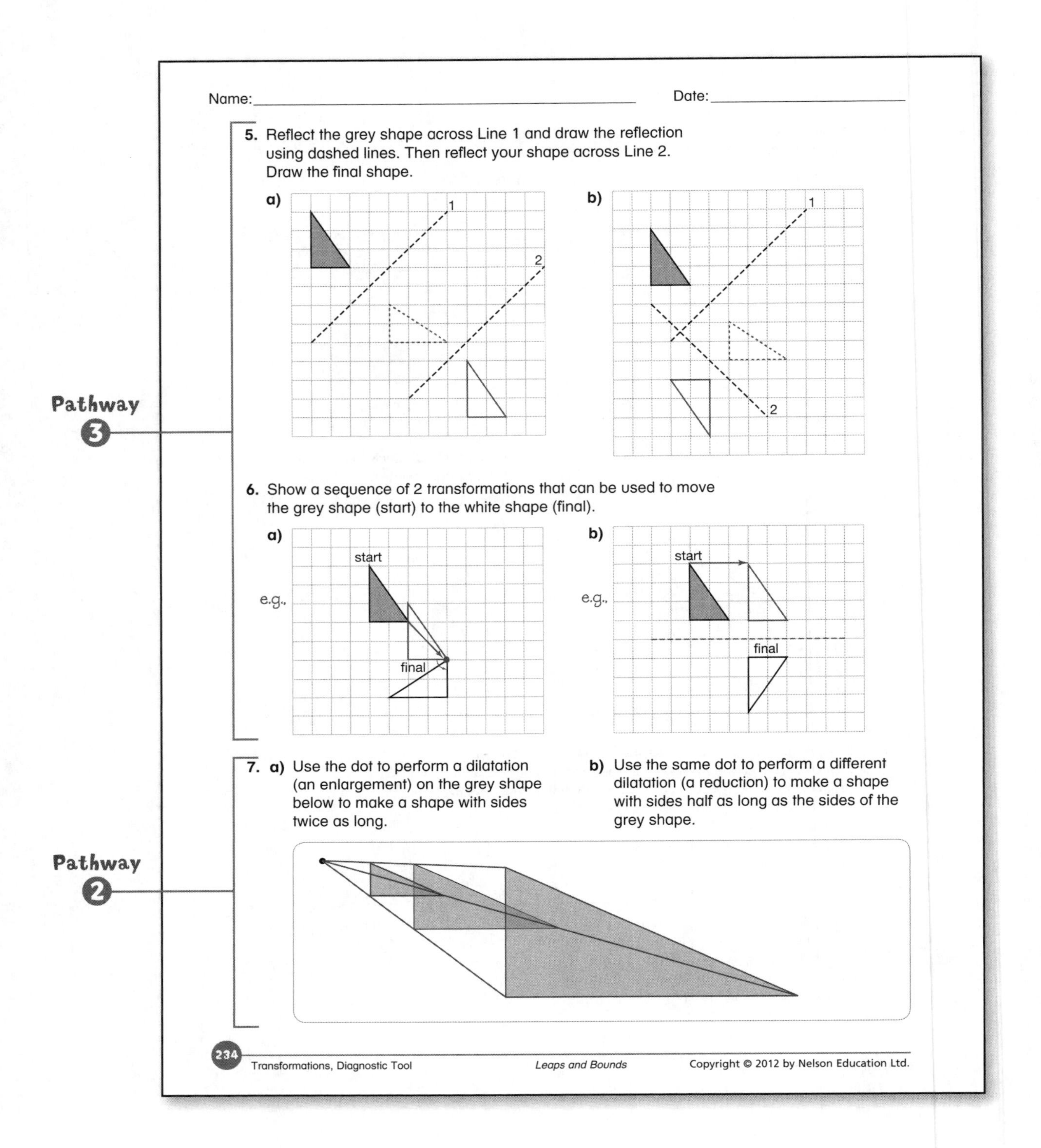

 Copyright © 2012 by Nelson Education Ltd.

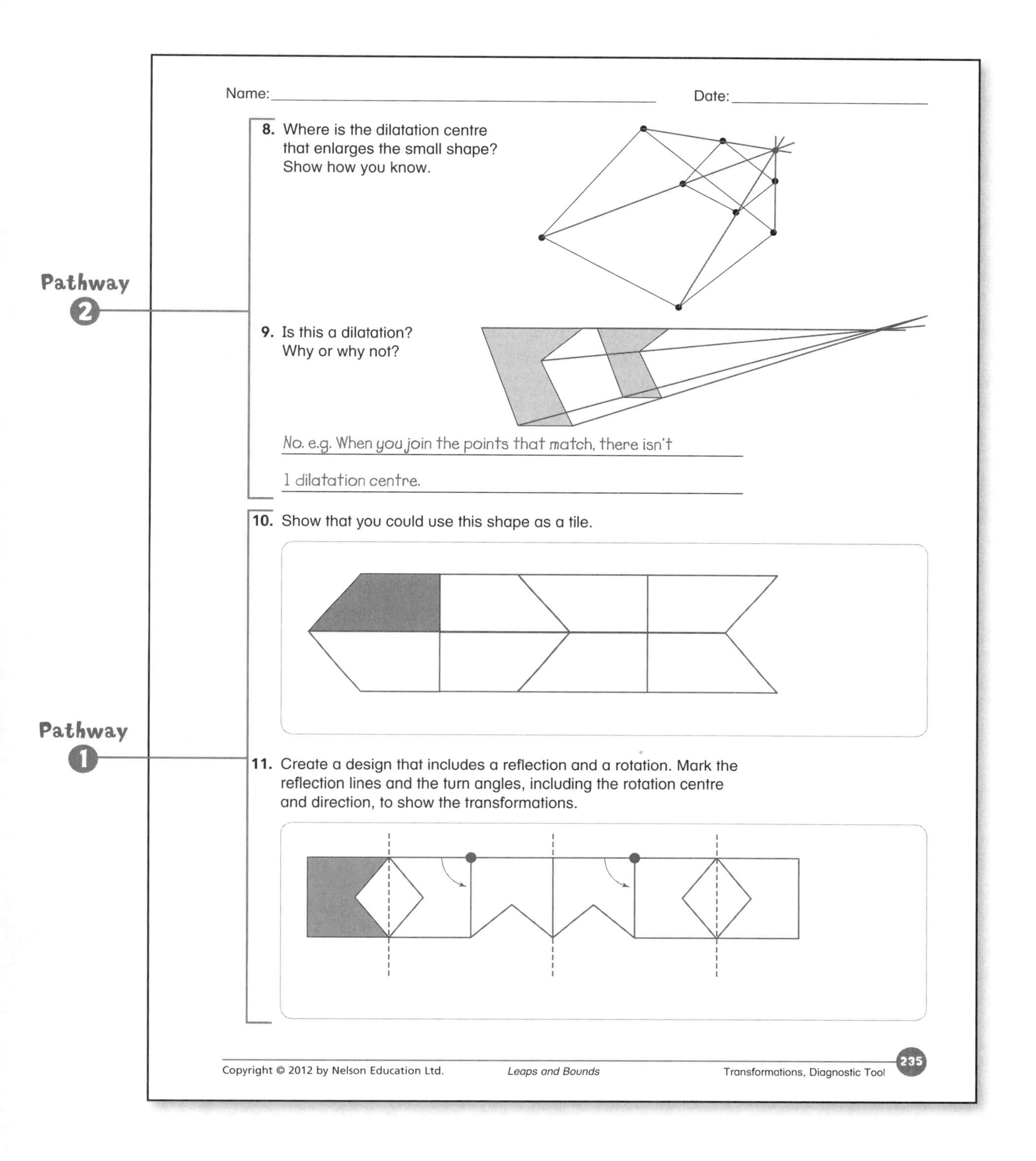

Name:____________________ Date:__________

Pathway 2

8. Where is the dilatation centre that enlarges the small shape? Show how you know.

9. Is this a dilatation? Why or why not?

No. e.g. When you join the points that match, there isn't 1 dilatation centre.

Pathway 1

10. Show that you could use this shape as a tile.

11. Create a design that includes a reflection and a rotation. Mark the reflection lines and the turn angles, including the rotation centre and direction, to show the transformations.

Copyright © 2012 by Nelson Education Ltd. *Leaps and Bounds* Transformations, Diagnostic Tool 235

Copyright © 2012 by Nelson Education Ltd.

Pathway 1 OPEN-ENDED

Using Transformations in Designs

You will need

- 2 congruent plastic or paper scalene right triangles
- pattern blocks (or BLMs 19 to 24)
- materials for drawing transformations (e.g., transparent mirrors such as Miras, rulers, protractors, tracing paper, or dynamic geometry software)
- Student Resource page 302

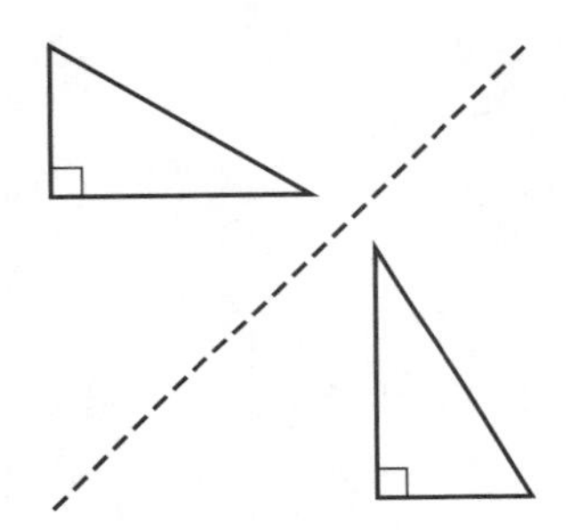

Open-Ended Intervention

Before Using the Open-Ended Intervention

Place a scalene right triangle on a piece of paper. Have students use the other (congruent) triangle to show a translation, a reflection, and then a 90° rotation. The reflection can be vertical or horizontal, and the turn can be clockwise or counterclockwise with a rotation centre inside, on, or outside the shape. Ask:

- ▸ What does a 180° rotation do to a shape? (e.g., *It turns it halfway around.*)

Mark a reflection line as shown in the margin art, using a transparent mirror or a ruler. Place one scalene right triangle on one side of the line. Ask:

- ▸ How do you reflect the triangle using this line? (e.g., *I look through the mirror to see where the corners go and then put the second triangle there.*)
- ▸ How would you perform a dilatation of the triangle? (e.g., *I would mark a point and draw a line from it to each corner. Then I would extend the lines by doubling or tripling all of them.*)

Display pattern block hexagons, trapezoids, rhombuses, and triangles. Ask:

- ▸ Choose one of the blocks. How can you show that it would make a good tile? (*I can place identical blocks edge to edge without gaps or overlaps.*)

Acquaint students with using transformation tools with technology, if available.

Using the Open-Ended Intervention **Student Resource page 302**

Read through the tasks on the student page together. Provide the materials and give students time to work, ideally in pairs.

Observe whether students

- can identify transformations in a design
- can create designs involving transformations, including dilatations
- can create a tessellation

Consolidating and Reflecting

Ensure understanding by asking questions such as these based on students' work:

- ▸ What did you do to make sure you reflected properly? (e.g., *I didn't use a mirror, but I checked that the distances from the reflection line matched.*)
- ▸ How did you perform the 180° rotation? (e.g., *I used drawing software, and I did $\frac{1}{4}$ turn ccw twice. I made sure that the turn centre didn't move.*)
- ▸ How did you perform the dilatation? (e.g., *I chose a dilatation point, and I joined each vertex to it. I extended the lines to make them twice as long.*)
- ▸ How can you tell if a 90° rotation has been performed? (e.g., *I would join the original vertex and the same vertex in the rotated shape to the turn centre to see if the 2 lines make a 90° angle.*)
- ▸ How did you know your shape would tessellate? (e.g., *I made sure that I kept turning it so that sides matched and there were no gaps or overlaps.*)

 Copyright © 2012 by Nelson Education Ltd.

Using Transformations in Designs

Pathway 1 GUIDED

Guided Intervention

Before Using the Guided Intervention

Place a scalene right triangle on a piece of paper. Have students use the other (congruent) triangle to show a translation, a reflection, and then a 90° rotation. Ask:

- ▸ What does a 180° rotation do to a shape? (e.g., *It turns it halfway around.*)

Mark a reflection line as shown in the margin art, using a transparent mirror or a ruler. Place one scalene right triangle on one side of the line. Ask:

- ▸ How do you reflect the triangle using this line? (e.g., *I look through the mirror to see where the corners go and then put the second triangle there.*)

Acquaint students with using transformation tools with technology, if available.

Using the Guided Intervention (Student Resource pages 303–306)

Provide the materials. Work through the instructional section of the student pages together and have students perform the transformations in the first design using the chevron on BLM 27. Make sure students understand the definitions for dilatations and tessellations.

Have students work through the **Try These** questions individually or in pairs.

Observe whether students

- identify transformations in a design, excluding dilatations (Questions 1, 2, 3, 6)
- can create designs involving transformations, including dilatations (Question 4)
- identify which shapes will tile (Questions 5, 7)
- can create a tiling design using polygons or combinations of polygons, and describe the transformation(s) involved (Question 8)

Consolidating and Reflecting

Ensure understanding by asking questions such as these based on students' work:

- ▸ How did you locate the rotations in Question 3? (e.g., *I saw that the first shape was turned halfway around by looking at the square corner. That's a 180° rotation.*)
- ▸ What reflections did you use in Question 4? (e.g., *I used 3 vertical reflections.*)
- ▸ What did you do to make sure you reflected properly? (e.g., *I checked the distances from the reflection line, and they matched.*)
- ▸ How did you perform the dilatation in Question 4? (e.g., *I put the dilatation point at the top left of the shape and drew a line from it to each other vertex. I extended the lines to make them twice as long. Then I joined the points at the end of the lines.*)
- ▸ How did you know which shapes would tessellate in Question 5? (e.g., *I tried them by moving them around different ways. My first way didn't always work.*)

You will need

- 2 congruent plastic or paper scalene right triangles
- materials for drawing transformations (e.g., transparent mirrors such as Miras, rulers, protractors, tracing paper, or dynamic geometry software)
- pattern blocks (or BLMs 19 to 24)
- polygon cutouts (e.g., Regular Polygons (BLM 26) and Polygons (BLM 27))
- Student Resource pages 303–306

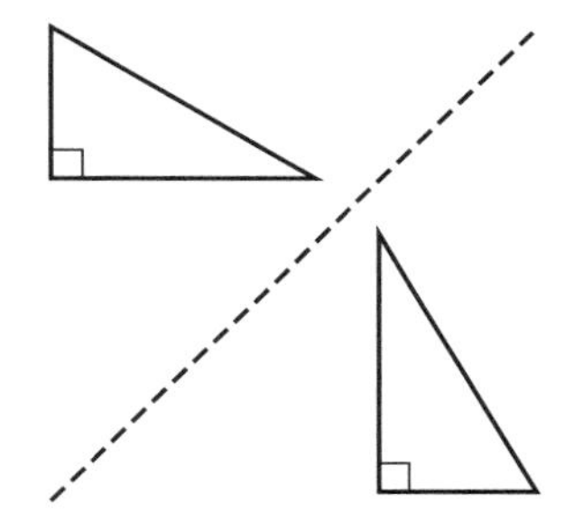

Pathway 2 OPEN-ENDED

Performing Dilatations

You will need

- pattern block triangles
- rulers
- compasses
- Star (BLM 28)
- a transparent 1 cm grid (made from BLM 12)
- dynamic geometry software (optional)
- Student Resource page 307

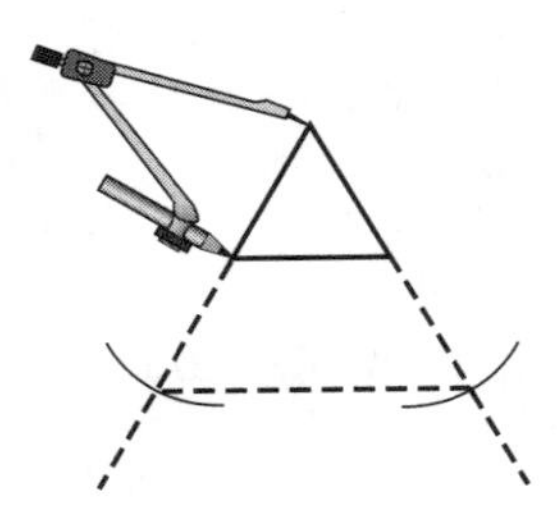

Open-Ended Intervention

Before Using the Open-Ended Intervention

Provide paper, pattern block triangles, rulers, and compasses. Have students trace around a triangle pattern block. Then have them extend 2 sides to double their lengths and draw a third line to join those 2 sides. (Demonstrate how to do this using a compass, where you set the radius to measure a given length and then mark that length on the extended line. See margin art.) Ask:

- ▸ What do you notice? (e.g., *I made a bigger triangle.*)
- ▸ How does the new bottom length compare to the original length? (*It's double.*)
- ▸ How many triangle blocks does it take to fill the bigger triangle? (*4*)

Have students trace the triangle again and extend 2 sides to 3 times their length.

- ▸ What do you notice? (e.g., *This triangle is bigger than the last triangle.*)
- ▸ Why are the 2 large triangles similar to the original one? (e.g., *They are enlargements, and the sides are proportional.*)

You might introduce the term *dilatation* by relating it to dilating pupils in the eye. Acquaint students with performing dilatations with technology, if available.

Using the Open-Ended Intervention (Student Resource page 307)

Provide the drawing materials and BLM 28. Read through the student page together. Some students might benefit from reading the instructional part of the guided intervention to clarify how to perform a dilatation, particularly on a Cartesian grid. Give students time to work, ideally in pairs.

Observe whether students

- can perform a dilatation to enlarge or reduce a shape
- can compare dilatations to other transformations

Consolidating and Reflecting

Ensure understanding by asking questions such as these based on students' work:

- ▸ Did your enlargements look the same when you used different dilatation centres? (e.g., *Yes and no. They were the same size, but were in different spots.*)
- ▸ How is enlarging a shape the same as and different from reducing a shape? (e.g., *You use a dilatation centre and measure distances to points on the sketch, but you create longer distances instead of shorter ones.*)
- ▸ If 2 shapes are similar, is one always a dilatation of the other? Explain. (*No.* e.g., *If you slide a dilatation, it wouldn't be a dilatation any more.*)

Place a transparent centimetre grid on top of the picture on BLM 28, making sure that the vertices correspond with grid intersection points. Ask:

- ▸ How might putting the original picture on a coordinate grid help you do the dilatation? (e.g., *I think you can just double or triple or halve the coordinates.*)

Copyright © 2012 by Nelson Education Ltd.

Performing Dilatations

Pathway 2
GUIDED

Guided Intervention

Before Using the Guided Intervention

Provide paper, pattern block triangles, rulers, and compasses. Have students trace around a triangle pattern block. Then have them extend 2 sides to double their lengths and draw a third line to join those 2 sides. (Demonstrate how to do this using a compass, where you set the radius to measure a given length and then mark that length on the extended line. See margin art.) Ask:

- What do you notice? (e.g., *I made a bigger triangle.*)

Point out that the larger triangle and smaller triangle share a vertex.

- How does the new bottom length compare to the original length? (*It's double.*)
- How many pattern block triangles would it take to fill it? (*4*)

Have students trace the triangle again and extend 2 sides to 3 times their length.

- What do you notice? (e.g., *It's even bigger than the last triangle.*)
- How many pattern block triangles would it take to fill it? (*9*)
- Why are the 2 large triangles similar to the original one? (e.g., *They are enlargements, and the sides are proportional.*)

You might introduce the term *dilatation* by relating it to dilating pupils in the eye. Acquaint students with performing dilatations with technology, if available.

You will need

- pattern block triangles
- rulers
- compasses
- 1 cm Grid Paper (BLM 12) or 0.5 cm Grid Paper (BLM 13)
- dynamic geometry software (optional)
- Student Resource pages 308–311

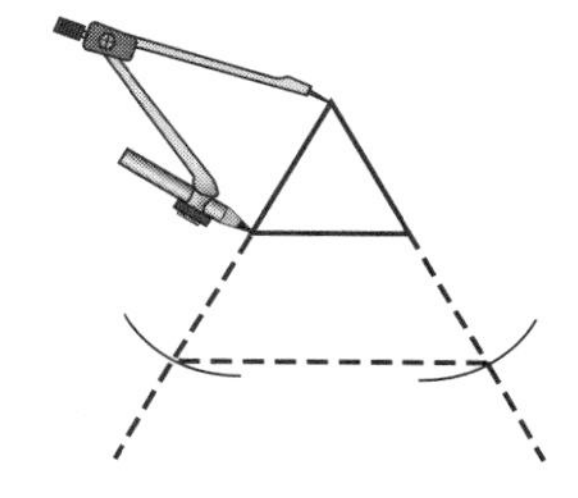

Using the Guided Intervention (Student Resource pages 308–311)

Work through the instructional section of the student pages together. Discuss how to perform dilatations using grid paper, and using technology if available.

Have students work through the **Try These** questions individually or in pairs.

Observe whether students

- can perform a dilatation to enlarge or reduce a shape (Questions 1, 2, 5)
- can locate a dilatation centre (Question 3)
- recognize whether an enlargement is or is not a dilatation (Questions 4, 5)
- can compare dilatations to other transformations (Question 6)

Consolidating and Reflecting

Ensure understanding by asking questions such as these based on students' work:

- How is enlarging a shape the same as and different from reducing a shape in Question 2? (e.g., *You still use a dilatation centre and measure distances to points on the shape, but you create longer distances instead of shorter ones.*)
- If 2 shapes are similar, is one always a dilatation of the other? Why? (*No.* e.g., *In Question 4, the largest triangle is similar to the original one, but not a dilatation of it because of where the dilatation centre is.*)
- How does putting a shape on a coordinate grid help you do the dilatation? (e.g., *If you double or triple or halve the coordinates, you make a dilatation.*)

Copyright © 2012 by Nelson Education Ltd.

Combining Transformations

You will need

- 1 cm Grid Paper (BLM 12) or 0.5 cm Grid Paper (BLM 13)
- transparent mirrors (e.g., Miras)
- materials for drawing transformations (e.g., rulers, protractors, tracing paper, or dynamic geometry software)
- Student Resource pages 312–313

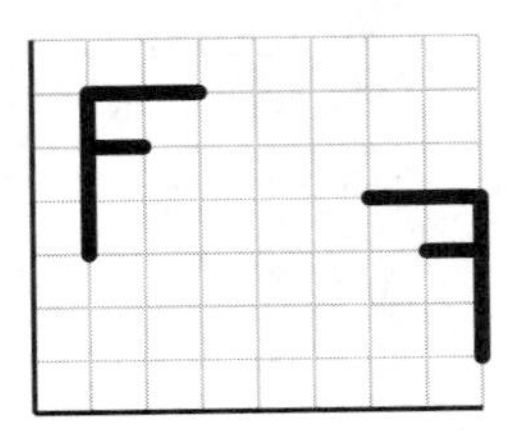

Open-Ended Intervention

Before Using the Open-Ended Intervention

Draw a capital letter *F* on a grid. Have a transparent mirror available, and ask:

- Suppose I reflect the *F* using a vertical line. What would the result look like? Check to see if you are right. (e.g., *a flipped over F*)
- What if I reflect the reflection in another vertical line? What would the result look like? Check to see if you are right. (e.g., *It looks like an F again.*)
- What if you rotated the *F* $\frac{1}{4}$ turn counterclockwise around the bottom of the letter? What will the result look like? (e.g., *It will be lying on its back to the left of the original letter.*)
- What if you did 2 translations, one right after the other? What will the result look like? (e.g., *I think it will look as if you did just 1 translation.*)

Draw a capital letter *F* and its image as shown in the margin art. Ask:

- Do you think you could use transformations to get from the *F* to the backward one? (e.g., *I know I would need a flip, but I might need other moves, too.*)

Acquaint students with using transformation tools, if the technology is available.

Using the Open-Ended Intervention (Student Resource pages 312–313)

Read through the tasks on the student pages together. Provide the materials for performing transformations. Make sure students can perform simple rotations, translations, and reflections, and can read coordinates on a Cartesian grid—this will help them describe the location of a turn centre, or a reflection line, or a slide arrow.

Give students time to work, ideally in pairs.

Observe whether students

- can perform a combination of transformations
- recognize that it is always possible to get from one shape to a congruent one using a combination of transformations

Consolidating and Reflecting

Ensure understanding by asking questions such as these based on students' work:

- Why did it make sense to use a rotation as one of the transformations? (e.g., *The end shape looked as if it had been turned 90°.*)
- Which reflections did you use? (e.g., *I reflected using a horizontal line once and a vertical line once.*)
- Why does it make sense that there was more than one set of transformations you could use to get from the start to the end? (e.g., *You can always show 1 translation using 2 translations, so if you use a translation, there is always more than one way to do it.*)

Copyright © 2012 by Nelson Education Ltd.

Combining Transformations

Pathway 3 GUIDED

Guided Intervention

You will need

- 1 cm Grid Paper (BLM 12) or 0.5 cm Grid Paper (BLM 13)
- transparent mirrors (e.g., Miras)
- materials for drawing transformations (e.g., rulers, protractors, tracing paper, or dynamic geometry software)
- Student Resource pages 314–317

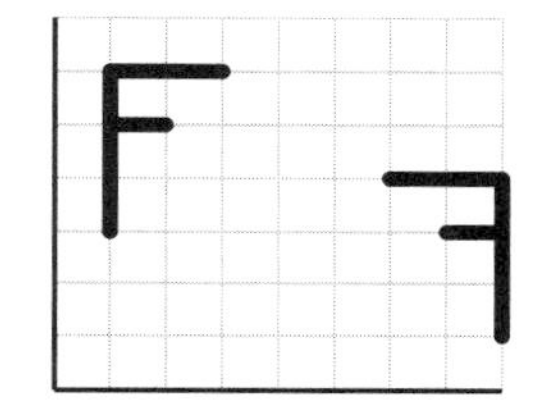

Before Using the Guided Intervention

Draw a capital letter *F* on a grid. Have a transparent mirror available, and ask:

- What if you rotated the *F* $\frac{1}{4}$ turn counterclockwise around the bottom of the letter? What will the result look like?
  (e.g., *It will be lying on its back to the left of the original letter.*)
- What if you then reflected it around the horizontal part of the image?
  (e.g., *It would point down instead.*)

Draw a capital letter *F* and its image as shown in the margin art. Ask:

- Do you think you could use transformations to get from the *F* to the backward one? (e.g., *I know I would need a flip, but I might need other moves, too.*)

Acquaint students with using transformation tools, if the technology is available.

Using the Guided Intervention (Student Resource pages 314–317)

Provide materials for drawing transformations and work through the instructional section of the student pages together.

Have students work through the **Try These** questions individually or in pairs. Students might like to do Question 8 on a separate piece of paper.

Observe whether students

- can perform a combination of transformations (Questions 1, 2, 5)
- relate double reflections to translations and rotations (Questions 3, 6, 7)
- recognize that a combination of translations can be performed as a single translation (Question 4)
- recognize that it is always possible to get from one shape to a congruent one using a combination of transformations (Question 8)

Consolidating and Reflecting

Ensure understanding by asking questions such as these based on students' work:

- When you compare the final shape with the original shape in Question 2, what happened and why? (e.g., *It is in the original direction, since you turned the same amount clockwise as counterclockwise.*)
- What did you notice in Question 3?
  (e.g., *that 2 reflections could end up looking like a translation or a rotation*)
- In Question 4, why does it make sense that 2 translations put together is the same as a single different translation? (e.g., *If you go right or left a certain amount, you could do it in 2 separate stages. The same is true for going up and down.*)
- In Question 8, you moved to a congruent shape using 2 or 3 transformations. How did you know whether you needed a reflection? (e.g., *It depended on whether the shape was facing the wrong way. If it was, I needed a reflection.*)

Pathway 4 OPEN-ENDED

Performing Single Transformations

You will need

- 1 cm Grid Paper (BLM 12) or 0.5 cm Grid Paper (BLM 13)
- materials for drawing transformations (e.g., rulers, protractors, tracing paper, or dynamic geometry software)
- Internet access
- Student Resource pages 318–319

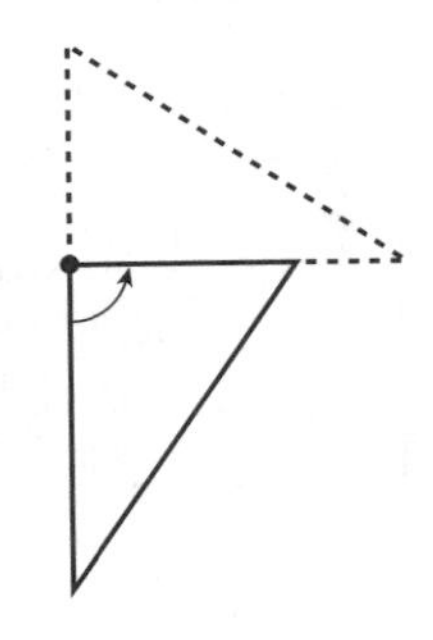

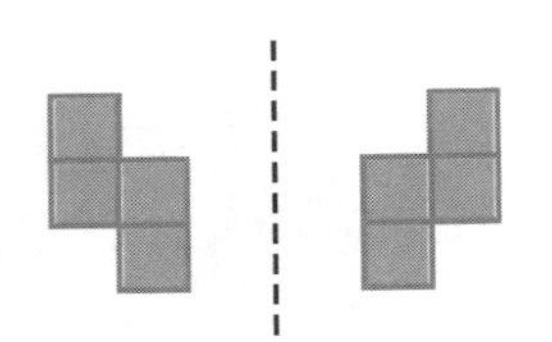

Open-Ended Intervention

Before Using the Open-Ended Intervention

On grid paper, draw a scalene right triangle. Ask:

- How will this look different if I reflect it across a vertical line? (e.g., *It will face the other way.*)
- How will it look different if I translate it? (e.g., *It won't, but it will be in a different place.*)
- How will it look different if I rotate it 90° counterclockwise? (See margin art.) (e.g., *First it's pointing down, but after the turn I think it would point to the right.*)

If possible, have students play an online version of the game Tetris.

Using the Open-Ended Intervention (Student Resource pages 318–319)

Read through the tasks on the student pages together. Provide the materials.

Give students time to work, ideally in pairs.

Observe whether students

- can perform translations, reflections, and rotations on a grid
- realize that some images can be achieved using more than one type of transformation

Consolidating and Reflecting

Ensure understanding by asking questions such as these based on students' work:

- Which Tetris shapes looked different when you did a 180° rotation? (*The square and the straight line didn't look different, but the others did.*)
- How did the zigzag one look when you reflected it across a vertical line? (See margin art.) (e.g., *In the reflection, the top part stuck out on the right instead of on the left.*)
- Which ones looked the same when you reflected across a vertical line? (e.g., *the straight line, the square, and the T*)
- How can you predict what a shape will look like after a rotation or reflection? (e.g., *After a rotation of 90°, what used to be vertical will be horizontal. After a reflection, things move to the other side of the reflection line; they move farther if they are farther from the line.*)

Copyright © 2012 by Nelson Education Ltd.

Performing Single Transformations

Pathway 4
GUIDED

Guided Intervention

Before Using the Guided Intervention

On grid paper, draw a scalene right triangle. Ask:

- ▸ How will this look different if I reflect it across a vertical line?
  (e.g., *It will face the other way.*)
- ▸ How will it look different if I translate it?
  (e.g., *It won't, but it will be in a different place.*)
- ▸ How will it look different if I rotate it 90° counterclockwise? (See margin art.)
  (e.g., *First it's pointing down, but after the turn I think it would point to the right.*)

You will need

- 1 cm Grid Paper (BLM 12) or 0.5 cm Grid Paper (BLM 13)
- transparent mirrors (e.g., Miras)
- materials for drawing transformations (e.g., rulers, protractors, tracing paper, or dynamic geometry software)
- Student Resource pages 320–323

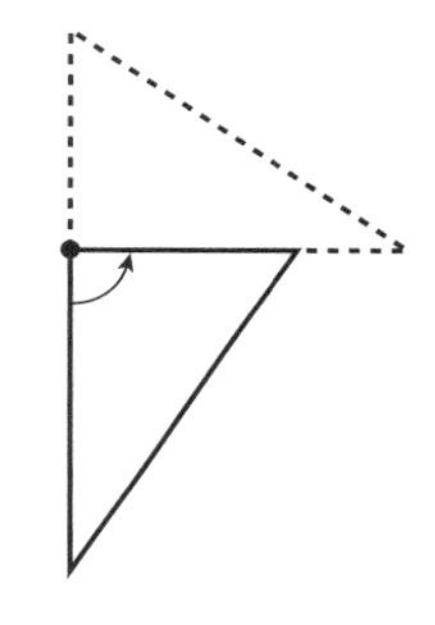

Using the Guided Intervention (Student Resource pages 320–323)

Work through the instructional section of the student pages together. Provide the materials, including BLM 12 so students can complete Question 8.

Have students work through the **Try These** questions individually or in pairs.

Observe whether students

- can identify transformations (Questions 1, 2)
- realize that some images can be achieved using more than one type of transformation (Questions 2, 7)
- can perform translations, reflections, and rotations on a grid (Questions 3, 4, 5, 6, 8)

Consolidating and Reflecting

Ensure understanding by asking questions such as these based on students' work:

- ▸ How did you tell whether a rotation was done in Question 1?
  (e.g., *When the turn centre is a vertex, one point doesn't change, and what used to be horizontal would be vertical if it was a 90° turn.*)
- ▸ Why do you need to show only 1 translation arrow to know where the whole shape moves?
  (e.g., *Every point moves exactly the same way, so showing it once is enough.*)
- ▸ How can you predict what a shape will look like after a rotation or reflection?
  (e.g., *After a rotation of 90°, what used to be vertical will be horizontal. After a reflection, things move to the other side of the reflection line; they move farther if they are farther from the line.*)
- ▸ Why is it sometimes hard to be sure which transformation has been performed?
  (e.g., *If a shape is symmetrical, then sometimes the effects of reflections, translations, and rotations look the same. So more than one transformation might have been performed.*)

Measurement Strand Overview

How were the measurement topics chosen?

This resource provides materials for assisting students with 4 measurement topics. These topics are drawn from the curriculum outcomes across the country for Grades 5 to 7. Topics are divided into distinct levels, called pathways, that address gaps in students' prerequisite skills and knowledge.

How were the pathways determined?

At the Grade 7/8 level, measurement topics include area, perimeter, volume, and angles, with an emphasis on the use of formulas. There is also an increased expectation that students will be able to convert measurement values from one metric unit to another, whether larger to smaller or smaller to larger.

The area and perimeter topic goes back as far as the formulas for the area and perimeter of rectangles. This topic includes pathways focusing both on formulas for areas of parallelograms and triangles and on areas of composite shapes. The pathway involving composite shapes specifically addresses the formula for the area of a trapezoid, treated as a composite shape, but also includes other composite shapes. Other pathways enable students to make sense of the formulas for the circumference and area of circles. The work with circles is not needed for remediation for Ontario students at this level, because circle formulas are introduced in Grade 8.

The volume and surface area topic involves 3-D shapes, again with an emphasis on formulas. The simplest pathway addresses the formula for the volume of a rectangular prism. Another pathway focuses on the surface area of other right prisms, and a third deals with the formula for the volume of right prisms.

The angle topic helps students who still need work with measuring angles with a protractor. It advances to angle construction and then to an understanding of why the sum of the interior angles of triangles and quadrilaterals is 180° and 360°, respectively.

The metric units topic is designed to assist students who need help in selecting an appropriate unit to use and progresses to renaming a measurement using a different metric unit. It includes the measurement attributes such as length, mass, area, and volume.

What measurement topics have been omitted?

The expectations from the Ontario curriculum related to measuring time were excluded because there is no follow-up on this material in Grades 7 and 8.

 Copyright © 2012 by Nelson Education Ltd.

The relationship between capacity and volume was addressed in the *Leaps and Bounds 5/6* materials, so it is not included here. The main idea needed—that 1 mL of water is displaced by 1 cm^3 of a solid—can be addressed, if needed, without an intervention.

Materials

The materials for assisting students with measurement topics are likely in the classroom or easily accessible. Blackline masters are provided at the back of this resource. Some materials are optional and are listed only in the Teacher's Resource.

- coloured markers
- a transparent centimetre grid (e.g., photocopy BLM 12 on a transparency) and a washable marker
- large pieces of paper
- rulers
- scissors
- calculators
- compasses
- skipping rope (optional)
- rulers and string or tape measures, marked in millimetres
- circular objects (e.g., plates, cans) of different sizes
- coloured pencils
- pattern blocks
- 1 cm Grid Paper (BLM 12)
- 1 cm Square Dot Paper (BLM 15)
- Fraction Circles (BLM 18)

Measurement Topics and Pathways

Topics and pathways in this strand are shown below.
Each pathway has an open-ended intervention and a guided intervention.

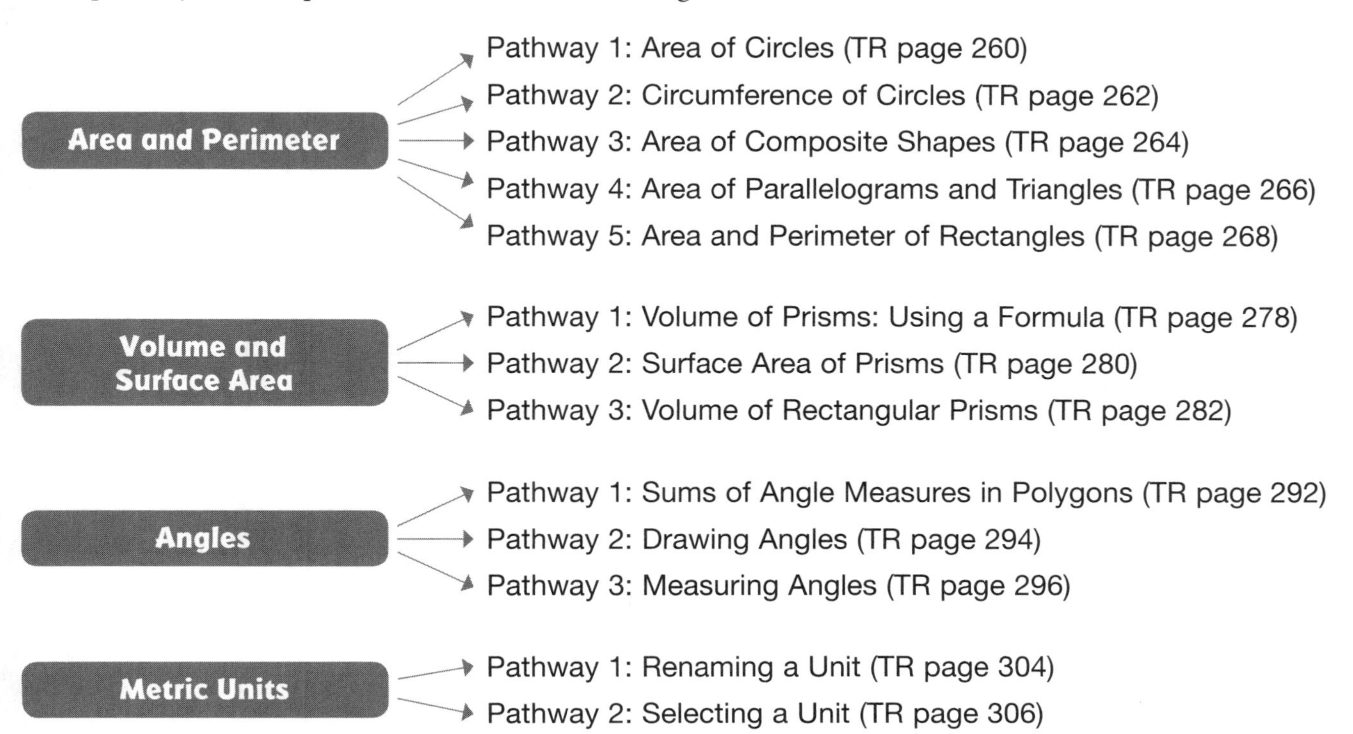

Strand: Measurement

Area and Perimeter

Planning For This Topic

Materials for assisting students with area and perimeter of 2-D shapes consist of a diagnostic tool and 5 intervention pathways. Pathway 1 focuses on area of circles. Pathway 2 deals with circumference of circles. Pathway 3 involves area of composite shapes, including trapezoids, regular polygons, and irregular polygons. Pathway 4 focuses on area of parallelograms and triangles. Pathway 5 deals with area and perimeter of rectangles.

Each pathway has an open-ended intervention and a guided intervention. Choose the type of intervention more suitable for your students' needs and your particular circumstances.

Professional Learning Connections

PRIME: Measurement, Background and Strategies (Nelson Education Ltd., 2010), pages 52–53, 55, 67–72, 75

Making Math Meaningful to Canadian Students K–8 (Nelson Education Ltd., 2008), pages 380–381, 384, 397–402, 405–407

Big Ideas from Dr. Small Grades 4–8 (Nelson Education Ltd., 2009), pages 137–139, 145–150

Good Questions (dist. by Nelson Education Ltd., 2009), pages 102–103, 105–107, 111, 115, 117–118

More Good Questions (dist. by Nelson Education Ltd., 2010), pages 124, 126, 128–129

Curriculum Connections

Grades 5 to 8 curriculum connections for this topic are provided online. See www.nelson.com/leapsandbounds. The Ontario curriculum does not cover area and circumference of circles until Grade 8, so Pathways 1 and 2 can be omitted for those students. The WNCP curriculum does not mention area of composite shapes, so users of this curriculum may omit Pathway 3. Pathways 4 and 5 are relevant to both curriculums.

Why might students struggle with area and perimeter?

Students might struggle with area and perimeter for any of the following reasons:

- They might mix up the terms *area* and *perimeter*.
- They might have difficulty using grids to determine areas of shapes with diagonal sides or correctly identifying side lengths on a grid.
- They might have difficulty identifying the correct value for the height of a parallelogram or triangle.
- They might be able to substitute linear dimensions into area or perimeter formulas to calculate those values, but not be able to start with those values and work backwards (e.g., given a triangle with an area of 12, it might be difficult to figure out that the base and height are numbers that multiply to 24).
- They might have difficulty deciding how to break up a composite shape into simpler shapes to calculate the total area.
- They might not recognize the potential of embedding a shape in another shape to determine its area. (See margin art.)

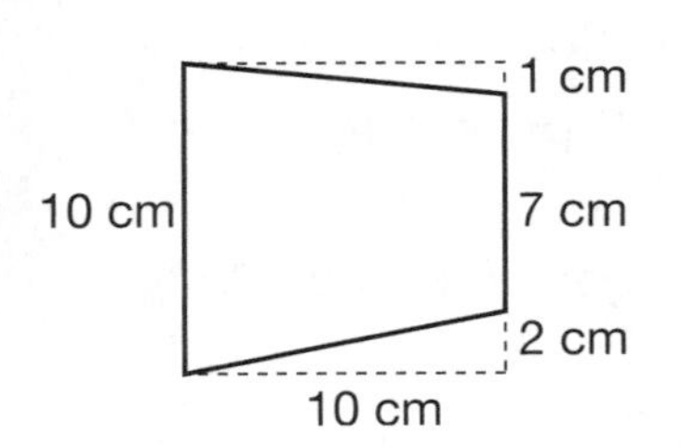

- They might not recognize the relationship between circumference and diameter of circles, even with a lot of data.
- They might mix up the formulas for area and circumference of circles, or not know when to use the radius or the diameter of a circle to calculate.

 Copyright © 2012 by Nelson Education Ltd.

Diagnostic Tool: Area and Perimeter

Use the diagnostic tool to determine the most suitable intervention pathway for area and perimeter. Provide Diagnostic Tool: Area and Perimeter, Teacher's Resource pages 252 to 255, and have students complete it. Provide rulers, coloured markers, and calculators.

If the diagnostic is too long for some students, it could be split into parts: use Questions 1 to 9 on one occasion to assess the need for Pathways 3, 4, and 5, and Questions 10 to 15 on another occasion to assess the need for Pathways 1 and 2.

See solutions on Teacher's Resource pages 256 to 259.

Intervention Pathways

The purpose of the intervention pathways is to help students see how measurements of shapes are interrelated. These concepts will be developed further as students work with 3-D measures: surface area and volume.

There are 5 pathways:

- Pathway 1: Area of Circles
- Pathway 2: Circumference of Circles
- Pathway 3: Area of Composite Shapes
- Pathway 4: Area of Parallelograms and Triangles
- Pathway 5: Area and Perimeter of Rectangles

Use the chart below (or the Key to Pathways on Teacher's Resource pages 256 to 259) to determine which pathway is most suitable for each student or group of students.

Diagnostic Tool Results	Intervention Pathway
If students struggle with Questions 13 to 15	use Pathway 1: Area of Circles *Teacher's Resource pages 260–261* *Student Resource pages 324–329*
If students struggle with Questions 10 to 12	use Pathway 2: Circumference of Circles *Teacher's Resource pages 262–263* *Student Resource pages 330–334*
If students struggle with Questions 7 to 9	use Pathway 3: Area of Composite Shapes *Teacher's Resource pages 264–265* *Student Resource pages 335–340*
If students struggle with Questions 4 to 6	use Pathway 4: Area of Parallelograms and Triangles *Teacher's Resource pages 266–267* *Student Resource pages 341–346*
If students struggle with Questions 1 to 3	use Pathway 5: Area and Perimeter of Rectangles *Teacher's Resource pages 268–269* *Student Resource pages 347–352*

Name: ______________________ Date: ______________

Area and Perimeter

Diagnostic Tool

1. Determine the area and perimeter of the shaded rectangle in each diagram.

a) area: ________ square units

perimeter: ________ units

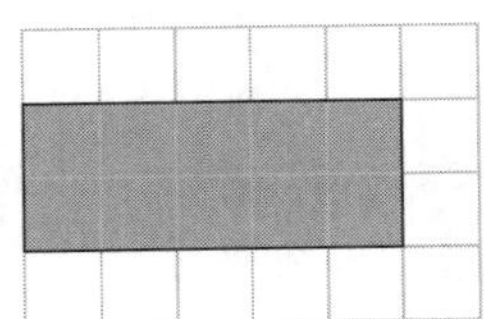

b) area: ____________ cm^2

perimeter: ____________ cm

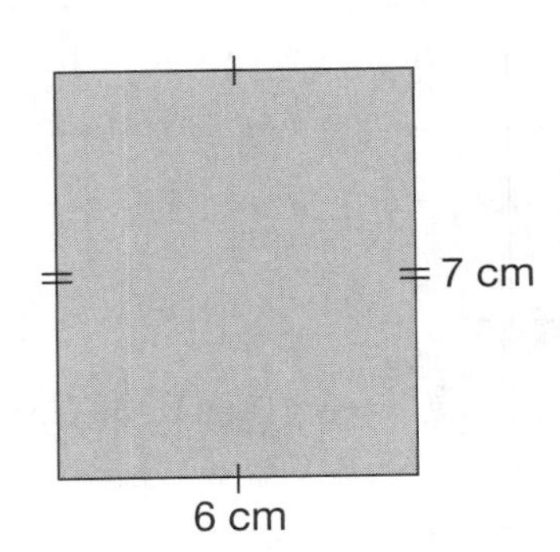

2. How could you prove that your answer for the area in Question 1b) is correct, without using the formula?

____

____

3. Sketch 2 different rectangles that have an area of 18 cm^2. Label the lengths and widths.

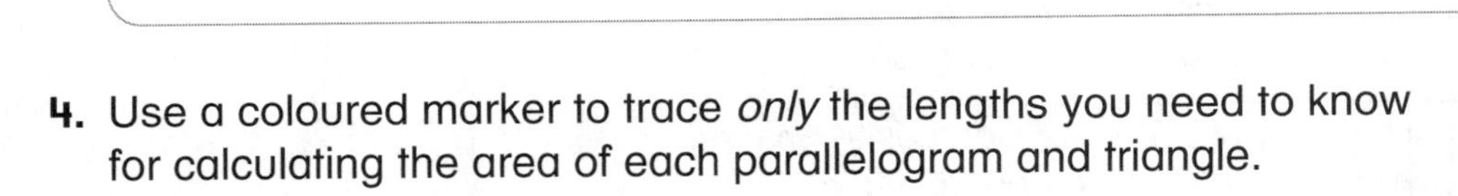

4. Use a coloured marker to trace *only* the lengths you need to know for calculating the area of each parallelogram and triangle.

a)

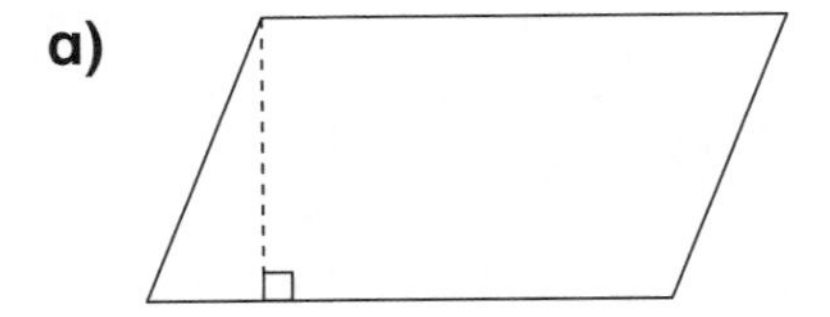

b)

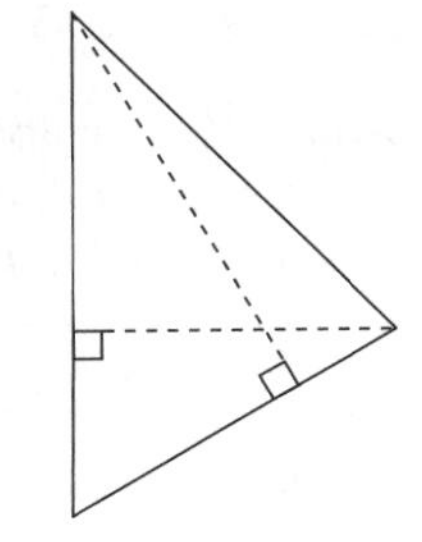

You will need

- a ruler
- a calculator
- a coloured marker

Remember

- Area of a parallelogram = base × height
- Area of a triangle = base × height ÷ 2

Copyright © 2012 by Nelson Education Ltd.

Name:________________________________ Date:________________

5. Tell how the areas of the 2 shapes below are related.

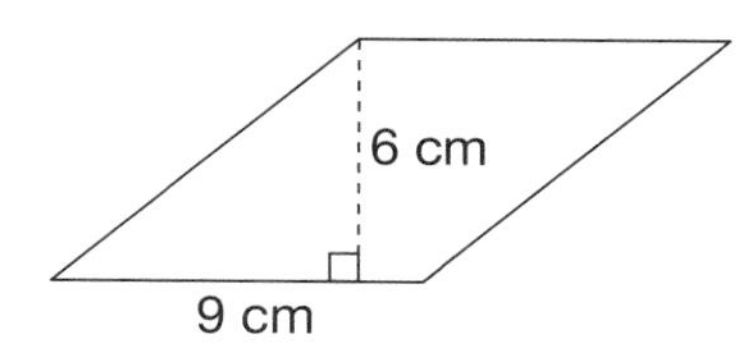

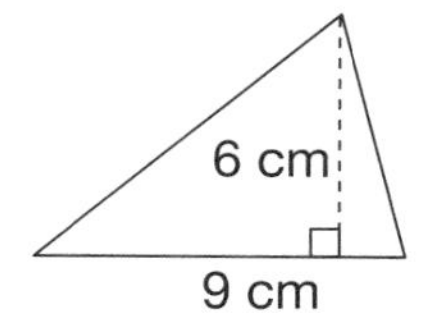

____

____

6. Calculate the area of each shape.

a)

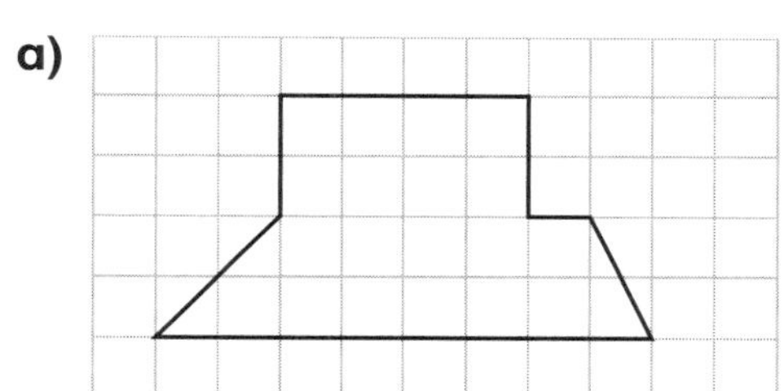

area: ________________

b)

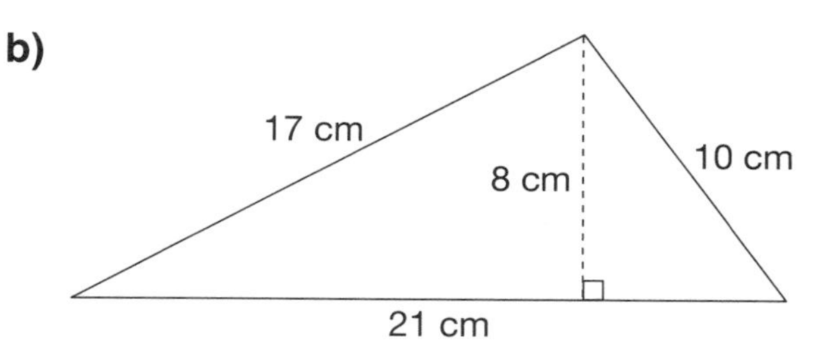

area: ________________

7. Show how to divide each shape into triangles, rectangles, or parallelograms to make it easier to figure out the total area. (You do *not* need to do the calculations.)

a)

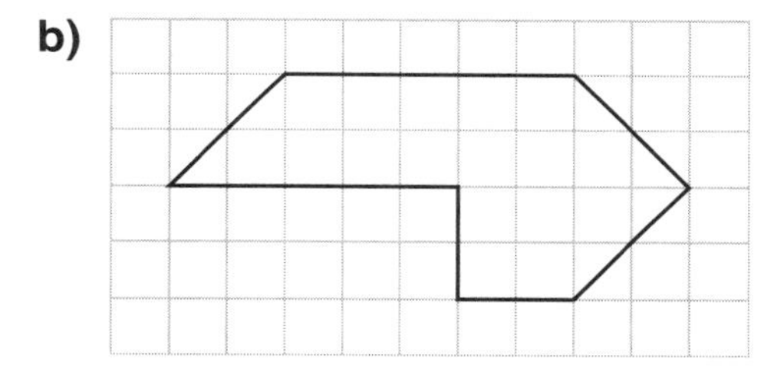

b)

8. Trish sketched these shapes using parallelograms and triangles. What is the area of the shaded part?

a)

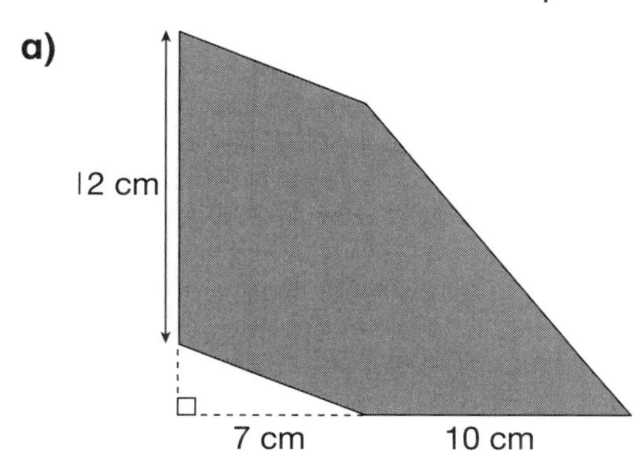

____

b)

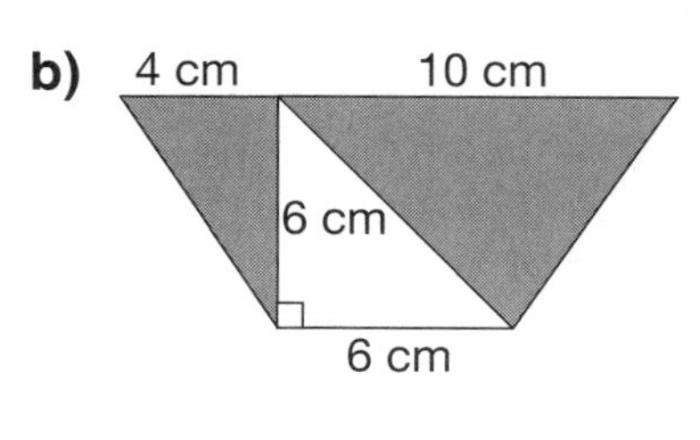

____

Copyright © 2012 by Nelson Education Ltd.

Name:________________________ Date:______________

9. How can you calculate the area of this trapezoid?

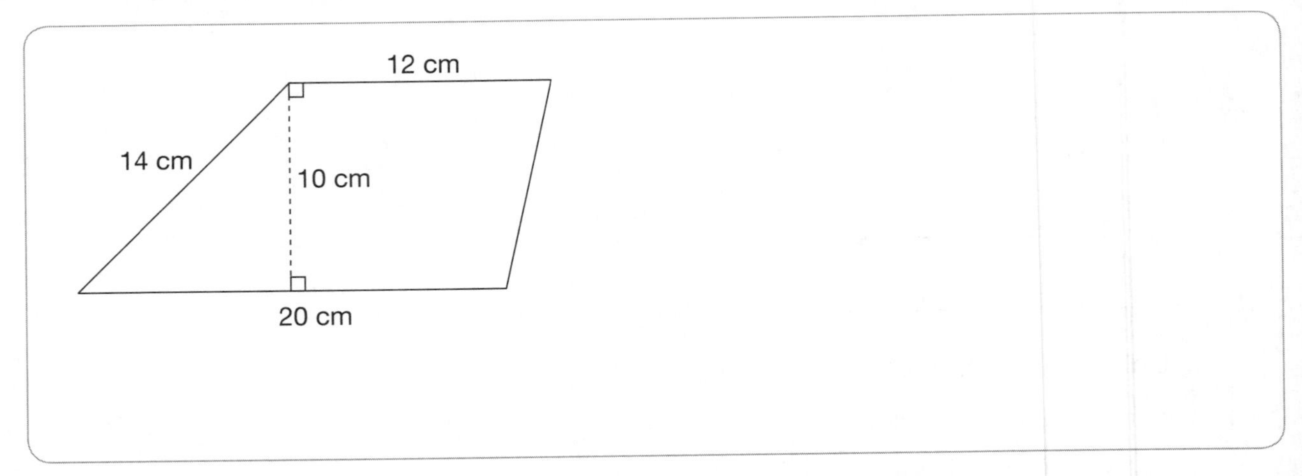

10. The circumference is the distance around a circle. What is the circumference of each circle, to the nearest whole number?

Remember

- Circumference = $\pi \times d$
  d is the diameter of a circle, and π is about 3.14.

a)

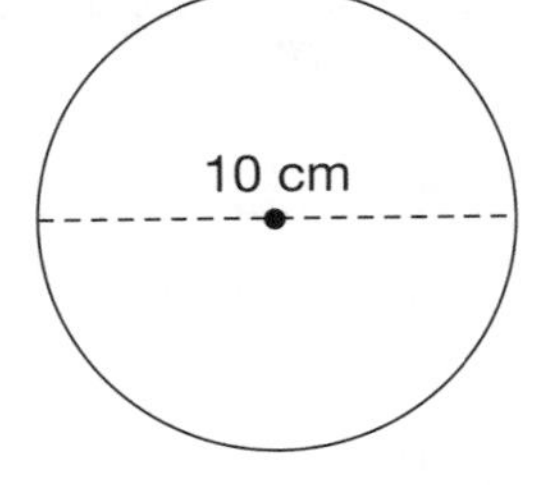

____

b)

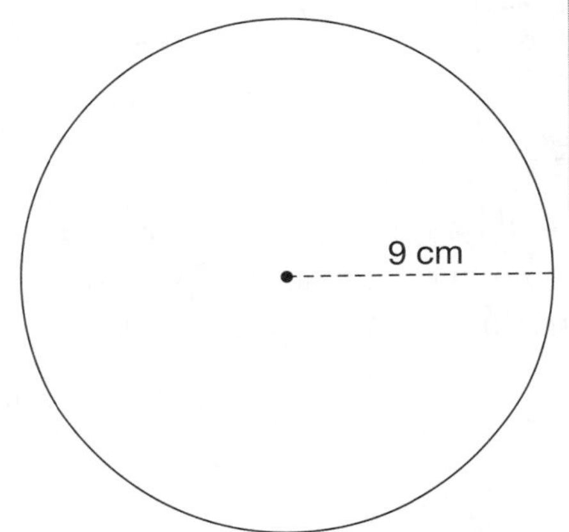

____

11. A string is about twice as long as the diameter of a circle.

Circle the statement below that is true.

- The string goes all the way around the circle.
- The string goes more than halfway around the circle, but not completely around.
- The string goes around the circle 3 times.
- The string goes around the circle about twice.

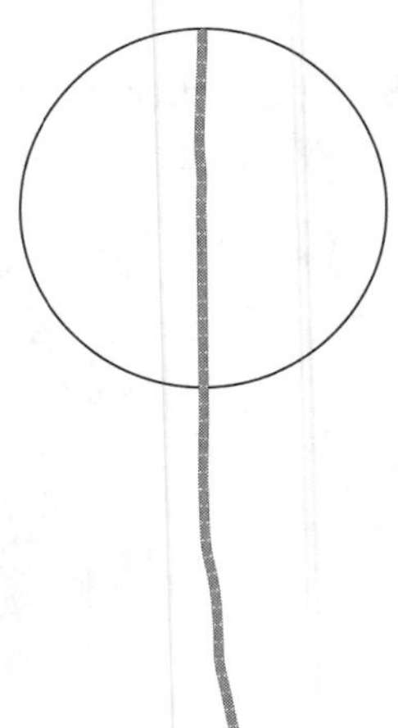

Copyright © 2012 by Nelson Education Ltd.

Name:______________________ Date:______________

12. Suppose the radius of a circle is 10 cm.

a) Estimate the circumference of the circle.

____

b) If a square has the same perimeter as your answer to part a), what is the side length of the square?

____

13. Estimate the area of each circle.

Remember

- Area of a circle = $\pi \times r \times r$

a)

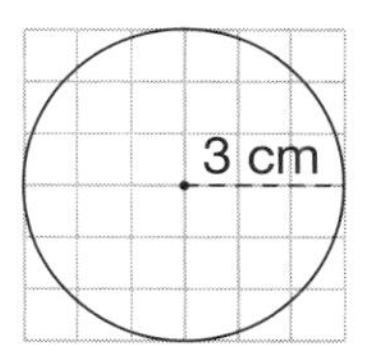

____

____

b)

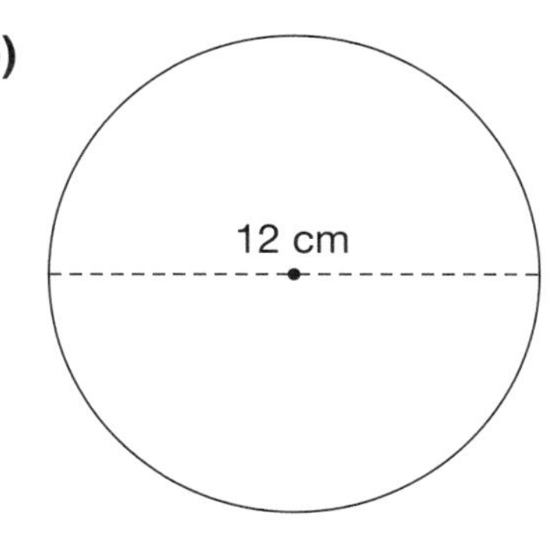

____

____

14. Estimate the area of the shaded shape.
Show your thinking.

15. A circle has an area of about 45 cm^2. Estimate the radius.
Show your thinking.

Copyright © 2012 by Nelson Education Ltd.

Solutions and Key to Pathways

Name: ______________________ Date: ______________

Area and Perimeter

Diagnostic Tool

You will need
- a ruler
- a calculator
- a coloured marker

Pathway 5

1. Determine the area and perimeter of the shaded rectangle in each diagram.

   a) area: 10 square units

   perimeter: 14 units

   b) area: $6 \times 7 = 42$ cm^2

   perimeter: $2 \times (6 + 7) = 26$ cm

   7 cm
   6 cm

2. How could you prove that your answer for the area in Question 1b) is correct, without using the formula?

   e.g., I could draw in 6 rows of 7 squares to fill the whole rectangle, and 6 sevens is 42.

3. Sketch 2 different rectangles that have an area of 18 cm^2. Label the lengths and widths.

   e.g.,
   9 cm, 2 cm, 2 cm, 9 cm
   6 cm, 3 cm, 3 cm, 6 cm

Pathway 4

4. Use a coloured marker to trace *only* the lengths you need to know for calculating the area of each parallelogram and triangle.

   a) e.g.,

   b) e.g.,

Remember
- Area of a parallelogram = base × height
- Area of a triangle = base × height ÷ 2

 Copyright © 2012 by Nelson Education Ltd.

 Copyright © 2012 by Nelson Education Ltd.

Name:________________ Date:________________

5. Tell how the areas of the 2 shapes below are related.

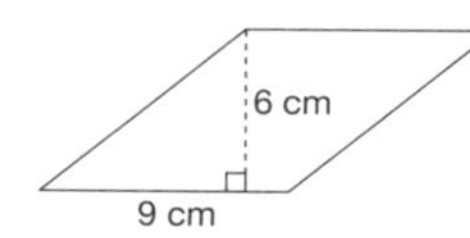

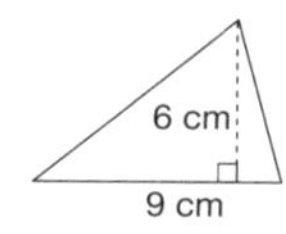

e.g., If you cut the parallelogram in half, you get the triangle, so the triangle should have half the area.

Pathway 4

6. Calculate the area of each shape.

a)

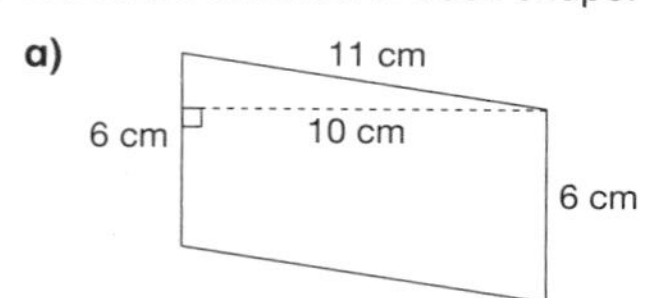

area: $6 \times 10 = 60\text{ cm}^2$

b)

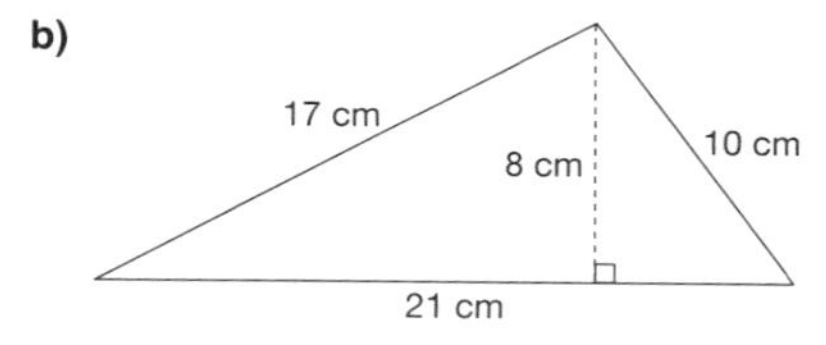

area: $21 \times 8 \div 2 = 84\text{ cm}^2$

7. Show how to divide each shape into triangles, rectangles, or parallelograms to make it easier to figure out the total area. (You do *not* need to do the calculations.)

a) e.g.,

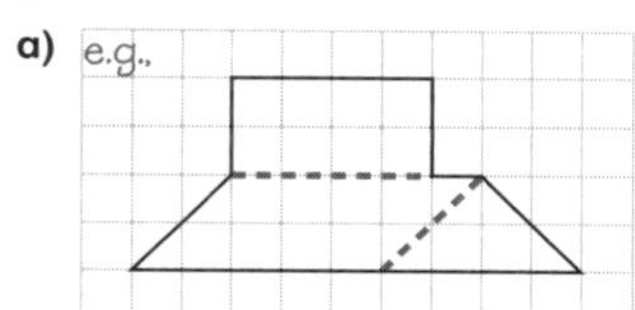

b) e.g.,

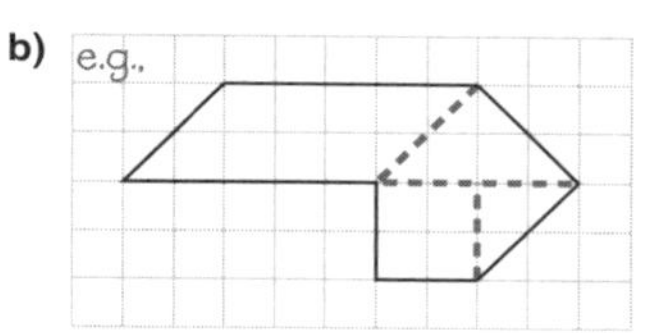

Pathway 3

8. Trish sketched these shapes using parallelograms and triangles. What is the area of the shaded part?

a)

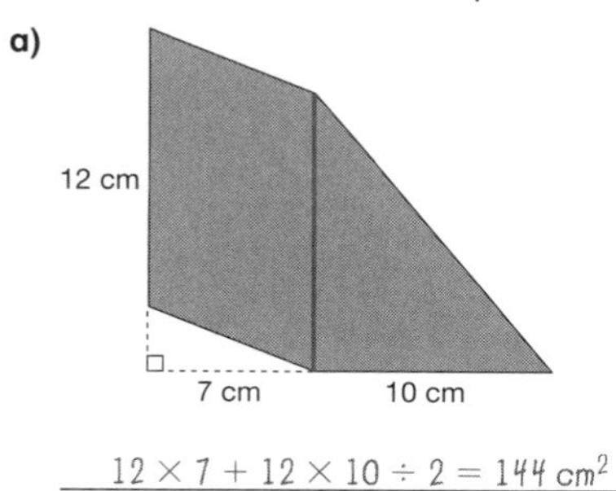

$12 \times 7 + 12 \times 10 \div 2 = 144\text{ cm}^2$

b)

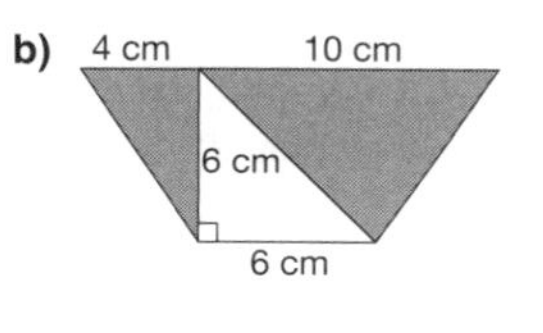

$4 \times 6 \div 2 + 10 \times 6 \div 2 = 42\text{ cm}^2$

Copyright © 2012 by Nelson Education Ltd.

Copyright © 2012 by Nelson Education Ltd.

Name:______________________ Date:______________

Pathway 3

9. How can you calculate the area of this trapezoid?

e.g., I drew a line to form 2 triangles.
The area of the larger triangle is
$20 \times 10 \div 2 = 100\ cm^2$.
The area of the smaller triangle is
$12 \times 10 \div 2 = 60$.
Area of trapezoid $= 100 + 60$
$= 160\ cm^2$

Pathway 2

10. The circumference is the distance around a circle. What is the circumference of each circle, to the nearest whole number?

Remember
- Circumference = $\pi \times d$
  d is the diameter of a circle, and π is about 3.14.

a)

e.g., 3.14×10 is about 31 cm.

b)

e.g., 3.14×18 is about 57 cm.

11. A string is about twice as long as the diameter of a circle.

Circle the statement below that is true.

- The string goes all the way around the circle.
- The string goes more than halfway around the circle, but not completely around. (circled)
- The string goes around the circle 3 times.
- The string goes around the circle about twice.

 Copyright © 2012 by Nelson Education Ltd.

 Copyright © 2012 by Nelson Education Ltd.

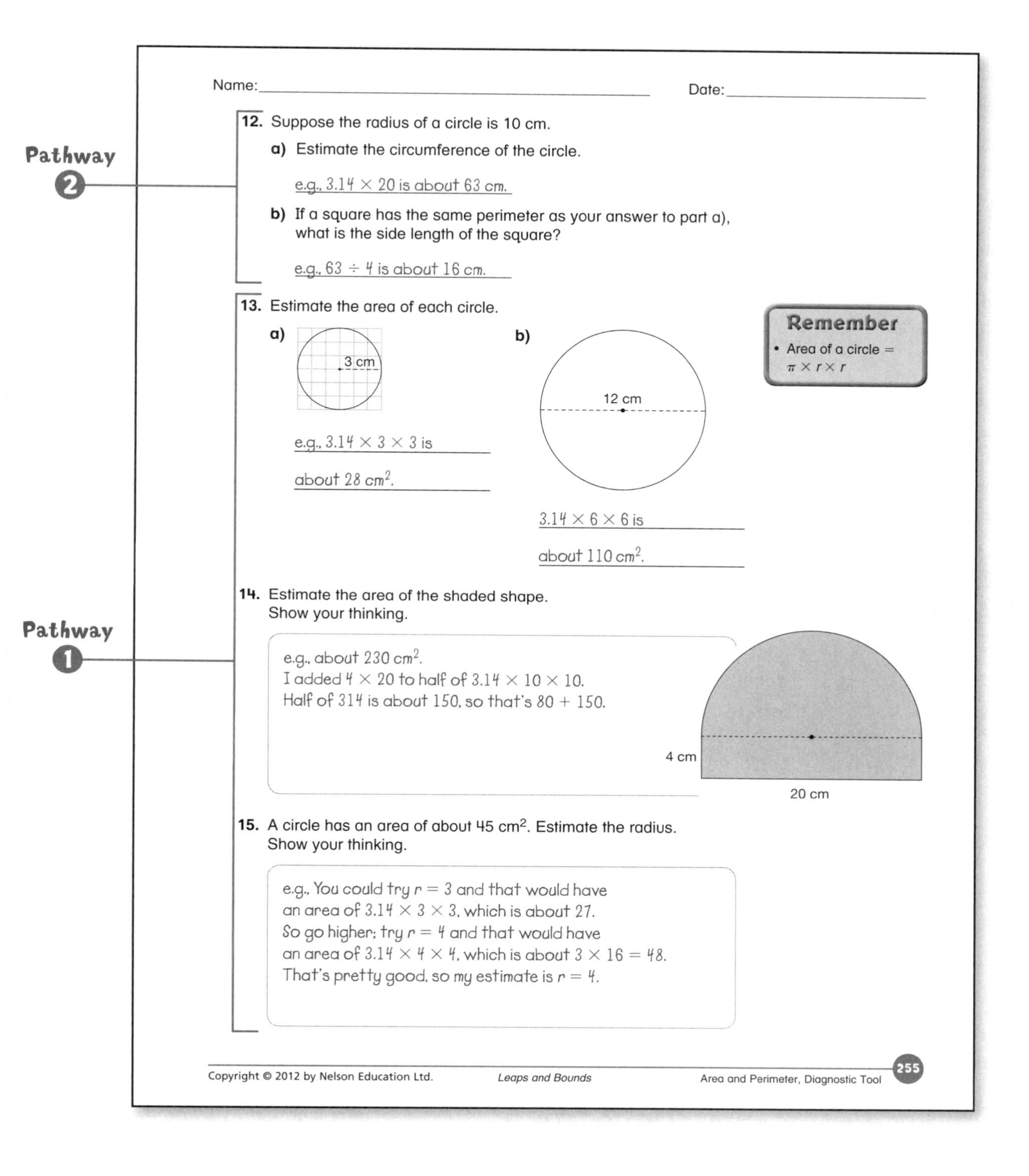

Pathway 2

Pathway 1

Name: ____________________ Date: ____________

12. Suppose the radius of a circle is 10 cm.

a) Estimate the circumference of the circle.

e.g., 3.14×20 is about 63 cm.

b) If a square has the same perimeter as your answer to part a), what is the side length of the square?

e.g., $63 \div 4$ is about 16 cm.

13. Estimate the area of each circle.

Remember

- Area of a circle = $\pi \times r \times r$

a)

e.g., $3.14 \times 3 \times 3$ is about 28 cm^2.

b)

$3.14 \times 6 \times 6$ is about 110 cm^2.

14. Estimate the area of the shaded shape. Show your thinking.

e.g., about 230 cm^2.
I added 4×20 to half of $3.14 \times 10 \times 10$.
Half of 314 is about 150, so that's $80 + 150$.

15. A circle has an area of about 45 cm^2. Estimate the radius. Show your thinking.

e.g., You could try $r = 3$ and that would have an area of $3.14 \times 3 \times 3$, which is about 27.
So go higher; try $r = 4$ and that would have an area of $3.14 \times 4 \times 4$, which is about $3 \times 16 = 48$.
That's pretty good, so my estimate is $r = 4$.

Copyright © 2012 by Nelson Education Ltd. *Leaps and Bounds* Area and Perimeter, Diagnostic Tool 255

Pathway 1 OPEN-ENDED

Area of Circles

You will need

- transparent centimetre grids (e.g., photocopy BLM 12 on a transparency)
- Fraction Circles (BLM 18)
- coloured markers
- scissors
- rulers
- calculators
- Student Resource pages 324–325

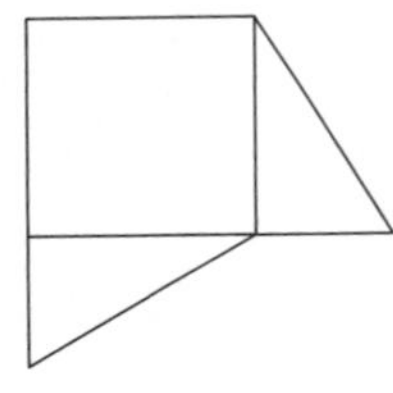

Open-Ended Intervention

Before Using the Open-Ended Intervention

Sketch a small irregular shape that covers full and partial squares on a grid, such as the first one in the margin art. Ask:

- How can you figure out the area of this shape?
  (e.g., *I would count the whole squares, and then estimate the partial amounts. You can count the squares that are more than half full and ignore the others.*)

Make a shape using pieces such as a square and 2 triangles (see margin art). Ask:

- Why might it be a good idea to move the pieces to figure out the area of this shape? (e.g., *I could match the bottom triangle with the one on the right and it would make a rectangle, so it would be whole squares instead of part squares.*)

Using the Open-Ended Intervention (Student Resource pages 324–325)

Read through the tasks on the student pages together. Provide the materials. Suggest that students outline the circumference with a coloured marker before they cut the circle into pieces. Give students time to work, ideally in pairs.

Observe whether students

- can estimate areas of circles
- relate areas of other shapes to those of a circle
- can use a formula for the area of a circle

Provide the formula for area of a circle ($A = \pi r^2$ or $\pi \times$ radius $\times$ radius) and help students relate it to what they did.

Consolidating and Reflecting

Ensure understanding by asking questions such as these based on students' work:

- How did you arrange the sections of a circle to form a parallelogram? (e.g., *I fit the pieces together so the coloured circumference was on the top and the bottom.*)
- Suppose you substituted the expression $2\pi r$ for C in your formula $A = \frac{1}{2}C \times r$. How would that get you the formula $A = \pi r \times r$?
  (e.g., *$C = 2\pi r$ and half of that is πr, so my formula is now $A = \pi r \times r$.*)
- Now that you know the formula is $A = \pi r \times r$, why does the relationship of the areas of the circle and the square that it just fits into make sense to you?
  (e.g., *The area of the square is $4 \times r \times r$ because it is made up of 4 small squares that are $r \times r$. The circle fits in the square and uses up most of it, but not all of it. Since π is about 3, the area is about $3 \times r \times r$, which is less than $4 \times r \times r$.*)
- Why is it useful to have a formula for the area of a circle instead of measuring the area with the grid?
  (e.g., *There are so many part squares, it's hard to use the grid. It's easy to measure the radius, though, with a ruler. Also, the formula gives the actual area, but the grid just gives an estimate.*)

 Copyright © 2012 by Nelson Education Ltd.

Area of Circles

Pathway 1
GUIDED

Guided Intervention

Before Using the Guided Intervention

Sketch a small shape that covers full and partial squares on a grid, such as the first one in the margin art. Ask:

- How can you figure out the area of this shape?
  (e.g., *I would count the whole squares, and then estimate the partial amounts. I think you can count the squares that are more than half full and ignore the others.*)

Make a shape using pieces such as a square and 2 triangles (see margin art).

- Why might it be a good idea to move the pieces to figure out the area of this shape? (e.g., *I could match the bottom triangle with the one on the right and it would make a rectangle, so it would be whole squares instead of part squares.*)

Using the Guided Intervention **Student Resource pages 326–329**

Provide the materials, and work through the instructional section of the student pages together. If necessary, help students fold the circle into 16 equal sections.

Have students work through the **Try These** questions individually or in pairs.

Observe whether students

- can relate areas of circles to areas of other shapes (Questions 1, 3, 6, 8)
- can estimate areas of circles (Questions 2, 3)
- can use the formula for the area of a circle (Questions 4, 8)
- recognize when it might be useful to apply the formula for the area of a circle (Questions 5, 7)

Consolidating and Reflecting

Ensure understanding by asking questions such as these based on students' work:

- Why might you estimate the radius of the circle as 5.7 cm in Question 3?
  (e.g., *The height of the triangle was 5.6 cm, and the radius is just a bit more than that.*)
- In Question 4, why didn't it matter if you were given the diameter instead of the radius?
  (e.g., *If you know the diameter, you just take half if you want the radius. The radius is what you need for the formula.*)
- In Question 8, how did you figure out the radius of the circle?
  (e.g., *I figured out that the area of the triangle was 50 cm². I tried different values for the radius until the area was close to 50.*)
- Why is it useful to have a formula for the area of a circle instead of measuring the area with the grid?
  (e.g., *There are so many part squares, it's hard to use the grid. It's easy to measure the radius, though, with a ruler. Also, the formula gives the actual area, but the grid just gives an estimate.*)

You will need

- a transparent centimetre grid (e.g., photocopy BLM 12 on a transparency) and a washable marker
- large pieces of paper
- compasses
- rulers
- scissors
- calculators
- Student Resource pages 326–329

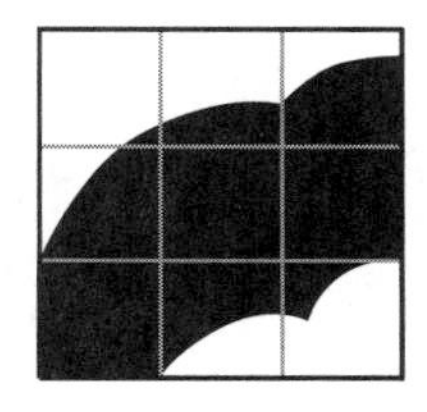

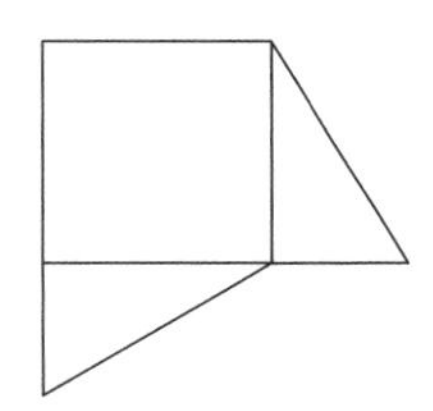

Pathway 2 OPEN-ENDED

Circumference of Circles

You will need

- circular objects of different sizes (e.g., plates, cans), or a compass
- rulers or tape measures
- string
- scissors
- skipping rope (optional)
- calculators (optional)
- Student Resource page 330

Open-Ended Intervention

Before Using the Open-Ended Intervention

Provide circular objects of 3 different sizes. Ask:

- Which circle has the greatest radius? How do you know? (e.g., *It is the biggest circle.*)
- How do you measure its radius? (e.g., *You use a ruler starting at the middle of the circle and going to the edge.*)
- Do you think that a circle with a bigger radius has a bigger distance around (circumference) than a circle with a smaller radius? Why? (e.g., *I am looking at these 3 circles, and it seems like if it has a bigger radius, the circle is wider, and so the distance around the edge would be bigger, too.*)

Have students cut a number of pieces of string the length of the diameter of one of the circles. Ask:

- About how many pieces of that string would it take to wrap around the circumference of that circle? (*about 3*)

Using the Open-Ended Intervention Student Resource page 330

Read through the tasks on the student pages together. Provide tape measures or rulers and string. You might also provide a long rope or a compass.

Give students time to work, ideally in pairs.

Observe whether students

- can determine a relationship between the circumference and diameter of a circle
- can determine a relationship between the circumference and radius of a circle
- can apply the formula for the circumference of a circle

Provide the formula $C = \pi \times d$, and explain that π is a number a little more than 3. Tell students that it is estimated as 3.14.

Consolidating and Reflecting

Ensure understanding by asking questions such as these based on students' work:

- What did you do that helped the most in figuring out the relationship between the diameter and the circumference? (e.g., *I cut the string the length of the diameter and counted how many times it fit around the circle.*)
- Why do you think the size of the circle did not matter when you compared the diameter lengths and the circumference lengths? (e.g., *If you have a bigger diameter, you have a bigger circle, so they go up together.*)
- Why is the circumference about 6 times the radius, but only about 3 times the diameter? (e.g., *The radius is only half as long as the diameter, so it takes twice as many of them to go around the circumference.*)

 Copyright © 2012 by Nelson Education Ltd.

Circumference of Circles

Pathway 2 GUIDED

Guided Intervention

You will need

- rulers or tape measures
- string
- circular objects (e.g., jar lids, plates) of different sizes
- calculators
- Student Resource pages 331–334

Before Using the Guided Intervention

Provide circular objects of 3 different sizes. Ask:

- Which circle has the greatest radius? How do you know? (e.g., *It is the biggest circle.*)
- How do you measure its radius? (e.g., *You use a ruler starting at the middle of the circle and going to the edge.*)
- Do you think that a circle with a bigger radius has a bigger distance around (circumference) than a circle with a smaller radius? Why? (e.g., *I am looking at these 3 circles, and it seems like if it has a bigger radius, the circle is wider, and so the distance around the edge would be bigger, too.*)

Using the Guided Intervention (Student Resource pages 331–334)

Work through the instructional section of the student pages together. Ensure students have tape measures or rulers and string to use.

Have students work through the **Try These** questions individually or in pairs.

Observe whether students

- recognize the direct relationship between the circumference and diameter of a circle (Questions 1, 5)
- can apply the formula for circumference of a circle to estimate the circumference based on the radius or diameter (Questions 2, 3, 6, 7)
- can apply the formula for circumference of a circle to estimate the diameter of a circle based on its circumference (Question 4)
- relate the circumference of a circle to the perimeters of relevant polygons (Questions 8, 9)

Consolidating and Reflecting

Ensure understanding by asking questions such as these based on students' work:

- Why do you think the size of the circle did not matter in Question 1 when you compared the diameter lengths and the circumference lengths? (e.g., *If you have a bigger diameter, you have a bigger circle, so they go up together.*)
- Why is the circumference about 6 times the radius, but only about 3 times the diameter? (e.g., *The radius is only half as long as the diameter, so it takes twice as many of them to go around the circumference.*)
- In Question 4, how did you use the formula to figure out the diameter if you knew the circumference? (e.g., *Just divide the circumference by π to get the diameter.*)
- How did the picture in Question 9 help you make more sense of the formula for circumference? (e.g., *The circumference was less than 4 times the diameter, but not a lot less.*)

Area of Composite Shapes

You will need

- pattern blocks
- coloured pencils
- 1 cm Square Dot Paper (BLM 15)
- rulers
- calculators
- Student Resource pages 335–336

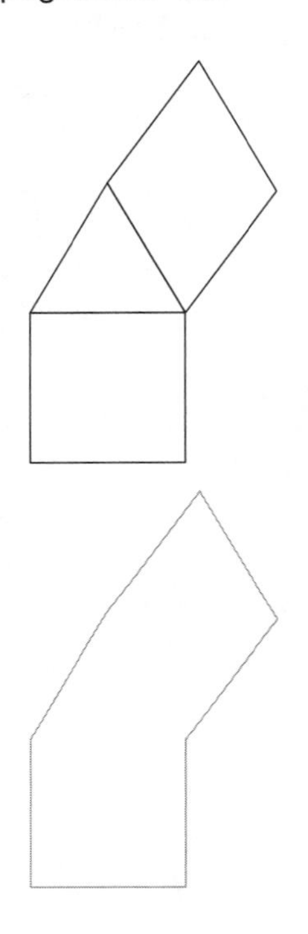

Open-Ended Intervention

Before Using the Open-Ended Intervention

Present 3 pattern blocks: blue rhombus, square, and triangle. Trace around each block on paper and show students the 3 shapes. Ask:

- ▸ How do you calculate the area of each?
  (*For the rhombus, I have to measure the base and the height and multiply. For the triangle, I do the same thing but then divide by 2. For the square, I just multiply the side length by itself.*)
- ▸ Suppose I made a new shape by putting together all 3 shapes (see margin art). How would I calculate the area?
  (*You would add the areas of all the parts.*)
- ▸ What if you couldn't see the parts and you could see only the outline?
  (e.g., *I would divide up the outline into parts myself.*)

Using the Open-Ended Intervention (Student Resource pages 335–336)

Read through the tasks on the student pages together. Provide the necessary materials. Make sure students realize that their explanations can involve the parts within a shape, or the shape fitting inside a larger rectangle.

Give students time to work, ideally in pairs.

Observe whether students

- can calculate the areas of pre-divided composite shapes
- can break up a composite shape into useful component parts
- can calculate areas of composite shapes by adding
- can calculate areas of composite shapes by subtracting
- can calculate the area of a trapezoid as a composite shape

Consolidating and Reflecting

Ensure understanding by asking questions such as these based on students' work:

- ▸ How did you calculate the area of the trapezoid? How did that help?
  (e.g., *I broke it into 2 triangles and a rectangle. Then I could use the formulas for calculating areas of triangles and rectangles.*)
- ▸ Why did the lengths of both bases as well as the height of the trapezoid matter?
  (e.g., *I needed to use the length of one base for the rectangle and the length of the other base for the rectangle plus the triangles. I needed the height to get all 3 areas.*)
- ▸ When was it useful to subtract areas to figure out the area of a shape?
  (e.g., *When my shape was most, but not all, of a square, I subtracted the extra pieces of the square from the area of the square.*)
- ▸ Do you think you could break up any polygon into smaller shapes to calculate its area using formulas?
  (e.g., *Yes, you can always cut it up into triangles and measure their bases and heights.*)

 Copyright © 2012 by Nelson Education Ltd.

Area of Composite Shapes

Pathway 3 GUIDED

Guided Intervention

You will need

- pattern blocks
- coloured pencils
- rulers
- calculators
- Student Resource pages 337–340

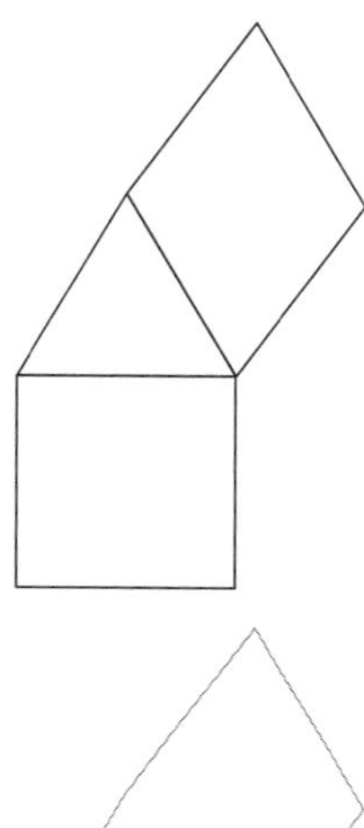

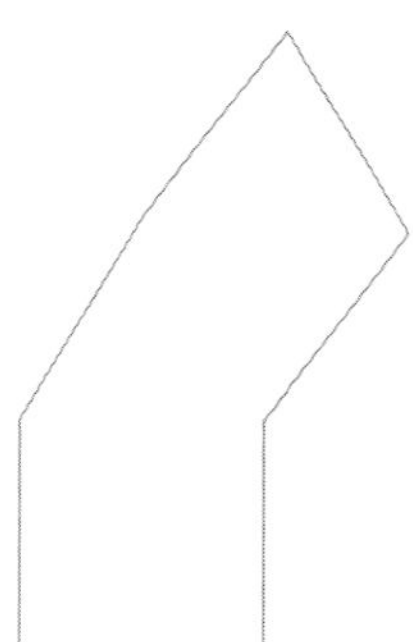

Before Using the Guided Intervention

Present 3 pattern blocks: blue rhombus, square, and triangle. Trace around each one on paper and show students the 3 shapes. Ask:

- How do you calculate the area of each?
  (*For the rhombus, I have to measure the base and the height and multiply. For the triangle, I do the same thing but then divide by 2. For the square, I just multiply the side length by itself.*)
- Suppose I made a new shape by putting together all 3 shapes (see margin art). How would I calculate the area?
  (*You would add the areas of all the parts.*)
- What if you couldn't see the parts and you could see only the outline?
  (e.g., *I would divide up the outline into parts myself.*)

Using the Guided Intervention Student Resource pages 337–340

Provide the necessary materials. Work through the instructional section of the student pages together.

Have students work through the **Try These** questions individually or in pairs.

Observe whether students

- can calculate the areas of pre-divided composite shapes (Question 1)
- can calculate areas of composite shapes by adding (Questions 1, 2, 3)
- can break up a composite shape into useful component parts (Questions 2, 3)
- can calculate areas of composite shapes by subtracting (Question 4)
- can create a composite shape with a given area (Questions 5, 6)
- can make sense of the formula for the area of a trapezoid (Question 7)

Consolidating and Reflecting

Ensure understanding by asking questions such as these based on students' work:

- How did you figure out the area of the trapezoid in Question 2? How did that help?
  (e.g., *I broke it up into 2 triangles. Then I could use the formulas for calculating areas of triangles.*)
- How did you figure out the area of the shape in Question 3c)?
  (e.g., *The shape looks like a hexagon pattern block made up of 6 triangle blocks.*)
- When was it useful to subtract areas to figure out the area of a shape?
  (e.g., *when my shape was most, but not all, of a square, as in Question 4b*))
- Do you think you could break up any polygon into smaller shapes to calculate its area using formulas?
  (e.g., *Yes, you can always cut it up into triangles and measure their bases and heights.*)

Area of Parallelograms and Triangles

You will need

- 1 cm Grid Paper (BLM 12)
- rulers
- scissors
- coloured pencils
- Student Resource pages 341–342

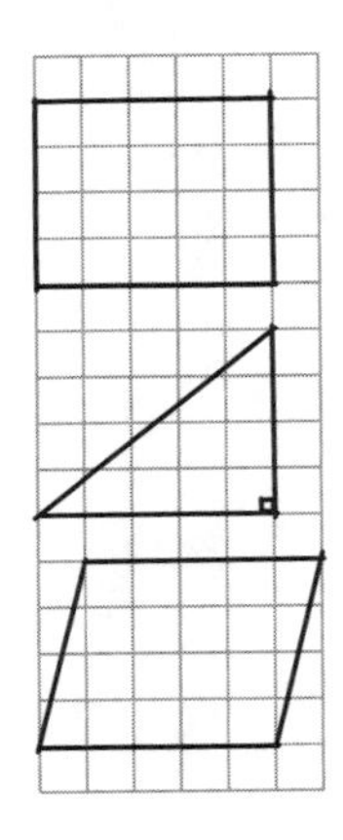

Note: *Ensure that students using the open-ended intervention are shown the conventional formulas for areas of parallelograms and triangles once they have completed their work. Help students relate these formulas to their own.*

Open-Ended Intervention

Before Using the Open-Ended Intervention

Have students sketch a rectangle with an area of 20 square units on grid paper. Then have them sketch and cut out 2 copies of a right triangle with the same base and height. (See margin art.) Ask:

- What would the area of this right triangle be? Why? (*It would be 10; I can use 2 of them to cover the rectangle, so each is half of 20.*)
- Why was it useful to compare the area of the triangle to the area of the rectangle? (e.g., *It's easy to get the area of the rectangle but not as easy for the triangle.*)

Help students sketch a parallelogram by shifting the top of the rectangle 1 unit. (See margin art.) Ask:

- How can you tell that the area of the parallelogram is the same as the area of the rectangle? (e.g., *The parallelogram is made up of a rectangle* (4×4) *and 2 triangles (each half of* 1×4*).* $16 + 2 + 2 = 20$)

Using the Open-Ended Intervention (Student Resource pages 341–342)

Read through the tasks on the student pages together. Provide extra centimetre grid paper if students require more space to sketch their shapes.

Give students time to work, ideally in pairs.

Observe whether students

- can develop and explain the formulas for areas of parallelograms and triangles
- realize that different parallelograms or triangles can have the same area
- can apply their formulas for areas of parallelograms and triangles

Consolidating and Reflecting

Ensure understanding by asking questions such as these based on students' work:

- What did you notice about how the area of the parallelogram relates to its base length? (e.g., *First I figured out the areas by counting squares. Then I noticed that if you multiplied the base by the height of the parallelogram, you got the number of square centimetres in the area.*)
- What did you notice about how the area of the right triangle relates to its base length? (e.g., *I made it half a rectangle, and I saw that you can multiply the base by the height, and then divide by 2 to get the number of square centimetres in the area.*) What about when the triangle was not a right triangle? (e.g., *When the triangle was not a right triangle, I built a rectangle around it and saw it was half the rectangle, so it was just like the right triangle, really.*)

Show a diagram like the bottom one in the margin. Then ask:

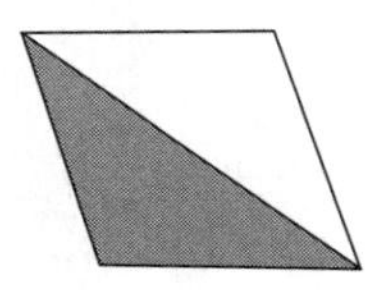

- How does this picture help you make sense of how the formula for the area of a triangle relates to the formula for the area of a parallelogram? (e.g., *The triangle is half of a parallelogram with the same base and same height.*)

 Copyright © 2012 by Nelson Education Ltd.

Area of Parallelograms and Triangles

Guided Intervention

Before Using the Guided Intervention

Have students sketch a rectangle with an area of 20 square units on grid paper. Then have them sketch and cut out 2 copies of a right triangle with the same base and height. (See margin art.) Ask:

- ▸ What would the area of this right triangle be? Why?
  (*It would be 10; I can use 2 of them to cover the rectangle, so each is half of 20.*)
- ▸ Why was it useful to compare the area of the triangle to the area of the rectangle?
  (e.g., *It's easy to get the area of the rectangle but not as easy for the triangle.*)

You will need

- 1 cm Grid Paper (BLM 12)
- scissors
- rulers
- calculators
- Student Resource pages 343–346

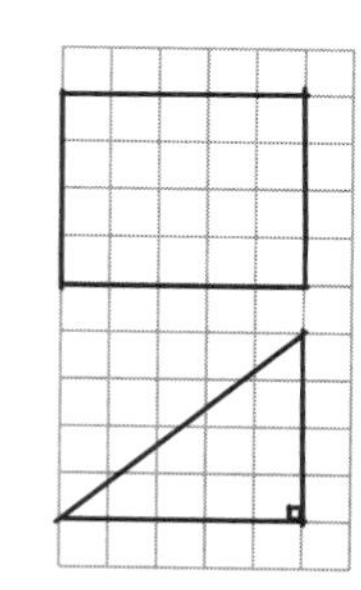

Using the Guided Intervention **Student Resource pages 343–346**

Work through the instructional section of the student pages together. Provide the necessary materials for students to investigate the area formulas.

Have students work through the **Try These** questions individually or in pairs.

Observe whether students

- can apply the formulas for areas of parallelograms and triangles (Questions 1, 2, 8)
- realize that different parallelograms or triangles can have the same area (Questions 3, 5, 6, 7)
- realize that any base–height combination can be used to determine area (Question 4)
- understand the formula for the area of a parallelogram (Question 9)

Consolidating and Reflecting

Ensure understanding by asking questions such as these based on students' work:

- ▸ How did you use the base and height of the parallelogram to determine its area in Question 2?
  (e.g., *I multiplied base by height to get the number of square centimetres for the area of the parallelograms.*)
- ▸ How did you use the base and height of the triangle to determine its area?
  (e.g., *I multiplied base by height and then divided by 2.*)
- ▸ Why might it be useful to know that you can use any base–height combination to calculate the area of a parallelogram or triangle?
  (e.g., *Maybe the numbers for one combination are a lot easier to use, or, as in Question 4, it might be nicer if the height were inside the triangle.*)
- ▸ In Question 6, why does it make sense that triangles with different bases and heights can have the same area?
  (e.g., *You could make the first one narrower, which would make the area less, but then you could make it taller, to make up the difference.*)

Copyright © 2012 by Nelson Education Ltd.

Pathway 5 OPEN-ENDED

Area and Perimeter of Rectangles

You will need

- 1 cm Grid Paper (BLM 12)
- string (about 25 cm)
- calculators (optional)
- Student Resource pages 347–348

Open-Ended Intervention

Before Using the Open-Ended Intervention

Shade 3 rectangles on grid paper (2-by-6, 2-by-8, and 3-by-8) and provide a piece of string. Have a student use the string to show the perimeter of one of the rectangles by going completely around the shape. Ask:

- Why did the length and width both affect the perimeter?
  (e.g., *When I went around, I used the length twice and the width twice.*)
- How can you use the 2 smaller rectangles (2-by-6 and 2-by-8) to show that the length affects the area?
  (e.g., *There are 4 extra squares when the length is longer.*)
- How can you use the 2 larger rectangles (2-by-8 and 3-by-8) to show that the width affects the area?
  (e.g., *There are 8 extra squares when there is 1 more unit of width.*)

Note: *Ensure that students using the open-ended intervention eventually see the formulas $A = l \times w$ and $P = 2 \times l + 2 \times w$ for a rectangle (or $P = 2 \times [l + w]$, depending on the direction a student took). Help students relate the formulas to what they did.*

Using the Open-Ended Intervention (Student Resource pages 347–348)

Read through the tasks on the student pages together. Provide centimetre grid paper, and provide calculators if students need them.

Give students time to work, ideally in pairs.

Observe students as they work, and use questions as necessary to see whether they

- can determine areas and perimeters of rectangles on a grid
- recognize when perimeter might not change even though dimensions do
- can articulate an understanding of the formulas for area and perimeter of rectangles

Consolidating and Reflecting

Ensure understanding by asking questions such as these based on students' work:

- What bread slice sizes did you use? (e.g., *a 13-by-13 slice and a 12-by-14 slice*)
- What does the perimeter tell you about the slice? What does the area tell you? (e.g., *The perimeter tells the crust length. The area tells how much space there is to spread butter on.*)
- What happened? (*The area got slightly smaller, but the perimeter did not change.*)
- Why didn't the perimeter change?
  (e.g., *If you make the width 1 less and the length 1 more, you are still adding the same total twice.*)
- What was your quick way to calculate area? (*Multiply length by width.*)
- Why might you write $A = l \times w$? (e.g., *It says the same thing but it's shorter.*)
- What was your quick way to calculate perimeter?
  (e.g., *I added the length and width, and then I doubled it.*)
- Why is it useful to have formulas for area and perimeter?
  (e.g., *so that you can calculate measurements you might need, without drawing*)

 Copyright © 2012 by Nelson Education Ltd.

Area and Perimeter of Rectangles

Pathway 5
GUIDED

Guided Intervention

Before Using the Guided Intervention

Shade 3 rectangles on grid paper (2-by-6, 2-by-8, and 3-by-8) and provide a piece of string. Have a student use the string to show the perimeter of one of the rectangles by going completely around the shape. Ask:

- ▸ Why did the length and width both affect the perimeter?
  (e.g., *When I went around, I used the length twice and the width twice.*)
- ▸ How can you use the 2 smaller rectangles (2-by-6 and 2-by-8) to show that the length affects the area? (e.g., *There are 4 extra squares when the length is longer.*)
- ▸ How can you use the 2 larger rectangles (2-by-8 and 3-by-8) to show that the width affects the area?
  (e.g., *There are 8 extra squares when there is 1 more unit of width.*)

You will need

- 1 cm Grid Paper (BLM 12)
- string (about 25 cm)
- rulers
- calculators (optional)
- Student Resource pages 349–352

Using the Guided Intervention **Student Resource pages 349–352**

Work through the instructional section of the student pages together. Provide centimetre grid paper and rulers for students to investigate area and perimeter. You might also provide calculators if students need them.

Have students work through the **Try These** questions individually or in pairs.

Observe whether students

- can determine areas and perimeters of rectangles (Questions 1, 2, 3, 4)
- can create rectangles with a given area or perimeter (Questions 5, 6)
- can describe situations where knowing area or perimeter is useful (Questions 7, 8)
- recognize when perimeter might not change, even though dimensions do (Question 9)
- can articulate an understanding of the formulas for area and perimeter of rectangles (Question 10)

Consolidating and Reflecting

Ensure understanding by asking questions such as these based on students' work:

- ▸ Why were you able to make 2 different rectangles with the same perimeter in Question 5? (e.g., *If you make the width a little less and the length the same amount more, you are still adding the same total twice.*)
- ▸ Why were you able to make 2 different rectangles with the same area in Question 6?
  (e.g., *You can find 2 different numbers that multiply to the same amount.*)
- ▸ What helped you think of ideas for Questions 7 and 8? (e.g., *I remembered that perimeter is about distance around and area is about space inside.*)
- ▸ Why is it useful to have formulas for area and perimeter?
  (e.g., *so that you can calculate measurements you might need, without drawing*)

Strand: Measurement

Volume and Surface Area

Planning For This Topic

Materials for assisting students with measurement of 3-D shapes consist of a diagnostic tool and 3 intervention pathways. Pathway 1 focuses on volume formulas for right prisms. Pathway 2 concentrates on surface area of right prisms. Pathway 3 deals with volume of right rectangular prisms.

Note: Throughout this resource, the term *prism* is associated with what is technically a *right prism*, which is a prism where the lateral faces are rectangles and the 2 congruent bases are directly opposite one another.

Each pathway has an open-ended intervention and a guided intervention. Choose the type of intervention more suitable for your students' needs and your particular circumstances.

Curriculum Connections

Grades 5 to 8 curriculum connections for this topic are provided online. See www.nelson.com/leapsandbounds. Pathway 1 and 2 topics are not covered in Grades 7 or below in the WNCP curriculum, so you may choose to omit them. Pathway 3 is relevant to both WNCP and Ontario curriculums.

Why might students struggle with volume or surface area?

Students might struggle with volume or surface area for any of the following reasons:

- They might mix up the terms *volume* and *surface area.*
- They might have difficulty identifying the base of a polygonal prism.
- They might not recognize that it is the area of the polygonal base and not the side lengths of the base that affect the volume of a non-rectangular prism.
- They might not remember how to calculate the area of non-rectangular bases.
- They might have difficulty visualizing the 3-D shape from a 2-D picture.
- They might confuse the height of the 2-D base with the height of the 3-D prism.
- They might not include all the faces when calculating surface area.
- They might assume that all of the rectangular faces of a prism have the same area when that might not be the case.

Professional Learning Connections

PRIME: *Measurement, Background and Strategies* (Nelson Education Ltd., 2010), pages 72–73, 84–91

Making Math Meaningful to Canadian Students K–8 (Nelson Education Ltd., 2008), pages 421–429

Big Ideas from Dr. Small Grades 4–8 (Nelson Education Ltd., 2009), pages 150–151, 153–158

Good Questions (dist. by Nelson Education Ltd., 2009), pages 103–104, 118–119

More Good Questions (dist. by Nelson Education Ltd., 2010), pages 127, 130

 Copyright © 2012 by Nelson Education Ltd.

Diagnostic Tool: Volume and Surface Area

Use the diagnostic tool to determine the most suitable intervention pathway for volume and surface area. Provide Diagnostic Tool: Volume and Surface Area, Teacher's Resource pages 272 to 274, and have students complete it in writing or orally. Provide centimetre cubes, rulers, 1 cm Grid Paper (BLM 12), models of a variety of prisms, and calculators. If you are using the WNCP curriculum, use only the questions related to Pathway 3.

See solutions on Teacher's Resource pages 275 to 277.

Intervention Pathways

The purpose of the intervention pathways is to help students see how measurements of shapes are interrelated. These concepts will be developed further as students work with volume and surface area of other shapes such as cylinders, cones, and pyramids.

There are 3 pathways:
- Pathway 1: Volume of Prisms: Using a Formula
- Pathway 2: Surface Area of Prisms
- Pathway 3: Volume of Rectangular Prisms

Use the chart below (or the Key to Pathways on Teacher's Resource pages 275 to 277) to determine which pathway is most suitable for each student or group of students.

Diagnostic Tool Results	Intervention Pathway
If students struggle with Questions 7 to 9	use Pathway 1: Volume of Prisms: Using a Formula *Teacher's Resource pages 278–279* *Student Resource pages 353–358*
If students struggle with Questions 4 to 6	use Pathway 2: Surface Area of Prisms *Teacher's Resource pages 280–281* *Student Resource pages 359–364*
If students struggle with Questions 1 to 3	use Pathway 3: Volume of Rectangular Prisms *Teacher's Resource pages 282–283* *Student Resource pages 365–370*

Name: ______________________ Date: ______________

Volume and Surface Area

Diagnostic Tool

You will need

- centimetre cubes
- a calculator
- a ruler (optional)
- 1 cm Grid Paper (BLM 12, optional)
- models of a variety of prisms (optional)

1. What is the volume of each prism?

a)

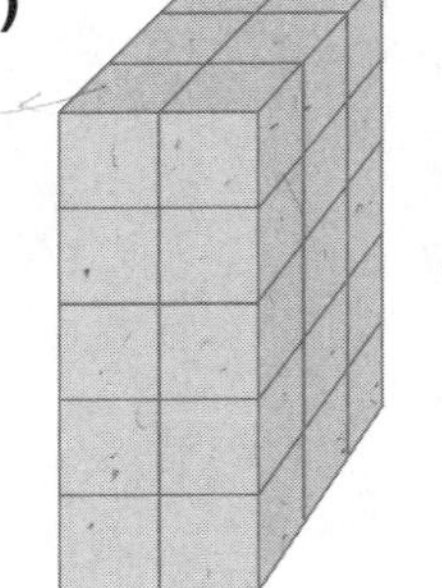

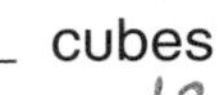

__________ cubes

c)

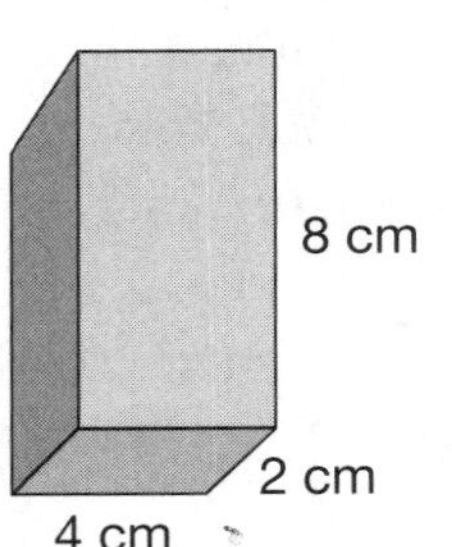

____

b)

__________ cubes

d)

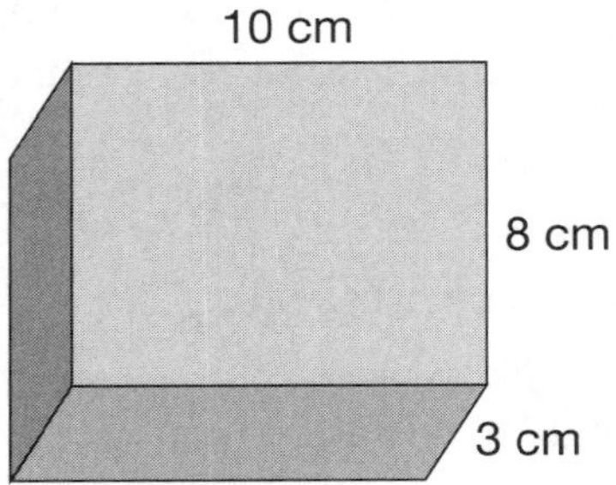

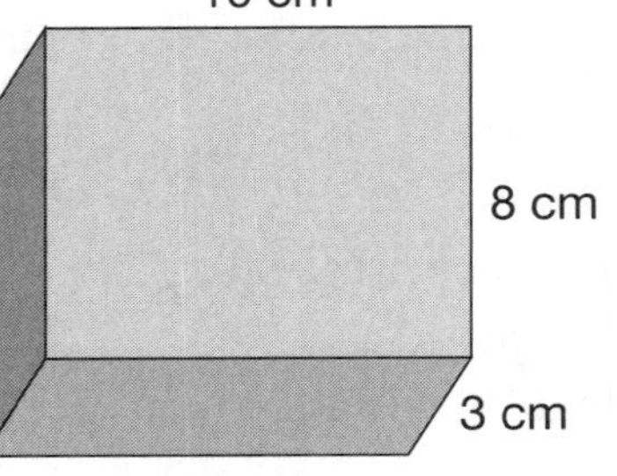

____

2. Sketch 2 different rectangular prisms so that the volume of each is 32 cm^3. Label the dimensions. (You can use centimetre cubes to model the prisms.)

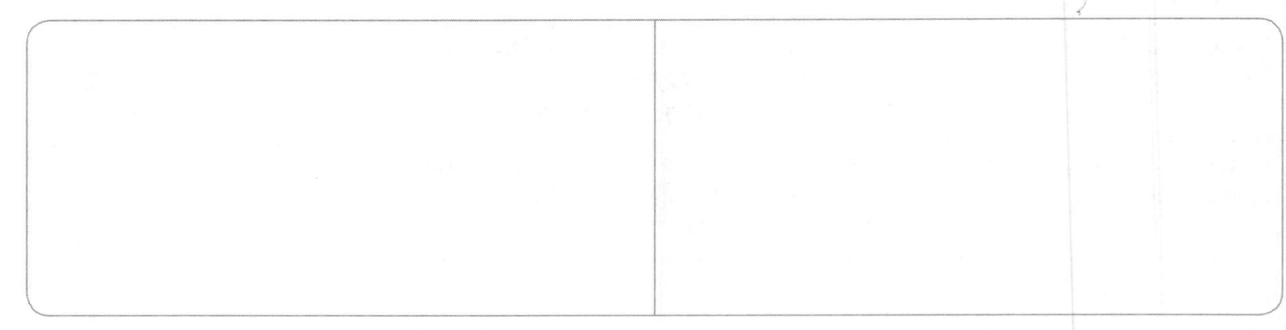

3. Why is it useful to know the length, width, and height when calculating the volume of a rectangular prism?

____

____

 Copyright © 2012 by Nelson Education Ltd.

Name:______________________ Date:______________

4. What is the surface area (the total area of the 6 faces) of the rectangular prism made from this net? Use square units.

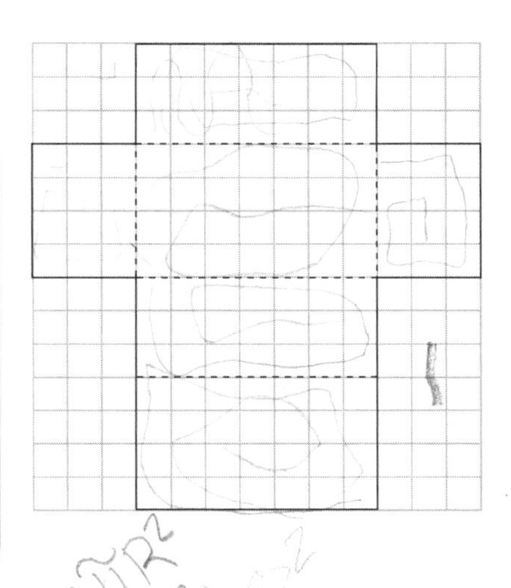

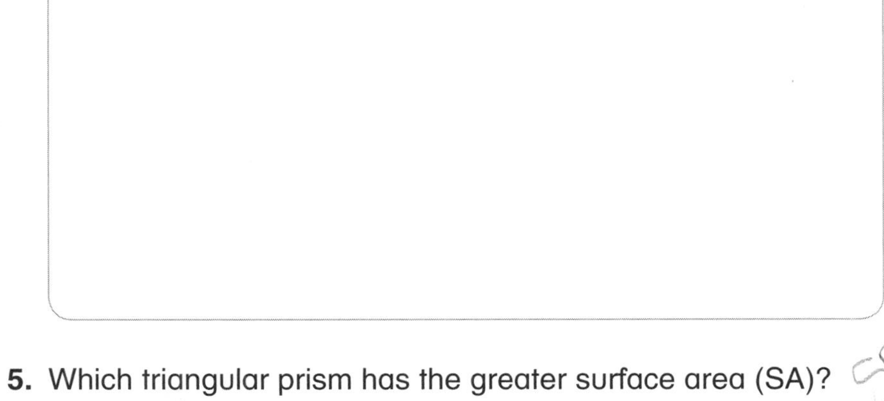

5. Which triangular prism has the greater surface area (SA)? How much greater is that surface area?

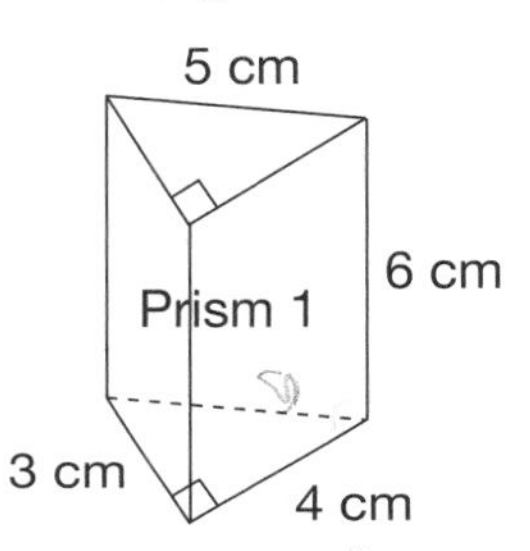

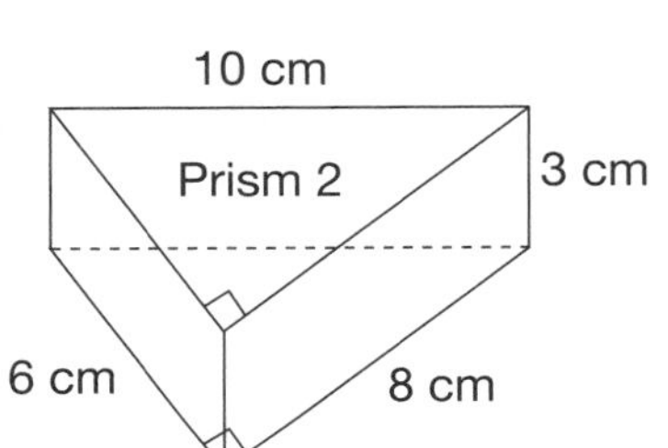

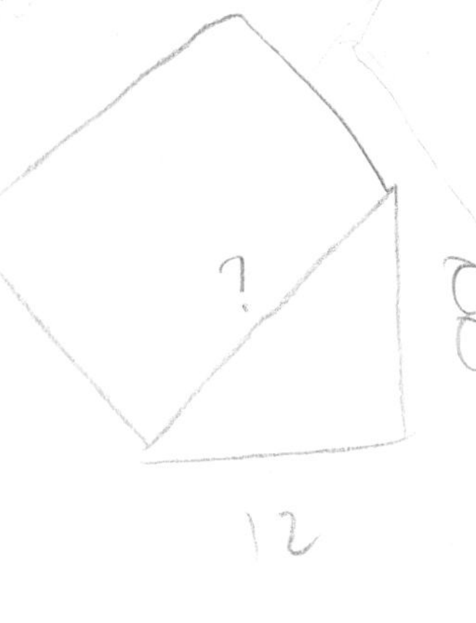

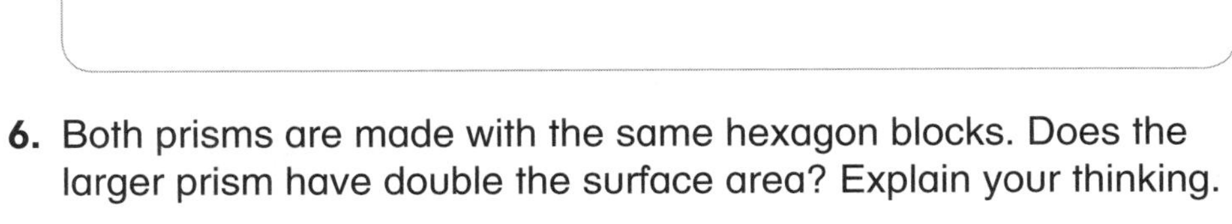

6. Both prisms are made with the same hexagon blocks. Does the larger prism have double the surface area? Explain your thinking.

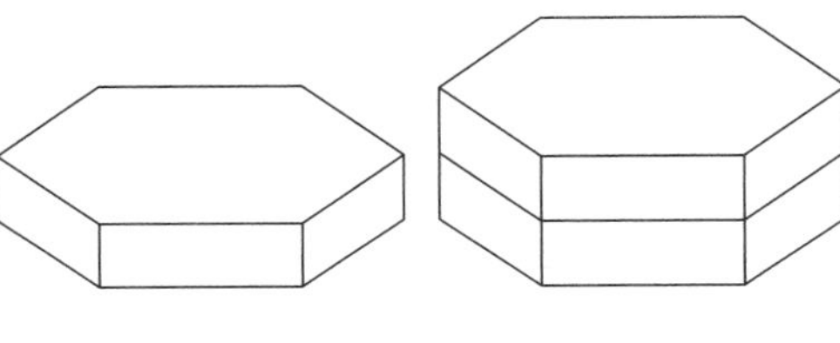

____

____

Copyright © 2012 by Nelson Education Ltd.

Name:________________________________ Date:________________

7. The shapes below are prisms. Calculate the volume of each.

a)

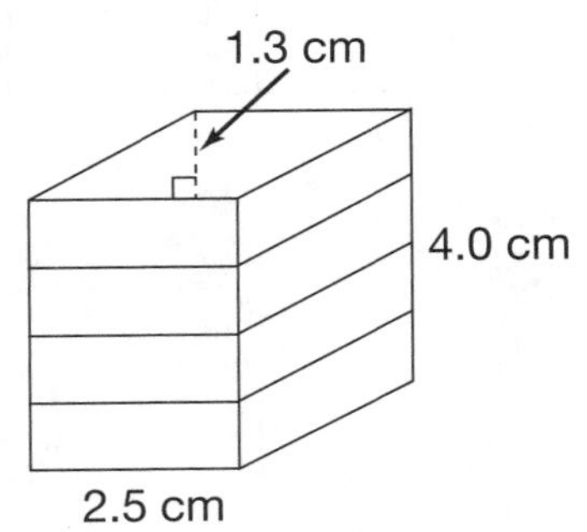

b)

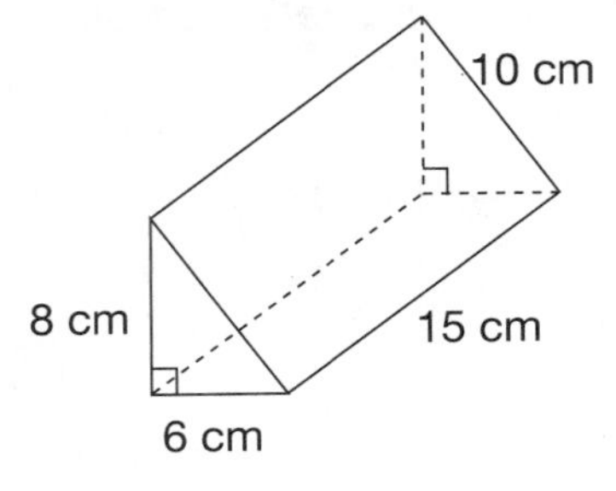

Remember

- The volume of a prism is the product of the area of its base and height.

8. Marty has 2 pads of notepaper in the shape of prisms. Each has a volume of 200 cm^3 and a height of 10 cm. The bases are not rectangular. Sketch 2 possible drawings for the bases and label the dimensions.

9. Two different parallelogram-based prisms have the same volume.

   a) Sketch 2 possible drawings for the bases and label the dimensions. Calculate the area of each base.

   b) Figure out the height of each prism so that they have the same volume.

sketch of base 1:	sketch of base 2:
area of base:	area of base:
height of prism:	height of prism:

 Copyright © 2012 by Nelson Education Ltd.

Solutions and Key to Pathways

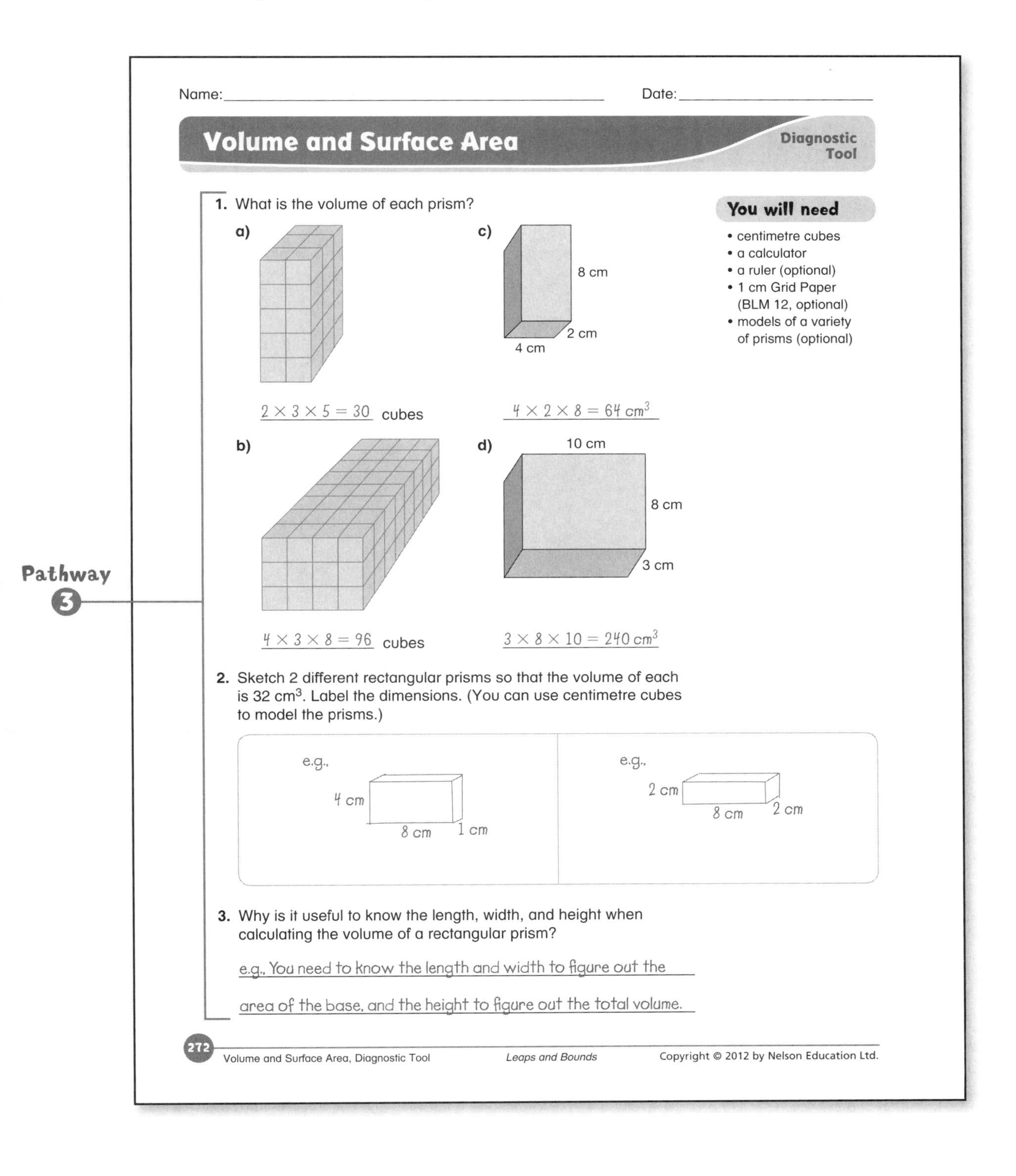

Name:______________________ Date:____________

Volume and Surface Area

Diagnostic Tool

You will need

- centimetre cubes
- a calculator
- a ruler (optional)
- 1 cm Grid Paper (BLM 12, optional)
- models of a variety of prisms (optional)

Pathway 3

1. What is the volume of each prism?

a) $2 \times 3 \times 5 = 30$ cubes

b) $4 \times 3 \times 8 = 96$ cubes

c) 8 cm, 2 cm, 4 cm — $4 \times 2 \times 8 = 64$ cm^3

d) 10 cm, 8 cm, 3 cm — $3 \times 8 \times 10 = 240$ cm^3

2. Sketch 2 different rectangular prisms so that the volume of each is 32 cm^3. Label the dimensions. (You can use centimetre cubes to model the prisms.)

e.g., 4 cm, 8 cm, 1 cm

e.g., 2 cm, 8 cm, 2 cm

3. Why is it useful to know the length, width, and height when calculating the volume of a rectangular prism?

e.g., You need to know the length and width to figure out the area of the base, and the height to figure out the total volume.

272 Volume and Surface Area, Diagnostic Tool *Leaps and Bounds* Copyright © 2012 by Nelson Education Ltd.

Name: ______________________ Date: ______________

4. What is the surface area (the total area of the 6 faces) of the rectangular prism made from this net? Use square units.

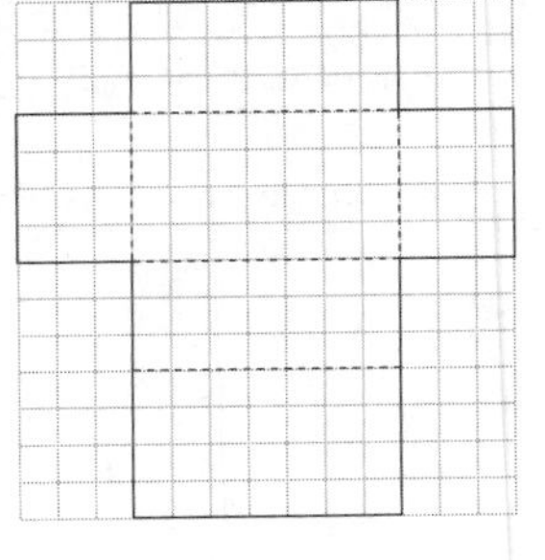

e.g., 2 × area of large rectangle
+ 2 × area of medium rectangle
+ 2 × area of small rectangle
$= 2 \times 7 \times 4 + 2 \times 7 \times 3 + 2 \times 3 \times 4$
$= 56 + 42 + 24$
$= 122$ square units

5. Which triangular prism has the greater surface area (SA)? How much greater is that surface area?

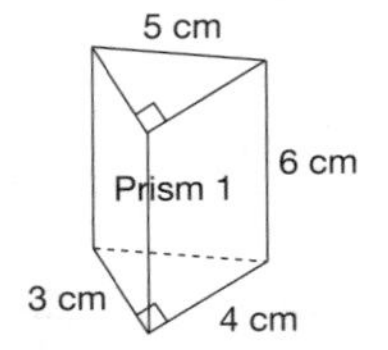

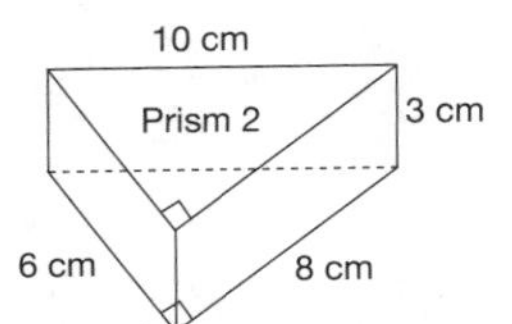

SA of Prism 1: $2 \times 3 \times 4 \div 2 + 3 \times 6 + 5 \times 6 + 4 \times 6$
$= 12 + 18 + 30 + 24$
$= 84\ cm^2$

SA of Prism 2: $2 \times 6 \times 8 \div 2 + 6 \times 3 + 10 \times 3 + 8 \times 3$
$= 48 + 18 + 30 + 24$
$= 120\ cm^2$

$120 - 84 = 36\ cm^2$

SA of Prism 2 is greater by $36\ cm^2$.

Pathway 2

6. Both prisms are made with the same hexagon blocks. Does the larger prism have double the surface area? Explain your thinking.

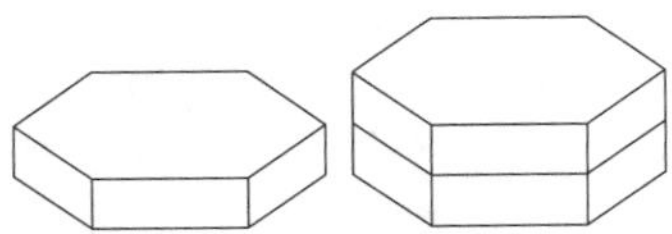

No, the side faces are double in area, but the top and bottom are not.

Copyright © 2012 by Nelson Education Ltd.

 Copyright © 2012 by Nelson Education Ltd.

Name:________________________ Date:____________

7. The shapes below are prisms. Calculate the volume of each.

a)

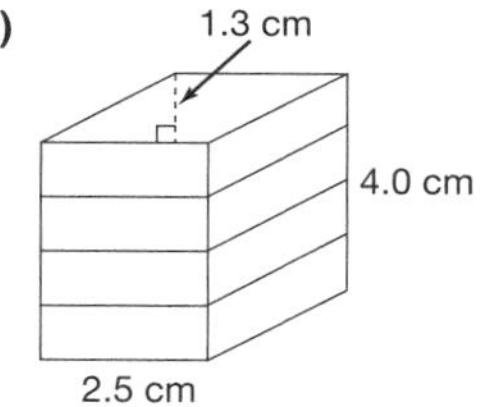

b)

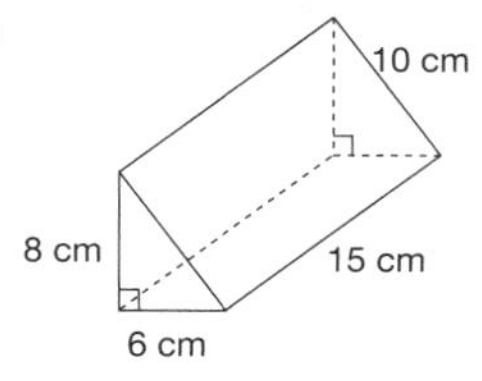

Remember
- The volume of a prism is the product of the area of its base and height.

V = area of base × height
$= 2.5 \times 1.3 \times 4$
$= 13 \text{ cm}^3$

V = area of base × height
$= 6 \times 8 \div 2 \times 15$
$= 360 \text{ cm}^3$

8. Marty has 2 pads of notepaper in the shape of prisms. Each has a volume of 200 cm^3 and a height of 10 cm. The bases are not rectangular. Sketch 2 possible drawings for the bases and label the dimensions.

e.g.,

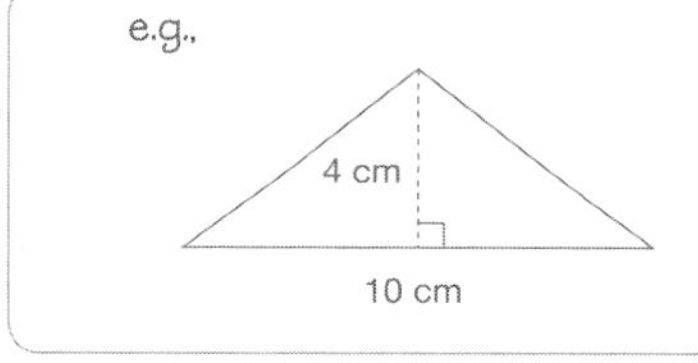

e.g.,

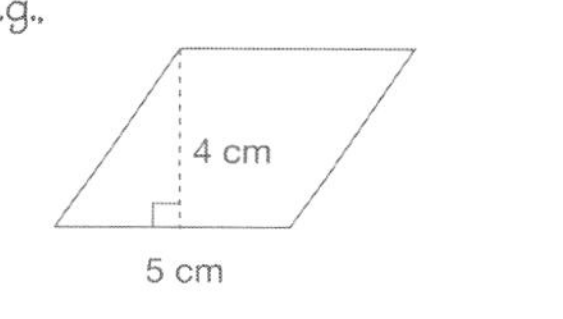

Pathway 1

9. Two different parallelogram-based prisms have the same volume.

a) Sketch 2 possible drawings for the bases and label the dimensions. Calculate the area of each base.
b) Figure out the height of each prism so that they have the same volume.

sketch of base 1:	sketch of base 2:
e.g., 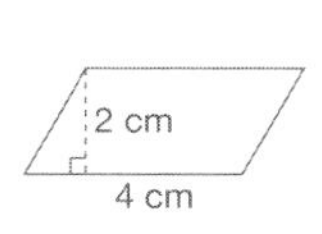	e.g., 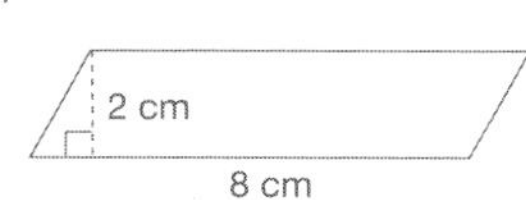
area of base: $4 \times 2 = 8 \text{ cm}^2$	area of base: $8 \times 2 = 16 \text{ cm}^2$
height of prism: 6 cm	height of prism: 3 cm

 Copyright © 2012 by Nelson Education Ltd.

Copyright © 2012 by Nelson Education Ltd.

Pathway 1 OPEN-ENDED

Volume of Prisms: Using a Formula

You will need

- linking cubes
- a camera (optional)
- pattern blocks
- centimetre cubes
- rulers
- calculators
- Triangle Dot Paper (BLM 16, optional)
- Student Resource pages 353–354

Open-Ended Intervention

Before Using the Open-Ended Intervention

Provide linking cubes. Ask:

- How can you build a rectangular prism with a volume of 10 cubic units? (e.g., *I would have 2 layers of rectangles that are 1 by 5.*)
- How do you know the volume is 10? (e.g., *I used 10 cubes.*)

Make a 4-by-3-by-3 prism and show only the 4-by-3 base. Ask:

- If I show you only the base and tell you the number of layers, could you still figure out the volume?
  (e.g., *Yes, I know how many cubes are in each layer, so I could just multiply.*)

Using the Open-Ended Intervention Student Resource pages 353–354

Read through the tasks on the student pages together. Provide the necessary materials.

Give students time to work, ideally in pairs. Students could use a camera to take pictures of their prisms, rather than sketch them. In Part B, they can sketch the area of the base of a pattern block on Triangle Dot Paper (BLM 16) and state the number of layers, rather than sketch the towers of pattern blocks.

Observe whether students

- can calculate the area of the base of a prism and identify the height
- recognize that the volume of a prism is the product of the area of its base and its height

Note: Ensure that students using the open-ended intervention eventually see the formula for the volume of a prism, Volume = area of base × height or $V = A_{base} \times h$.

Consolidating and Reflecting

Ensure understanding by asking questions such as these based on students' work:

- Why did it make sense to measure volume in units that were cubes in Part A, but triangle block units in Part B?
  (e.g., *The shapes in Part A were made of cubes, but the shapes in Part B were easier to build with triangle blocks than cubes.*)
- What did you discover about the formula for the volume of any prism?
  (e.g., *You multiply the area of the base by the height of the prism.*)
- How did you calculate the area of your trapezoid base? (e.g., *I know the area of a green triangle, so I multiplied that by 3, since 3 triangles make a red trapezoid.*)
- Why did it make sense that your hexagon prism had 4 times the volume of your trapezoid prism? (e.g., *The hexagon area is double the trapezoid area, and there were twice as many layers, so I had to double twice.*)

 Copyright © 2012 by Nelson Education Ltd.

Volume of Prisms: Using a Formula

Pathway 1 GUIDED

Guided Intervention

You will need

- linking cubes
- pattern blocks
- rulers
- calculators
- Student Resource pages 355–358

Before Using the Guided Intervention

Provide linking cubes. Ask:

- How can you build a rectangular prism with a volume of 10 cubic units? (e.g., *I would have 2 layers of rectangles that are 1 by 5.*)
- How do you know the volume is 10? (e.g., *I used 10 cubes.*)

Make a 4-by-3-by-3 prism and show only the 4-by-3 base. Ask:

- If I show you only the base and tell you the number of layers, could you still figure out the volume? (e.g., *Yes, I know how many cubes are in each layer, so I could just multiply.*)

Using the Guided Intervention (Student Resource pages 355–358)

Provide the materials. Work through the instructional section of the student pages together. Encourage students to model the prisms using pattern blocks.

Have students work through the **Try These** questions individually or in pairs.

Observe whether students

- can calculate the area of the base of a prism and identify the height (Questions 1, 2, 3)
- recognize that the volume of a prism is the product of the area of its base and its height (Questions 1, 2, 3, 5, 7, 8, 9)
- can relate the volumes of 2 prisms (Question 3)
- can estimate the height of the base and the height of a prism (Question 4)
- recognize that different prisms can have the same volume (Questions 5, 9)
- choose which units to use appropriately (Questions 6, 7)

Consolidating and Reflecting

Ensure understanding by asking questions such as these based on students' work:

- In Question 1, why did it make sense that your hexagon-based prism had the same volume as your trapezoid-based prism? (e.g., *The hexagon area is double the trapezoid area but there are half as many layers. So you multiply by 2 but then divide by 2, and it has no effect.*)
- Why did it make sense in Questions 5 and 9 that 2 different prisms can have the same volume? (e.g., *You could switch the values for the height and the area of the base.*)
- Suppose some of your length measurements are given in centimetres and some in metres. What would you do when calculating volume? (e.g., *You could either convert the centimetres to metres or the metres to centimetres; otherwise, the number you get wouldn't mean anything.*)
- Why is it useful to know the formula for the volume of a prism? (e.g., *It's quicker to multiply some numbers than to count cubic units every time.*)

Pathway 2 OPEN-ENDED

Surface Area of Prisms

You will need

- linking cubes
- a transparent centimetre grid (e.g., photocopy BLM 12 on a transparency)
- prisms from a set of geometric solids
- connecting faces (e.g., Polydrons or cardboard polygons)
- rulers
- calculators
- 1 cm Grid Paper (BLM 12, optional)
- Student Resource pages 359–360

Open-Ended Intervention

Before Using the Open-Ended Intervention

Show a 3-by-5-by-2 prism made of linking cubes and provide a transparent centimetre grid. Ask:

- ▸ Which face is 3 cubes by 5 cubes? (*the 3-by-5 face*)
- ▸ How would you get the area of the other faces in linking cube square units? (e.g., *I remember that the area of a rectangle is the product of its length and width.*)
- ▸ What if you want the area in square centimetres? (e.g., *Figure out the area of 1 face of a linking cube in square centimetres and multiply by that number.*)

Using the Open-Ended Intervention Student Resource pages 359–360

Read through the tasks on the student pages together. If students are unsure about what surface area is, they might read through the instructional section of the guided intervention first. Provide the necessary materials. Students might use connecting faces or centimetre grid paper to create nets, and/or sketch what the prisms look like in 3-D. Their sketches may not look accurate, but the dimensions should be labelled correctly.

Give students time to work, ideally in pairs.

Observe whether students

- can calculate the area of the bases and lateral (side) faces of a prism
- recognize that the surface area of a prism is the sum of the areas of all faces
- recognize that areas can be added more efficiently by doubling, tripling, and so on, the area of one face when possible
- can determine some dimensions of a prism when the surface area is known

Consolidating and Reflecting

Ensure understanding by asking questions such as these based on students' work:

- ▸ How did you decide how many numbers you needed to add to calculate the surface area? (e.g., *I counted the faces.*)
- ▸ How did you decide whether you had to add a lot of individual numbers or whether you could use multiplication to make things quicker? (e.g., *If you have equal faces, you can double or triple the area. You can always double the area of one base to get the area of both bases.*)
- ▸ How could you have known that the length, width, and height could not be too big? (e.g., *200 is the sum of at least 5 faces and so none can be all that big, since if one is big, there are probably other big areas, too.*)
- ▸ What strategies did you use to solve the problem? (e.g., *I chose a height first and didn't make it too big. I realized that I would be multiplying that height by each side length of the base and adding them without using the top and bottom base. So I tried a few numbers and then adjusted if it was too big or too small.*)

 Copyright © 2012 by Nelson Education Ltd.

Surface Area of Prisms

Pathway 2 GUIDED

Guided Intervention

Before Using the Guided Intervention

Show a 3-by-5-by-2 rectangular prism made of linking cubes and provide a transparent centimetre grid. Ask:

- Which face is 3 cubes by 5 cubes? (*the 3-by-5 face*)
- How would you get the area of the other faces in linking cube square units? (e.g., *I remember that the area of a rectangle is the product of its length and width.*)
- What if you wanted the area in square centimetres? (e.g., *I would have to figure out the area of one face of each linking cube in square centimetres and then multiply by that number.*)

Using the Guided Intervention (Student Resource pages 361–364)

Work through the instructional section of the student pages together. Help students see that in a rectangular prism, opposite sides are identical rectangles, so any of them could be the base if you turned the shape.

Have students work through the **Try These** questions individually or in pairs. Provide tape measures or metre sticks for Question 5 and possibly linking cubes for Question 9.

Observe whether students

- can calculate the area of the bases and lateral (side) faces of a prism (Questions 1, 2)
- recognize that the surface area of a prism is the sum of the areas of all faces (Questions 3 to 9)
- use multiplication to add face areas efficiently (Questions 6, 7)
- realize that shapes with the same volume can have different surface areas (Question 8)
- can determine some dimensions of a prism when the surface area is known (Question 9)

Consolidating and Reflecting

Ensure understanding by asking questions such as these based on students' work:

- How did you decide how many numbers to add to calculate the surface areas in Question 3? (e.g., *I counted the faces.*)
- How did you figure out your answer to Question 7a)? (e.g., *If there are 6 faces with the same area, you must have had a cube.*)
- Why do you think the longer prism had the greater surface area in Question 8? (e.g., *There are some really long faces that make the surface area big.*)
- What strategies did you use to solve Question 9? (e.g., *I knew the numbers couldn't be that big, so I just tried combinations of small numbers. I knew that I had to add 3 areas and double them to get the total.*)

You will need

- linking cubes
- a transparent centimetre grid (e.g., photocopy BLM 12 on a transparency)
- prisms from a set of geometric solids
- calculators
- a tape measure or a metre stick
- connecting faces (e.g., Polydrons or cardboard polygons, optional)
- Student Resource pages 361–364

Copyright © 2012 by Nelson Education Ltd.

Volume of Rectangular Prisms

You will need

- 50 linking cubes (2 cm on each side)
- centimetre cubes
- calculators
- 3 classroom objects in the shape of rectangular prisms
- rulers
- a camera (optional)
- Student Resource pages 365–366

Open-Ended Intervention

Before Using the Open-Ended Intervention

Provide linking cubes and centimetre cubes. Make a 1-layer prism with 4 rows of 3 linking cubes. Ask:

- Why would you say the volume of your prism is 12 cubic units? (e.g., *It took 12 cubes to make it.*)
- Why would it take more centimetre cubes than linking cubes to build this size of prism? (e.g., *The linking cubes are bigger.*)
- Does that mean the volume has changed? (e.g., *No, it's just different units.*)
- What other prism could you build with 12 linking cubes? (e.g., *a 2-layer prism where each layer is 3 by 2*)

Using the Open-Ended Intervention (Student Resource pages 365–366)

Read through the tasks on the student pages together. Provide the necessary materials. Students might sketch their prisms on paper and label the dimensions, or use a camera to take pictures of them.

Give students time to work, ideally in pairs.

Observe whether students

- recognize that the volume of a prism is the product of the area of its base and its height
- notice that there are alternative possible bases and that the height used must be relative to the base used
- can calculate dimensions of a rectangular prism, given the volume
- recognize that different rectangular prisms can have the same volume

Note: Ensure that students using the open-ended intervention eventually see the formula Volume = area of base × height for the volume of a prism. Also, the formula can be shown as $V = A_{base} \times h$.

Consolidating and Reflecting

Ensure understanding by asking questions such as these based on students' work:

- How did you make sure you didn't use more than 50 cubes? (e.g., *I built 2 small prisms and then figured out how many cubes I had left. I tried to figure out what prisms I could build with that many, or almost that many, cubes.*)
- Did you always count all the cubes, or did you use shortcuts? (e.g., *I used shortcuts. Once I had one layer, I just multiplied by the number of layers.*)
- Why did making linking-cube prisms to match the sizes of the classroom objects help? (e.g., *It took too many centimetre cubes to build the objects, so I just used the measurements and calculated. But using the linking cubes helped me be sure.*)
- What did you discover about the formula for the volume of a prism? (e.g., *You multiply the area of the base by the height of the prism.*)

 Copyright © 2012 by Nelson Education Ltd.

Volume of Rectangular Prisms

Pathway 3
GUIDED

Guided Intervention

Before Using the Guided Intervention

Provide linking cubes and centimetre cubes. Make a 1-layer prism with 4 rows of 3 linking cubes. Ask:

- Why would you say the volume of your prism is 12 cubic units? (e.g., *It took 12 cubes to make it.*)
- Why would it take more centimetre cubes than linking cubes to build this size of prism? (e.g., *The linking cubes are bigger.*)
- Does that mean the volume has changed? (e.g., *No, it's just different units.*)
- What other prism could you build with 12 linking cubes? (e.g., *a 2-layer prism where each layer is 3 by 2*)

You will need

- linking cubes (2 cm on each side)
- centimetre cubes
- calculators
- rulers
- Student Resource pages 367–370

Using the Guided Intervention Student Resource pages 367–370

Work through the instructional section of the student pages together. (If students use linking cubes to model a stack of books, help them see that a linking cube has the same volume as a 2-by-2-by-2 prism of centimetre cubes, so its volume is 8 cm^3.)

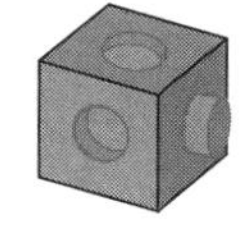
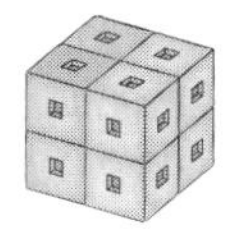

Have students work through the **Try These** questions individually or in pairs.

Observe whether students

- can calculate the area of the base of a prism and identify the height of the prism (Question 1)
- recognize that the volume of a prism is the product of the area of its base and its height (Questions 1 to 9)
- recognize that alternative bases can be used and the height is relative to the base used (Question 2)
- can calculate dimensions of a rectangular prism, given the volume (Questions 3, 4, 8)

Consolidating and Reflecting

Ensure understanding by asking questions such as these based on students' work:

- The formula says that you multiply the height by the base area to get the volume. What did you learn in Question 2? (e.g., *You can choose your base, but you have to use the height that goes with that base.*)
- Did you always count all the cubes, or did you use shortcuts? (e.g., *I used shortcuts. Once I had 1 layer, I just multiplied by the number of layers.*)
- When you knew the total volume, how did you calculate the length, width, and height? (e.g., *I picked a number that was a factor of the volume to be the height and saw what was left when I divided. Then I found 2 numbers to multiply to get that number.*)

Copyright © 2012 by Nelson Education Ltd.

Angles

Planning For This Topic

Materials for assisting students with angles consist of a diagnostic tool and 3 intervention pathways. Pathway 1 establishes an understanding of the sum of the angles in triangles and in quadrilaterals. Pathway 2 focuses on drawing angles using a protractor. Pathway 3 concentrates on measuring angles.

Each pathway has an open-ended intervention and a guided intervention. Choose the type of intervention more suitable for your students' needs and your particular circumstances.

Professional Learning Connections

PRIME: Measurement, Background and Strategies (Nelson Education Ltd., 2010), pages 119–126
PRIME: Geometry, Background and Strategies (Nelson Education Ltd., 2007), pages 78, 109
Making Math Meaningful to Canadian Students K–8 (Nelson Education Ltd., 2008), pages 295–296, 315, 460–465
Big Ideas from Dr. Small Grades 4–8 (Nelson Education Ltd., 2009), pages 97, 99, 111, 128, 163–165
Good Questions (dist. by Nelson Education Ltd., 2009), page 100
More Good Questions (dist. by Nelson Education Ltd., 2010), page 109

Curriculum Connections

Grades 5 to 8 curriculum connections for this topic are provided online. See www.nelson.com/leapsandbounds. Pathway 1 is appropriate for students working with the WNCP curriculum; the topic is not addressed in Ontario until Grades 8 and 9. Pathway 2 and 3 topics are both addressed in the Ontario and the WNCP curriculums.

Why might students struggle with angles?

Students might struggle with angles for any of the following reasons:

- They might confuse angle measure with the arm length of an angle (e.g., they might mistakenly think that the measure of an angle with longer arms has to be greater than one with shorter arms).
- They might use the wrong scale on a protractor, particularly when measuring an obtuse angle (e.g., measuring a 120° angle as a 60° angle).
- They might not line up the 0° mark properly on the protractor.
- They might have difficulty measuring an angle with arms that are too short to reach the scale on the protractor.
- They might be able to measure an angle but have more difficulty drawing one. They might have particular difficulty with obtuse angles.
- They might not correctly identify all of the interior angles of a polygon.
- They might not recognize that the measures of angles in polygons might not be exact and, therefore, not reach appropriate conclusions about the sums of angles in polygons.
- They might not understand the relationship between the sum of the angles in a quadrilateral and the sum of the angles in a triangle.

 Copyright © 2012 by Nelson Education Ltd.

Diagnostic Tool: Angles

Use the diagnostic tool to determine the most suitable intervention pathway for angles. Provide Diagnostic Tool: Angles, Teacher's Resource pages 286 to 288, and have students complete it in writing or orally. Provide straightedges or rulers, and protractors and coloured pencils. If you are using the Ontario curriculum, use only the questions related to Pathways 2 and 3.

See solutions on Teacher's Resource pages 289 to 291.

Intervention Pathways

The purpose of the intervention pathways is to help students measure and construct angles to 180° and to recognize angle relationships in triangles and quadrilaterals. These concepts will be developed further as students investigate geometric properties of shapes, including angle sums in general polygons.

There are 3 pathways:

- Pathway 1: Sums of Angle Measures in Polygons
- Pathway 2: Drawing Angles
- Pathway 3: Measuring Angles

Use the chart below (or the Key to Pathways on Teacher's Resource pages 289 to 291) to determine which pathway is most suitable for each student or group of students.

Diagnostic Tool Results	Intervention Pathway
If students struggle with Questions 8 to 10	use Pathway 1: Sums of Angle Measures in Polygons *Teacher's Resource pages 292–293* *Student Resource pages 371–375*
If students struggle with Questions 5 to 7	use Pathway 2: Drawing Angles *Teacher's Resource pages 294–295* *Student Resource pages 376–380*
If students struggle with Questions 1 to 4	use Pathway 3: Measuring Angles *Teacher's Resource pages 296–297* *Student Resource pages 381–386*

Copyright © 2012 by Nelson Education Ltd.

Name: ______________________ Date: ______________

Angles

Diagnostic Tool

You will need

- a straightedge or a ruler
- a protractor
- coloured pencils

1. Estimate the size of each angle.

a)

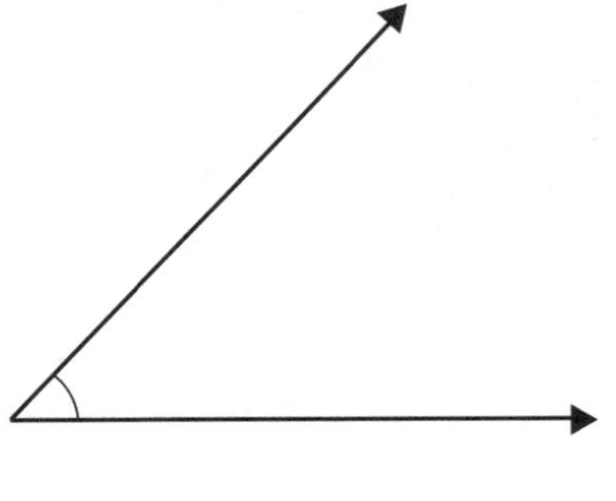

about ________

b)

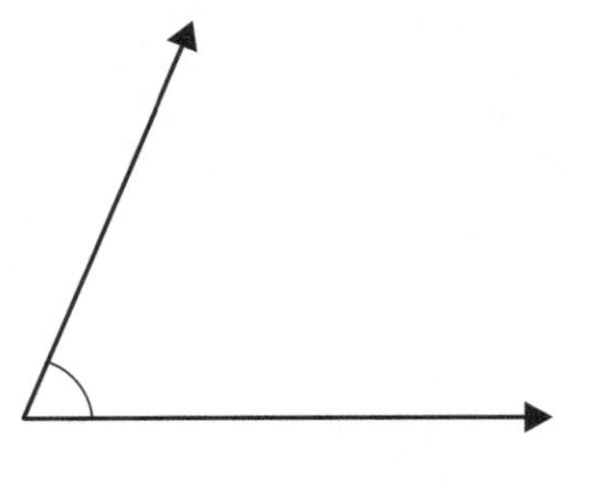

about ________

c)

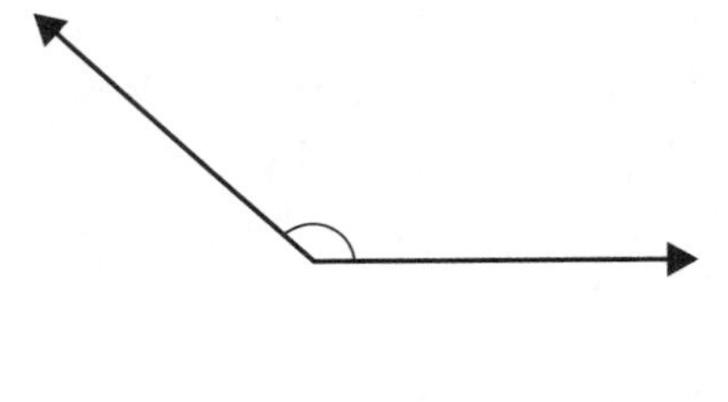

about ________

d)

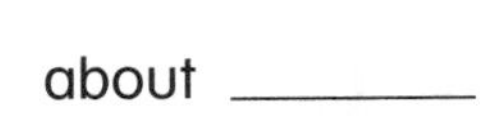

about ________

2. What is the measure of each angle?

a)

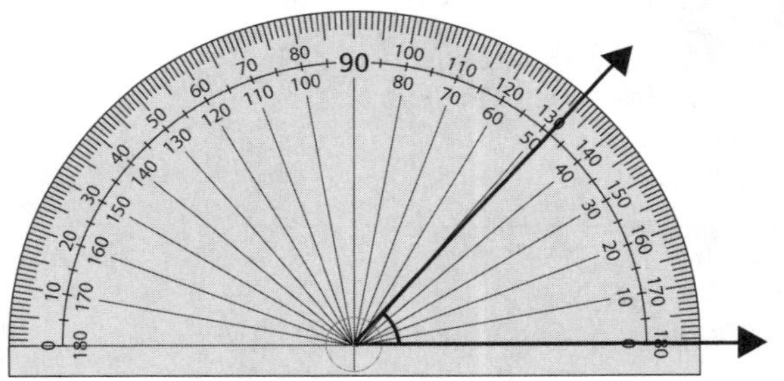

____

b)

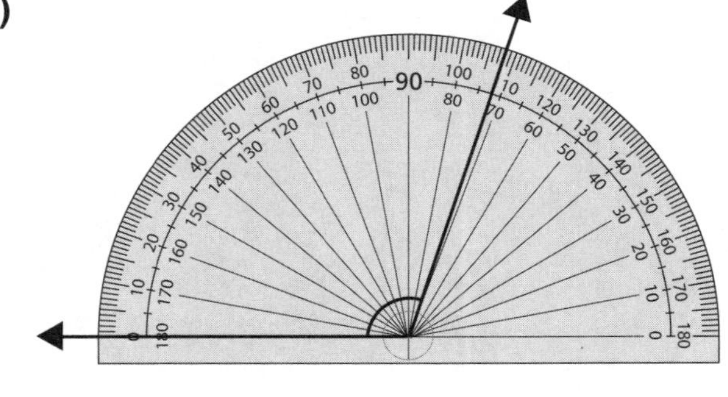

____

c)

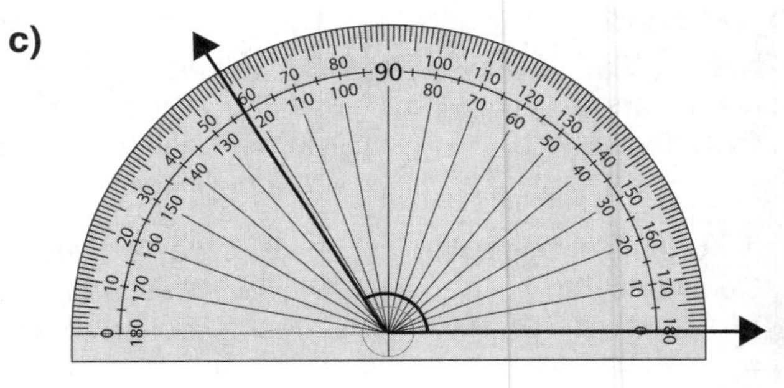

____

d)

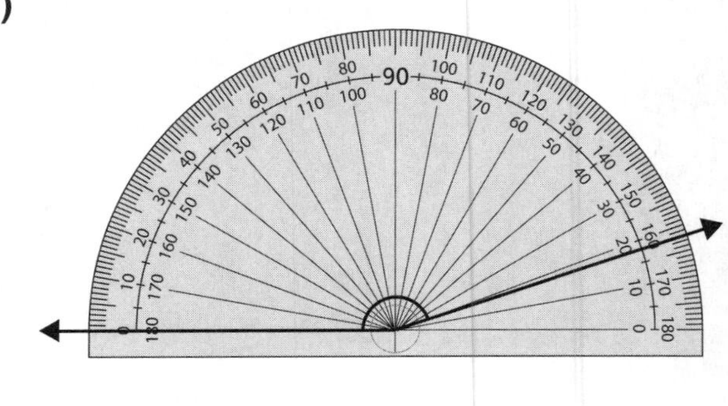

____

 Copyright © 2012 by Nelson Education Ltd.

Name: ______________________________ Date: ____________________

3. Mason says the protractor shows that the angle is 35°.
Do you agree? Explain your thinking.

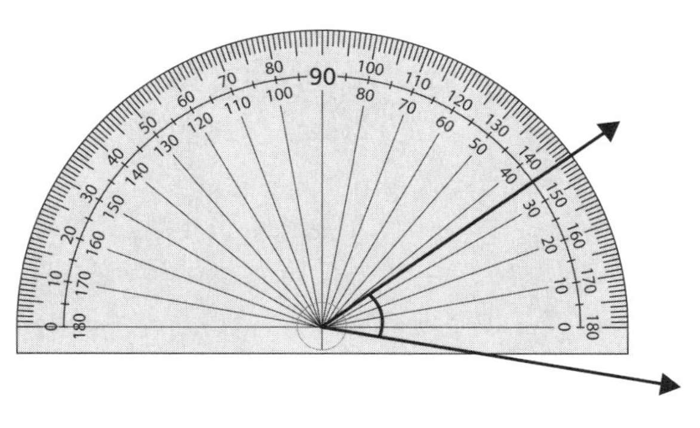

____

4. Use coloured pencils to mark one acute angle and one obtuse angle in each picture. Then measure and label each angle.

a)

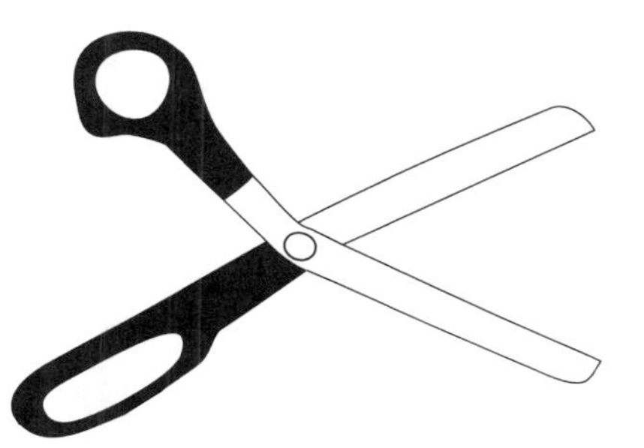

b)

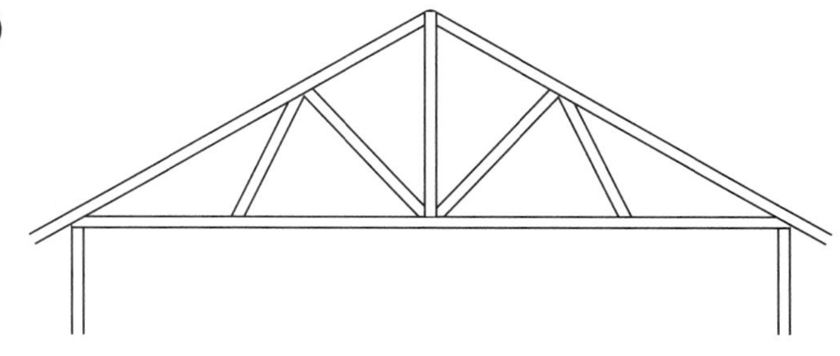

5. Draw each angle. Classify it as right, acute, or obtuse.

a) a 125° angle: ____________________

b) a 19° angle: ____________________

6. Draw a triangle with one angle that measures 150°. Tell what you notice about the other angles and how the triangle looks.

Copyright © 2012 by Nelson Education Ltd.

Name:______________________________ Date:____________________

7. Draw a polygon (a shape with straight sides) with at least 3 sides. Include one angle that is 80°, one angle that is 53°, and one angle that is 29°. Label each angle.

8. Complete each sentence.

a) The sum of the angles in any isosceles triangle is ________.

b) The sum of the angles in any obtuse triangle is ________.

c) The sum of the angles in any parallelogram is ________.

d) The sum of the angles in this grey shape is ________.

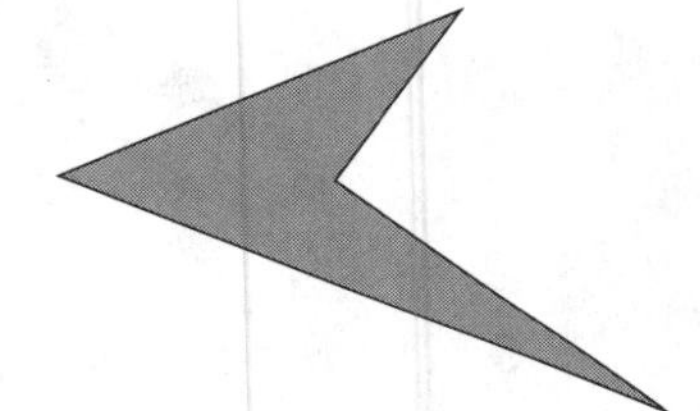

9. Figure out the measure of the unmarked angle, without using a protractor.

a)

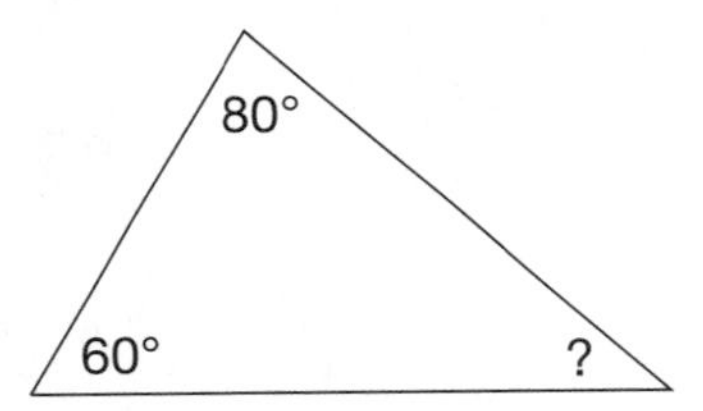

____

b)

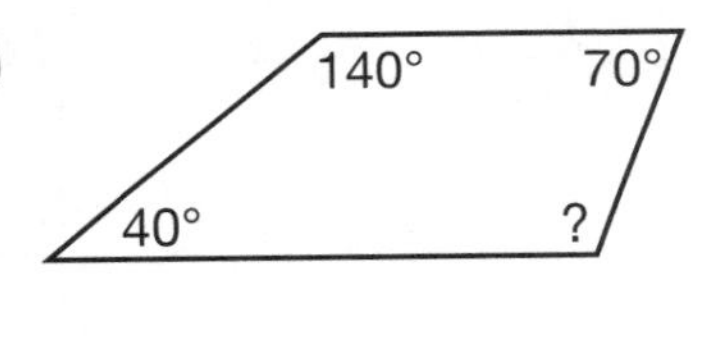

____

____

10. Explain why an obtuse triangle can have only one obtuse angle.

____

____

Copyright © 2012 by Nelson Education Ltd.

Solutions and Key to Pathways

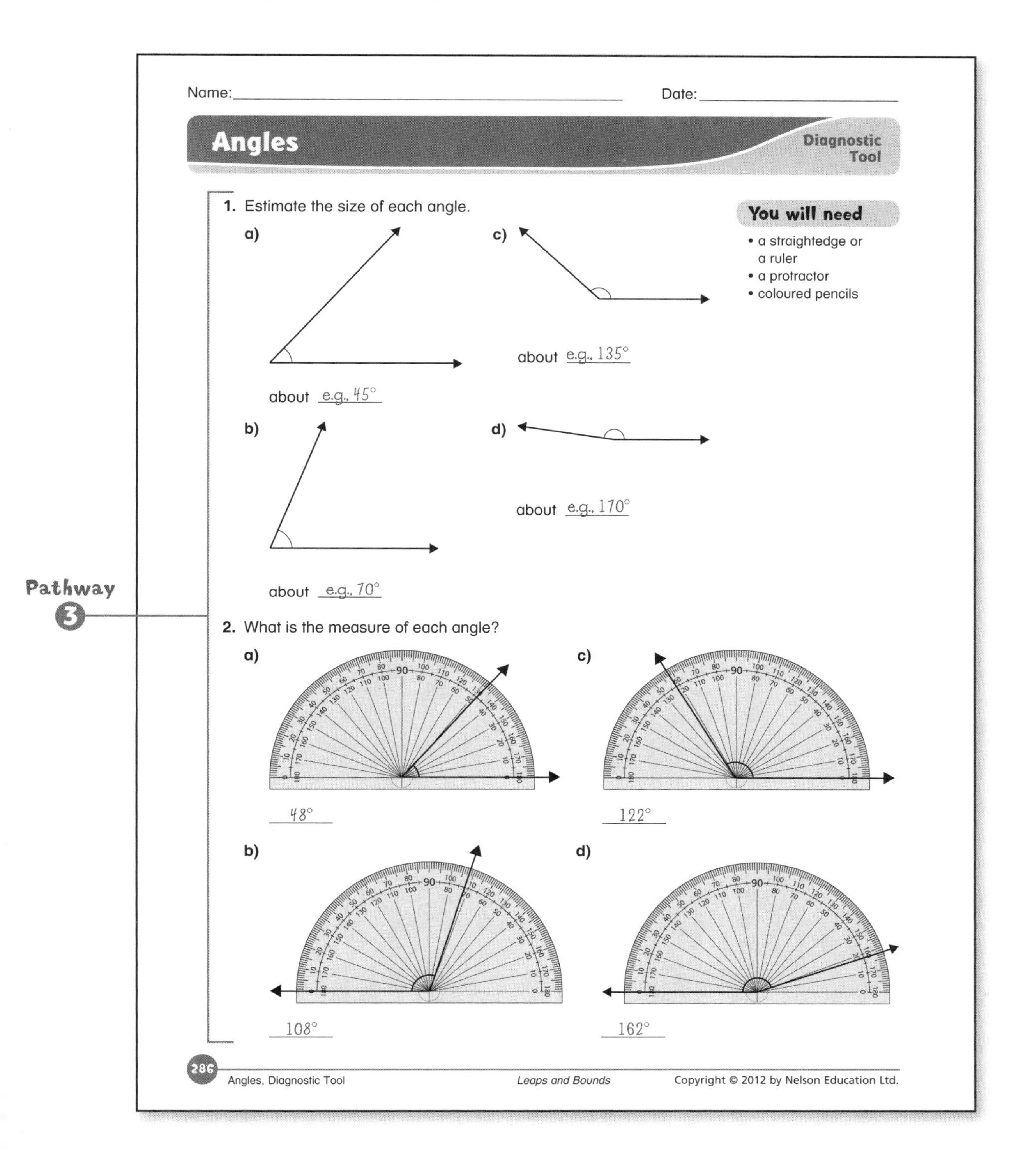

Name: ______________________ Date: ______________

3. Mason says the protractor shows that the angle is 35°. Do you agree? Explain your thinking.

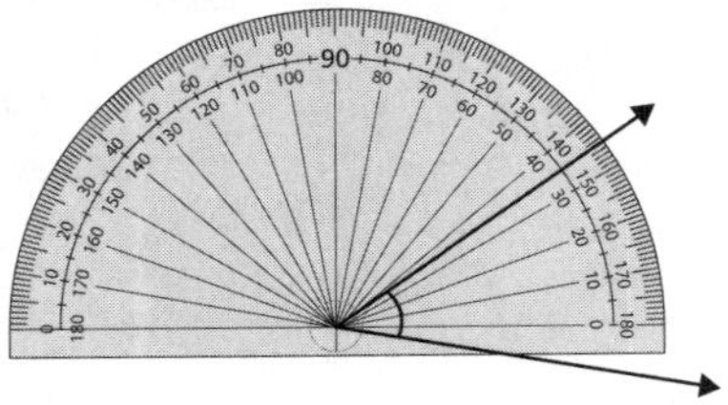

e.g., No, the protractor doesn't line up along the 0 line.

4. Use coloured pencils to mark one acute angle and one obtuse angle in each picture. Then measure and label each angle.

a) e.g., 50° 118°

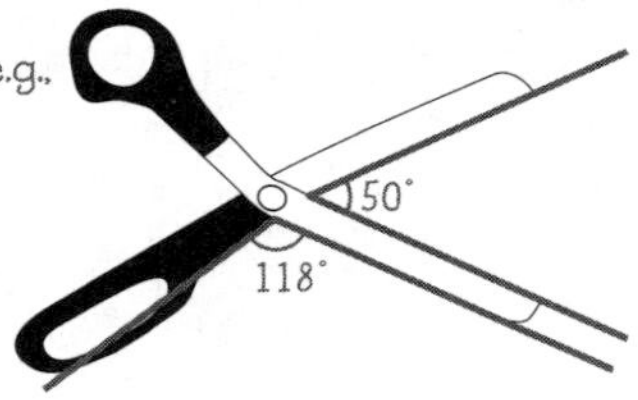

b) e.g., 118° 62°

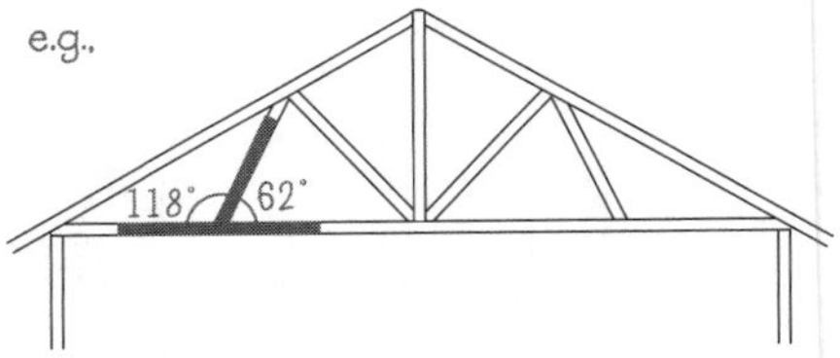

Pathway 3

5. Draw each angle. Classify it as right, acute, or obtuse.

a) a 125° angle: obtuse

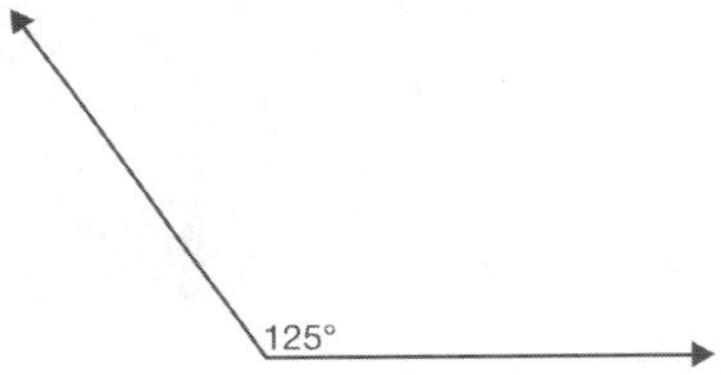

125°

b) a 19° angle: acute

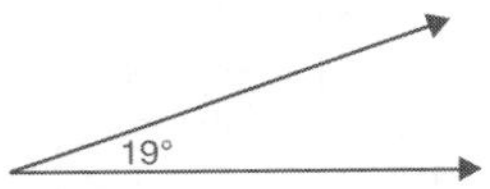

19°

6. Draw a triangle with one angle that measures 150°. Tell what you notice about the other angles and how the triangle looks.

e.g.,

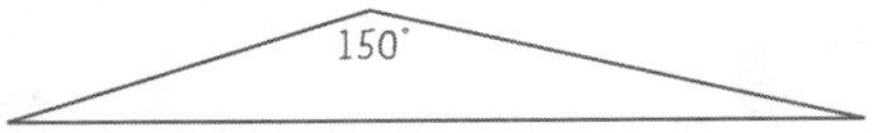

150°

The other angles are small. It's kind of wide and skinny.

Pathway 2

Name:______________________ Date:______________

Pathway 2

7. Draw a polygon (a shape with straight sides) with at least 3 sides. Include one angle that is 80°, one angle that is 53°, and one angle that is 29°. Label each angle.

e.g., 80°, 53°, 29°

Pathway 1

8. Complete each sentence.

a) The sum of the angles in any isosceles triangle is 180°.

b) The sum of the angles in any obtuse triangle is 180°.

c) The sum of the angles in any parallelogram is 360°.

d) The sum of the angles in this grey shape is 360°.

9. Figure out the measure of the unmarked angle, without using a protractor.

a) 80°, 60°, ?

180 − 140 = 40°

b) 140°, 70°, 40°, ?

40 + 140 + 70 = 250

360 − 250 = 110°

10. Explain why an obtuse triangle can have only one obtuse angle.

e.g., If there is more than one obtuse angle, the sum of the angles would be more than 180° and the sides wouldn't meet.

 Copyright © 2012 by Nelson Education Ltd.

Copyright © 2012 by Nelson Education Ltd.

Pathway 1 OPEN-ENDED

Sums of Angle Measures in Polygons

You will need

- straightedges or rulers
- protractors
- extra paper
- scissors
- glue (optional)
- dynamic geometry software (optional)
- Student Resource page 371

Open-Ended Intervention

Before Using the Open-Ended Intervention

Provide rulers and protractors, and ask students to draw an angle with a degree measure that is fairly small (e.g., 1° to 30°). Ask:

- How do you measure the angle with a protractor?
  (e.g., *I line up the 0° line of the protractor on one arm and see what number the other arm goes through.*)
- How did you know your angle would have a small measure?
  (e.g., *The sides are really close together.*)

Ask students to sketch polygons to show what these terms mean: scalene triangle, right triangle, parallelogram, quadrilateral, obtuse angle, and acute angle. Ask:

- How many angles are there to measure in each shape?
  (*3 for the triangles, but 4 for the parallelogram and the other quadrilateral*)

Using the Open-Ended Intervention Student Resource page 371

Read through the tasks on the student page together. Ensure that students have the necessary materials and clarify the descriptions as needed. If dynamic geometry software is available, it can be used.

Give students time to work, ideally in pairs.

Observe whether students

- can measure angles in a polygon and add the measures
- recognize why a quadrilateral has a greater total angle measure than a triangle

Consolidating and Reflecting

Ensure understanding by asking questions such as these based on students' work:

- Did the sum of the measures of the angles in a triangle depend on the look of the triangle? (*No.* e.g., *It was always close to 180°.*)
- How did tearing off the corners and putting them together help you see why the total would be 180° for a triangle? (e.g., *They made a straight angle.*)
- Draw a quadrilateral. How could you cut it up to show why the sum would have to be twice the sum of the measures of the angles in a triangle?
  (e.g., *I could draw a diagonal. Then there are 2 triangles, and their angles make up the angles of the quadrilateral.*)
- Suppose you knew 3 angle measures in a quadrilateral. How would you figure out the measure of the fourth angle?
  (e.g., *I would add the 3 I know and take it away from 360°.*)
- How does knowing the angle sums of triangles and quadrilaterals help you figure out whether a certain combination of angles in each of those shapes is possible?
  (e.g., *I would have to make sure that the sum of the angles was the right number.*)

 Copyright © 2012 by Nelson Education Ltd.

Sums of Angle Measures in Polygons

Pathway 1 GUIDED

Guided Intervention

Before Using the Guided Intervention

Provide straightedges or rulers and protractors, and ask students to draw an angle with a degree measure that is fairly small (e.g., 1° to 30°). Ask:

- How do you measure the angle with a protractor?
  (e.g., *I line up the 0° line of the protractor on one arm and see what number the other arm goes through.*)
- How did you know your angle would have a small measure?
  (e.g., *The sides are really close together.*)

Using the Guided Intervention Student Resource pages 372–375

Work through the instructional section of the student pages together. Provide straightedges or rulers, protractors, and scissors.

Have students work through the **Try These** questions individually or in pairs. Provide extra paper or dynamic geometry software for Question 9.

Observe whether students

- can measure angles in a shape and add the measures (Questions 1, 2)
- can use the total angle measure in a triangle or quadrilateral to determine a missing angle measure (Questions 2, 4)
- relate the total angle measure in a quadrilateral to that in a triangle by seeing the visual relationship (Question 3)
- use information about angle sums in a triangle or quadrilateral to determine what angle combinations are possible or not possible (Questions 5, 6, 7, 8, 9, 10)

Consolidating and Reflecting

Ensure understanding by asking questions such as these based on students' work:

- Did the side lengths of the triangle affect the sum of the angles in Question 1?
  (e.g., *The sum was always 180°, no matter what the side lengths were.*)
- Why might your total have not been exactly 180°?
  (e.g., *It's hard to be exact with a protractor.*)
- Why was what you did in Question 3 useful to help you figure out the sum of the angles of quadrilaterals?
  (e.g., *I could draw a diagonal. Then there are 2 triangles and the sum of the angle measures in each is 180°. Together that makes 360°.*)
- How could you use the idea in Question 6 to see that no triangle can have 2 right angles?
  (e.g., *The same thing would happen. When you draw 2 right angles, the lines don't meet. Since the sum of the angle measures is already 180°, there is no room for another angle.*)
- What are 4 angle measures that would not be possible for a quadrilateral?
  (e.g., *4 angles of 100°*)

You will need

- straightedges or rulers
- protractors
- scissors
- extra paper
- dynamic geometry software (optional)
- Student Resource pages 372–375

Copyright © 2012 by Nelson Education Ltd.

Drawing Angles

You will need

- Roof Angles (BLM 29)
- straightedges or rulers
- protractors
- 1 cm Square Dot Paper (BLM 15) or Triangle Dot Paper (BLM 16)
- dynamic geometry software (optional)
- Student Resource page 376

Open-Ended Intervention

Before Using the Open-Ended Intervention

Show students a variety of roof pictures, such as those on Roof Angles, BLM 29. Ask:

- Which roof has an angle of about 90°? (*the roof on the birdhouse*)
- Which roof has an angle of about 140°? (*the one on the dog house*)
- How can you tell which roof has an angle of about 140°?
  (e.g., *140° angles are pretty flat, and that roof is the flattest.*)
- How could you use a protractor to check your estimate?
  (e.g., *First, I'd line up the 0° line on the right side of the roof. Then I'd read the number that was more than 90 where the left side of the roof hit the protractor.*)
- What would you do if the picture was so small that the angle arms didn't extend to the edge of the protractor?
  (e.g., *I would trace over the sides and make them longer.*)

Using the Open-Ended Intervention Student Resource page 376

Read through the tasks on the student page together. Provide the necessary materials. Students might use Triangle Dot Paper (BLM 16) to draw on (allowing them to sketch regular triangles and hexagons in their diagrams), or they might use dynamic geometry software.

Give students time to work, ideally in pairs.

Observe whether students

- can draw both acute and obtuse angles
- can identify right, acute, and obtuse angles
- can describe how to draw an obtuse angle

Consolidating and Reflecting

Ensure understanding by asking questions such as these based on students' work:

- How did you know your roof would not look very traditional?
  (e.g., *Most roofs are symmetrical, and there would be fewer angles and equal ones.*)
- When you drew your angles, what were your steps after you drew the first side of the angle and marked the vertex?
  (e.g., *I lined up the side I had already drawn with the 0° line and put the point in the middle of the protractor at the vertex of the angle. I marked a dot there and then joined that dot to the point I had already marked.*)
- When you put your point at the number, how did you decide which of the 2 numbers to use?
  (e.g., *I looked to see if I was measuring an angle more or less than 90° and figured out which scale to use to make that happen.*)

 Copyright © 2012 by Nelson Education Ltd.

Drawing Angles

Pathway 2
GUIDED

Guided Intervention

Before Using the Guided Intervention

Show students a variety of roof pictures, such as those on Roof Angles, BLM 29. Ask:

- Which roof has an angle of about 90°? (*the roof on the birdhouse*)
- Which roof has an angle of about 140°? (*the one on the dog house*)
- How can you tell which roof has an angle of about 140°?
  (e.g., *140° angles are pretty flat, and that roof is the flattest.*)
- How could you use a protractor to check your estimate?
  (e.g., *First, I'd line up the 0° line on the right side of the roof. Then I'd read the number that was more than 90 where the left side of the roof hit the protractor.*)
- What would you do if the picture was so small that the angle arms didn't extend to the edge of the protractor?
  (e.g., *I would trace over the sides and make them longer.*)

Using the Guided Intervention Student Resource pages 377–380

Provide the necessary materials and/or dynamic geometry software. Work through the instructional section of the student pages together. Suggest that students do a quick sketch first and then a more accurate drawing.

Have students work through the **Try These** questions individually or in pairs.

Observe whether students

- recognize that it is necessary to decide which scale on the protractor to use to draw an angle (Question 1)
- can construct both acute and obtuse angles (Questions 2, 3, 4, 5, 6)
- can identify right, acute, and obtuse angles (Question 3)
- notice relationships in the sizes of angles in shapes (Questions 4, 6)

Consolidating and Reflecting

Ensure understanding by asking questions such as these based on students' work:

- In Question 1, would your actions be a lot different in the 2 cases?
  (e.g., *Not really, but both times I had to decide which scale to use.*)
- When you drew your angles in Question 2, what were your steps after you drew the first side of the angle and marked the vertex?
  (e.g., *I lined up the side I had already drawn with the 0° line and put the point in the middle of the protractor at the vertex of the angle. I marked a dot there and then joined that dot to the point I had already marked.*)
- Look at your work in Question 4. Why might someone say that once you've constructed a 145° angle, you've automatically constructed a 35° angle?
  (e.g., *Each time, the angle on the other side of the obtuse angle was the acute angle that matched it on the protractor. When I looked at the protractor, I noticed that those numbers added to 180.*)

You will need

- Roof Angles (BLM 29)
- rulers
- protractors
- Regular Polygons (BLM 26)
- dynamic geometry software (optional)
- Student Resource pages 377–380

Copyright © 2012 by Nelson Education Ltd.

Measuring Angles

You will need

- a geoboard and geobands (e.g., a square or circular model can be used), or 2 cm Square Dot Paper (BLM 14), or Circular Geoboard Paper (BLM 17)
- straightedges or rulers
- protractors
- classroom objects to measure angles (e.g., pattern blocks, BLMs 19 to 24)
- a camera or magazine pictures or Internet access (optional)
- Student Resource pages 381–382

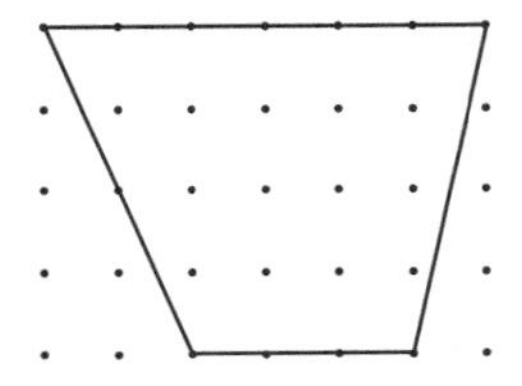

Open-Ended Intervention

Before Using the Open-Ended Intervention

On a geoboard or a copy of BLM 14 or 17, create a polygon with a number of angles, including both acute and obtuse angles. (See art in margin.) Ask:

- Which angles are acute angles? (*the ones at the top*)
- What type are the other angles? (*obtuse angles*)
- How could you use a protractor to check?
  (e.g., *I would put the 0° line on one of the edges of the shape and figure out the number where the other edge hits. I would have to make sure to use the right scale.*)
- What would you do if you just had a small picture and the angle arms didn't reach the edge of the protractor? (e.g., *I would make the arms longer.*)
- Would that change the measurement? (*no*)

Note: If a student does not know how to use the protractor, you may demonstrate or you may refer him or her to the instructional part of the guided intervention. Show how to measure an acute angle and an obtuse angle starting from either side of the protractor, using the appropriate scale.

Using the Open-Ended Intervention **Student Resource pages 381–382**

Read through the tasks on the student pages together. Make sure students have protractors and straightedges or rulers. Students can locate real objects in the classroom and use images from the Internet or magazines. They can sketch or take photos of those items, or cut them out. Also make sure students realize that when we measure 3-D shapes, we are measuring the flat face, as if it were 2-D.

Give students time to work, ideally in pairs.

Observe whether students

- can estimate angle measures
- can measure acute and obtuse angles with a protractor
- observe relationships between angles
- can distinguish between acute and obtuse angles in a variety of orientations
- recognize angles in their environment

Consolidating and Reflecting

Ensure understanding by asking questions such as these based on students' work:

- How did you estimate the angle measures? (e.g., *I knew that a 45° angle is half of a 90° angle, and I know what a 90° and a 180° angle look like, so I compared to those.*)
- When you measure an angle with a protractor, what are the important things you need to remember? (e.g., *Line up one arm with the 0° line, put the point in the middle of the protractor at the vertex of the angle, and use the right scale.*)
- How do you decide what scale to use? (e.g., *First, I estimate the measure of the angle. Then, from the 2 possible numbers, I choose the one that makes sense.*)

 Copyright © 2012 by Nelson Education Ltd.

Measuring Angles

Pathway 3 GUIDED

Guided Intervention

Before Using the Guided Intervention

On a geoboard or a copy of BLM 14 or 17, create a polygon with a number of angles, including both acute and obtuse angles. (See art in margin.) Ask:

- Which angles are acute angles?
  (*the ones at the bottom*)
- What do you call the other angles?
  (*obtuse angles*)
- Do you think the angle at the bottom right is closer to a right angle or closer to half a right angle?
  (*a right angle*)
- About how big do you think the obtuse angle at the top right is?
  (e.g., *about 100°*)

Using the Guided Intervention Student Resource pages 383–386

Work through the instructional section of the student pages together. To help students read their protractors correctly when measuring angles, encourage them to estimate the measure first by comparing it to a right angle, half a right angle, or a straight angle. Students can use copies of BLM 26 to help them with Question 7.

Have students work through the **Try These** questions individually or in pairs.

Observe whether students

- can estimate angle measures (Questions 1, 2, 4)
- can measure acute and obtuse angles with a protractor (Questions 1, 2, 3, 4, 5, 6)
- can distinguish between acute and obtuse angles (Questions 1, 2, 6)
- observe relationships between angles in a polygon (Questions 4, 6, 7, 8)
- recognize angles in their environment (Question 8)

Consolidating and Reflecting

Ensure understanding by asking questions such as these based on students' work:

- How did you estimate the angle measures in Question 4?
  (e.g., *I know that a 45° angle is half of a 90° angle, and I know what a 90° and a 180° angle look like, so I compared to those.*)
- When you measure an angle with a protractor, what are the important things you need to remember?
  (e.g., *Line up one arm with the 0° line, put the point in the middle of the protractor at the vertex of the angle, and use the right scale.*)
- How do you decide what scale to use?
  (e.g., *First, I estimate the measure of the angle. Then, from the 2 possible numbers, I choose the one that makes sense.*)

You will need

- a geoboard and geobands (e.g., a square or circular model can be used), or 2 cm Square Dot Paper (BLM 14), or Circular Geoboard Paper (BLM 17)
- straightedges or rulers
- protractors
- Regular Polygons (BLM 26)
- Student Resource pages 383–386

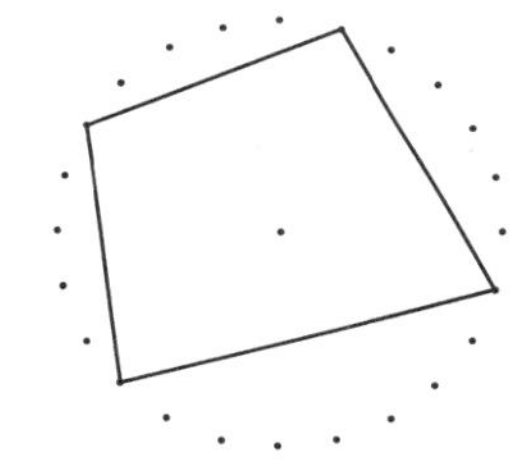

Metric Units

Planning For This Topic

Materials for assisting students with metric units consist of a diagnostic tool and 2 intervention pathways. Pathway 1 focuses on converting from one unit to another. Pathway 2 involves selecting measurement units.

Each pathway has an open-ended intervention and a guided intervention. Choose the type of intervention more suitable for your students' needs and your particular circumstances.

Curriculum Connections

Grades 5 to 8 curriculum connections for this topic are provided online. See www.nelson.com/leapsandbounds. Students studying under the WNCP curriculum use the ideas from Pathway 2 only in Grade 5, but only for the attributes of length and capacity. In grades before that, they did meet the other types of measurement conversions (e.g., kilograms to grams). The ideas in Pathway 1 are required in both the WNCP and Ontario curriculums.

Professional Learning Connections

PRIME: Measurement, Background and Strategies (Nelson Education Ltd., 2010), pages 43–44

Making Math Meaningful to Canadian Students K–8 (Nelson Education Ltd., 2008), pages 369–370

Big Ideas from Dr. Small Grades 4–8 (Nelson Education Ltd., 2009), pages 133–135

Why might students struggle with metric units?

Students might struggle with selecting appropriate metric units or renaming units for any of the following reasons:

- They might not have good personal benchmarks for some of the units, so it might be difficult to determine whether a measurement seems reasonable.
- They might not be familiar with the relationships among metric prefixes.
- They might have difficulties with place value that interfere when they are converting metric measures (e.g., they might not realize that 1000 = 100 000 hundredths when converting kilometres to centimetres).
- They might confuse whether a converted measure should become greater or less (e.g., converting 3 cm to 300 m instead of 0.03 m).
- They might convert linear measures correctly, such as litres to millilitres or metres to centimetres, but not area measures, such as square metres to square centimetres, or volume measures, such as cubic metres to cubic centimetres (e.g., they might think that 1 m^2 = 100 cm^2 rather than 10 000 cm^2).

Copyright © 2012 by Nelson Education Ltd.

Diagnostic Tool: Metric Units

Use the diagnostic tool to determine the most suitable intervention pathway for metric units. Provide Diagnostic Tool: Metric Units, Teacher's Resource pages 300 and 301, and have students complete it in writing or orally. Students should have access to calculators to help, if necessary, with their calculations.

See solutions on Teacher's Resource pages 302 and 303.

Intervention Pathways

The purpose of the intervention pathways is to help students recognize how we select metric units to use, and why and how we can rename one unit as another. These concepts will be used in later grades as students work with a broader range of metric units.

There are 2 pathways:
- Pathway 1: Renaming Units
- Pathway 2: Selecting a Unit

Use the chart below (or the Key to Pathways on Teacher's Resource pages 302 and 303) to determine which pathway is most suitable for each student or group of students.

Diagnostic Tool Results	Intervention Pathway
If students struggle with Questions 4 to 6	use Pathway 1: Renaming Units *Teacher's Resource pages 304–305* *Student Resource pages 387–391*
If students struggle with Questions 1 to 3	use Pathway 2: Selecting a Unit *Teacher's Resource pages 306–307* *Student Resource pages 392–396*

Name: ______________________________ Date: ____________________

Metric Units

Diagnostic Tool

You will need

- a calculator

1. Circle the unit that makes the most sense for measuring the given item.

a) the length of a classroom

centimetres

metres

kilometres

b) the amount of liquid in a punch bowl

millilitres

litres

kilolitres

c) the area of a living room rug

square centimetres

square metres

d) the volume of a cellphone

cubic centimetres

cubic metres

e) the thickness of a candle wick

millimetres

centimetres

metres

2. Think of an item that would be appropriate to measure using each unit below. What part of the item would be measured?

a) millimetres

____

b) cubic metres

____

c) square centimetres

____

d) kilometres

____

Copyright © 2012 by Nelson Education Ltd.

Name:________________________________ Date:__________________

3. Explain why it might make more sense to measure the amount of water in a drinking glass in millilitres rather than litres.

____

____

____

4. Rename each measurement using the given unit.

a) 5 kg = __________ g

b) 3.12 km = __________ m

c) 0.46 m = __________ cm

d) 15.23 L = __________ mL

e) 812 mm = __________ m

f) 4123 cm = __________ m

g) 62 133 g = __________ kg

h) 1 270 000 mm = __________ km

5. Explain why the 2 missing numbers below are not the same.

4.12 m = ■ cm   4.12 m^2 = ■ cm^2

____

____

____

6. Fill in the blanks with unit symbols to make each statement true.

a) 400 ________ = 4 ________

b) 15 000 ________ = 15 ________

c) 1 230 000 ________ = 1.23 ________

Copyright © 2012 by Nelson Education Ltd.

Solutions and Key to Pathways

Name: ____________________ Date: ____________

Metric Units

Diagnostic Tool

You will need

- a calculator

Pathway 2

1. Circle the unit that makes the most sense for measuring the given item.
   - **a)** the length of a classroom — centimetres / (metres) / kilometres
   - **b)** the amount of liquid in a punch bowl — millilitres / (litres) / kilolitres
   - **c)** the area of a living room rug — square centimetres / (square metres)
   - **d)** the volume of a cellphone — (cubic centimetres) / cubic metres
   - **e)** the thickness of a candle wick — (millimetres) / centimetres / metres

2. Think of an item that would be appropriate to measure using each unit below. What part of the item would be measured?
   - **a)** millimetres

     e.g., the thickness of a penny
   - **b)** cubic metres

     e.g., the volume of a room
   - **c)** square centimetres

     e.g., the area of a book cover
   - **d)** kilometres

     e.g., the distance that would take a while to drive

300 Metric Units, Diagnostic Tool *Leaps and Bounds* Copyright © 2012 by Nelson Education Ltd.

 Copyright © 2012 by Nelson Education Ltd.

Name: ____________________ Date: ____________

Pathway 2

3. Explain why it might make more sense to measure the amount of water in a drinking glass in millilitres rather than litres.

e.g., There are probably only 200 or 300 mL in a drinking glass, and a litre is 1000 mL and that is too much.

Pathway 1

4. Rename each measurement using the given unit.

a) 5 kg = 5000 g

b) 3.12 km = 3120 m

c) 0.46 m = 46 cm

d) 15.23 L = 15 230 mL

e) 812 mm = 0.812 m

f) 4123 cm = 41.23 m

g) 62 133 g = 62.133 kg

h) 1 270 000 mm = 1.27 km

5. Explain why the 2 missing numbers below are not the same.

4.12 m = ■ cm $\quad$ $4.12\ m^2$ = ■ cm^2

e.g., There are 100 cm in 1 m, but there are $100 \times 100\ cm^2$ in $1\ m^2$.
OR 4.12 m is 412 cm; and $4.12\ m^2$ is $41\ 200\ cm^2$.

6. Fill in the blanks with unit symbols to make each statement true.

e.g.,
a) 400 cm = 4 m

b) 15 000 m = 15 km

c) 1 230 000 cm^3 = 1.23 m^3

Renaming Units

You will need

- calculators
- Student Resource page 387

Open-Ended Intervention

Before Using the Open-Ended Intervention

Ask:

- What metric units might you use to measure length? (e.g., *metres, centimetres, kilometres, and millimetres*)
- What does the *kilo* in kilometres mean? (*that 1 km is 1000 m*)
- What does the *centi* in centimetres mean? (*that it takes 100 to make 1 m*)
- How can you tell how many centimetres long something is if you know how many metres long it is? (*You multiply by 100.*)
- Would the number of millimetres in a metre be more or less than 100? Why? (*more,* e.g., *since millimetres are smaller than centimetres*)
- Would you use the same units to measure area as to measure length? (*No, you would use square centimetres and square metres.*)

Using the Open-Ended Intervention Student Resource page 387

Read through the task on the student page together. Discuss whether the statement about the car is reasonable. (e.g., *Yes, 4120 mm = 4.12 m, which is a little more than 2 people long—that seems right for a car.*) Review the formulas for the area of a rectangle and the volume of a rectangular prism, and the rules for multiplying and dividing with decimals.

Give students time to work, ideally in pairs.

Observe whether students

- relate the metric prefixes to the place value system
- can rename larger (or smaller) metric units as smaller (or larger) ones
- realize that conversions for linear units are different from conversions for area and volume units involving the same base units

Consolidating and Reflecting

Ensure understanding by asking questions such as these based on students' work:

- Choose a statement in Group 1 where the number was large. How did you check if it was reasonable? (e.g., *For the pitcher holding 22 000 mL, I know that 1000 mL = 1 L, so that means there are 22 L, which is a lot.*)
- Choose a statement in Group 2 where the number was a decimal. How did you check whether it was reasonable? (e.g., *For the jewellery box, I know that 1 m^3 = 1 000 000 cm^3 and 0.01 m^3 is 100th the size, so that's 10 000 cm^3. That could be 10 cm by 100 cm by 10 cm. And 0.02 m^3 is even more. That's too much.*)
- How does knowing the relationship between metres and millimetres tell you the relationship between square metres and square millimetres? (e.g., *1 metre is 1000 millimetres, so you multiply 1000 by 1000 to go from square metres to square millimetres.*)

 Copyright © 2012 by Nelson Education Ltd.

Renaming Units

Pathway 1 GUIDED

Guided Intervention

You will need

- calculators
- Student Resource pages 388–391

Before Using the Guided Intervention

Ask:

- ▸ What metric units might you use to measure length? (e.g., *metres, centimetres, kilometres, and millimetres*)
- ▸ What does the *kilo* in kilometres mean? (*that 1 km is 1000 m*)
- ▸ What does the *centi* in centimetres mean? (*that it takes 100 to make 1 m*)
- ▸ How can you tell how many centimetres long something is if you know how many metres long it is? (*You multiply by 100.*)
- ▸ Would the number of millimetres in a metre be more or less than 100? Why? (*more,* e.g., *since millimetres are smaller than centimetres*)
- ▸ Would you use the same units to measure area as to measure length? (*No, you would use square centimetres and square metres.*)

Using the Guided Intervention Student Resource pages 388–391

Work through the instructional section of the student pages together.

Have students work through the **Try These** questions individually or in pairs.

Observe whether students

- relate the metric prefixes to the place value system (Questions 1, 4)
- can rename larger metric units as smaller ones (Questions 2, 5, 6, 7, 8)
- can rename smaller metric units as larger ones (Questions 3, 6, 7, 8)
- realize that conversions for linear units are different from conversions for area and volume units involving the same base units (Question 5)
- recognize that when the value increases, the unit is smaller (Question 9)

Consolidating and Reflecting

Ensure understanding by asking questions such as these based on students' work:

- ▸ How did you rename the 0.96 cm in Question 2h)? (e.g., *There are 10 times as many millimetres as centimetres, so I multiplied by 10.*)
- ▸ How did you explain Question 5c)? (e.g., *1 m^3 = 100 cm × 100 cm × 100 cm and 100 × 100 × 100 = 1 000 000, so I multiplied 4.2 by 1 million.*)
- ▸ What choices did you have for solving Question 6a)? (e.g., *I could have written 23 cm as 0.23 m and multiplied by 0.6, or written 0.6 m as 60 cm and multiplied by 23. Once I had one answer, I could have multiplied or divided by 10 000 to get the other one.*)
- ▸ In Question 8c), how did you rename 2 423 000 mg? (e.g., *I used the idea in Question 1 and I divided by 10 six times. That got me to kilograms.*)
- ▸ How does knowing the relationship between metres and millimetres tell you the relationship between cubic metres and cubic millimetres? (e.g., *1 m = 1000 mm, so you multiply 1000 by 1000 to go from square metres to square millimetres.*)

Pathway 2
OPEN-ENDED

Selecting a Unit

You will need

- a balance scale with 10 g and 1 kg masses
- a ruler
- a metre stick
- base ten blocks
- a teaspoon
- a 1 L container
- Student Resource pages 392–393

Open-Ended Intervention

Before Using the Open-Ended Intervention

Display the measuring materials. Ask:

- Which of the tools would you use to measure length? What would the units be? (*a ruler for centimetres and millimetres; a metre stick for metres and centimetres*)
- How could you measure kilometres? (e.g., *Use the metre stick 1000 times.*)
- Which tools are for mass, and what are the units? (*the balance scale; the units are grams and kilograms*)
- What mass might be hard to measure with the balance scale? (e.g., *a gram, since the balance scale doesn't notice things that are super light*)
- How could you measure 1 square metre? (e.g., *You could make a line 1 m long and then turn and make another line 1 m wide, and outline the full square.*)
- What are the container and spoon used to measure, and what units would you use? (e.g., *capacity, in millilitres or litres*)

Using the Open-Ended Intervention Student Resource pages 392–393

Read through the task on the student pages together and ensure students understand it. Have the measuring tools available.

Give students time to work, ideally in pairs.

Observe whether students

- suggest appropriate items when given measurement units
- can justify unit choices
- choose to relate the use of small units to precision
- note that a variety of units can be used to measure the same thing

Consolidating and Reflecting

Ensure understanding by asking questions such as these based on students' work:

- Choose one of your length units. How did you choose your items? (e.g., *For millimetres, I chose something really small where even centimetres are too big for the measurement.*)
- How did you distinguish the items you used for square metres from those you used for square centimetres? (e.g., *I imagined they were rectangular and decided whether I would probably measure the length and width with centimetres or metres.*)
- What helps you to decide which unit to use? (e.g., *First, I think about whether I'm measuring length or area or capacity or mass or volume, and then I think about whether the item is big or little.*)
- Is there always just one possible answer? (*No.* e.g., *You can always measure in another unit of the same type, but you might end up with too many of those units or maybe not even a whole unit, and the numbers are not nice.*)

 Copyright © 2012 by Nelson Education Ltd.

Selecting a Unit

Pathway 2 GUIDED

Guided Intervention

You will need

- a balance scale with 10 g and 1 kg masses
- a ruler
- a metre stick
- base ten blocks
- a teaspoon
- a 1 L container
- Student Resource pages 394–396

Before Using the Guided Intervention

Display the measuring materials. Ask:

- Which of the tools would you use to measure length? What would the units be? (*a ruler for centimetres and millimetres; a metre stick for metres and centimetres*)
- How could you measure kilometres? (e.g., *Use the metre stick 1000 times.*)
- What mass might be hard to measure with the balance scale? (e.g., *a gram, since the balance scale doesn't notice things that are super light*)
- How could you measure 1 square metre? (e.g., *You could make a line 1 m long and then turn and make another line 1 m wide, and outline the full square.*)
- What are the container and spoon used to measure, and what units would you use? (e.g., *capacity, in millilitres or litres*)

Using the Guided Intervention Student Resource pages 394–396

Work through the ways of measuring pennies in the instructional section of the student pages together. Have the measuring tools available.

Have students work through the **Try These** questions individually or in pairs.

Observe whether students

- suggest appropriate units when given items to measure (Questions 1, 5, 7)
- can justify unit choices (Questions 2, 3, 4, 9)
- relate the use of small units to precision (Question 3)
- recognize that a variety of units can be used to measure the same thing (Questions 6, 8)
- realize that sometimes one unit is preferable to another unit (Question 9)

Consolidating and Reflecting

Ensure understanding by asking questions such as these based on students' work:

- How did you decide on your answer to Question 3? (e.g., *I could see that the lengths were almost the same, so a really small unit would be best.*)
- How did you decide on your answer to Question 4? (e.g., *I thought about how to measure the depth and the sides of the pool, and it would be with metres and not centimetres since the sides are long and the pool is deep.*)
- What helps you decide which unit to use? (e.g., *First I think about whether I'm measuring length or area or capacity or mass or volume, and then I think about whether the item is big or little.*)
- Is there always just one possible answer? (*No.* e.g., *You can always measure in another unit of the same type, but you might end up with too many of those units or maybe not even a whole unit, and that is more awkward.*)

Data and Probability Strand Overview

How were the data and probability topics chosen?

This resource provides materials for assisting students with 3 data and probability topics. These topics are drawn from curriculum outcomes across the country for Grades 5 to 7. Topics are divided into levels, called pathways, that address gaps in students' prerequisite skills and knowledge.

How were the pathways determined?

At the Grade 7/8 level, the focus is on 3 topics: collecting, displaying, and interpreting data; summarizing data; and probability.

The collecting, displaying, and interpreting data topic (entitled Displaying Data) addresses the creation and interpretation of circle graphs and line graphs. It looks at issues related to bias and sampling in collecting data, and to interpreting graphs (mostly types that are familiar from previous grades, such as bar graphs) and tables. Some attention is given to misleading graphs.

The summarizing data topic addresses the effect of data changes (e.g., removing outliers, adding the same amount to each piece of data, etc.) on measures of central tendency. It also looks at the use and interpretation and comparison of measures of central tendency (mean, median, and mode), and the meaning and calculation of the mean of a set of data.

The probability topic addresses complex issues such as calculating the theoretical probability of 2 independent events and simpler topics such as calculating simple theoretical probabilities and calculating simple experimental probabilities.

What data and probability topics have been omitted?

These topics involve a minimal amount of actual data collection and organization, and focus more on interpreting and displaying data. However, some attention is given to how to collect the data, particularly in the pathway on bias and sampling.

There is no emphasis on curriculum outcomes that focus on specific vocabulary, such as correctly using the word *census* versus *sample*, the term *discrete data* versus *continuous data*, or the terms *first-hand (primary) data* and *second-hand (secondary) data*. The specific vocabulary was not a focus, since conceptual gaps, rather than vocabulary, are the focus of these materials.

 Copyright © 2012 by Nelson Education Ltd.

Materials

The materials listed below for assisting students with data and probability are likely in the classroom or easily accessible. Blackline masters are provided at the back of this resource. Many of these materials are optional and are listed only in the Teacher's Resource.

- straightedges or rulers
- coloured pencils
- materials for graphing (e.g., a spreadsheet or graphing software, a compass and protractor, percent circle, grid paper)
- calculators
- access to the Internet
- linking cubes
- small blank cards or pieces of paper
- dice
- coins
- paper bags
- identical coloured objects such as square tiles, counters, or bingo chips
- pencils and paper clips
- centimetre cubes
- standard deck of playing cards (optional)

BLM 12: 1 cm Grid Paper
BLM 13: 0.5 cm Grid Paper
BLM 18: Fraction Circles/Spinners
BLM 30: Types of Graphs
BLM 31: Percent Circles

Data and Probability Topics and Pathways

Topics and pathways in this strand are shown below.
Each pathway has an open-ended intervention and a guided intervention.

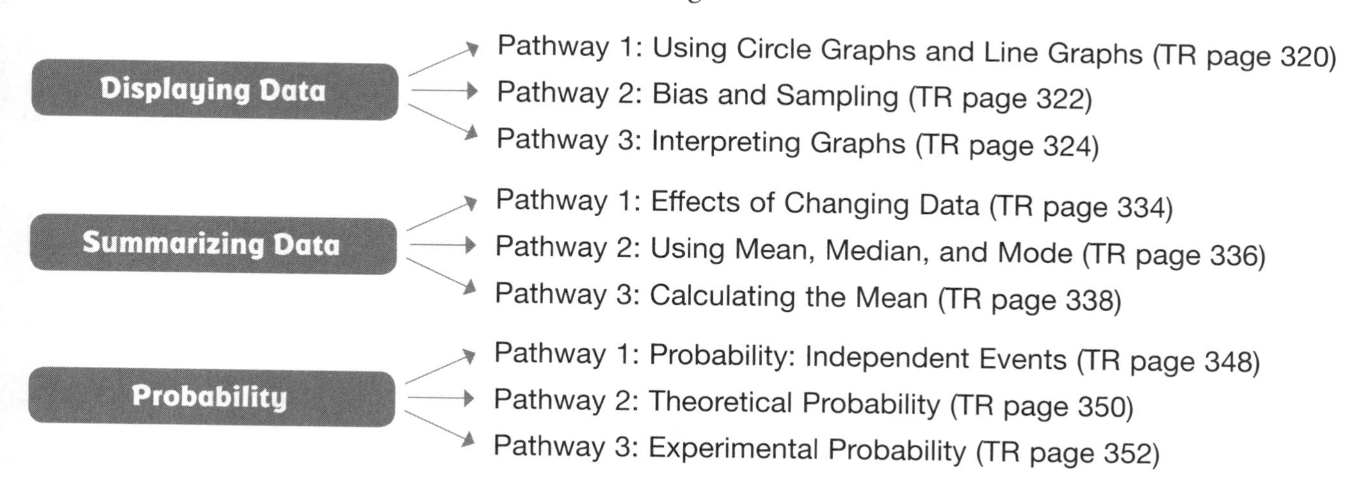

Copyright © 2012 by Nelson Education Ltd.

Displaying Data

Planning For This Topic

Materials for assisting students with displaying and interpreting data consist of a diagnostic tool and 3 intervention pathways. Pathway 1 focuses on constructing circle graphs and line graphs, and includes interpretation of these graphs. Pathway 2 addresses bias and sampling. Pathway 3 involves interpreting tables of data, bar graphs, and line graphs—including misleading graphs. It also includes creating graphs from tables of data.

Each pathway has an open-ended intervention and a guided intervention. Choose the type of intervention more suitable for your students' needs and for your particular circumstances.

Curriculum Connections

Grades 5 to 8 curriculum connections for this topic are provided online. See www.nelson.com/leapsandbounds. In these pathways, a limited amount of data collection and organization (e.g., collecting survey data) is included. The WNCP curriculum does not cover bias until Grade 8, so Pathway 2 may be omitted for those students.

Pathways 1 and 3 are relevant to both the WNCP and Ontario curriculums. Note that the term *broken-line graph* is not used in the WNCP curriculum, but simply *line graph*, which involves continuous data that can be plotted and joined by line segments, regardless of whether the line segments fall in a straight line or not.

Professional Learning Connections

PRIME: Data Management and Probability, Background and Strategies (Nelson Education Ltd., 2006), pages 54–56, 64–68, 71–72, 74–78, 80, 87–90

Making Math Meaningful to Canadian Students K–8 (Nelson Education Ltd., 2008), pages 487–491, 493–494, 497–498, 505–508, 525–531

Big Ideas from Dr. Small Grades 4–8 (Nelson Education Ltd., 2009), pages 169–173, 181–183, 187–194

More Good Questions (dist. by Nelson Education Ltd., 2010), pages 155–159, 172–175

Why might students struggle with data?

Students might struggle with displaying and interpreting data for any of the following reasons:

- They might not understand that different graphs are used for different purposes (e.g., line graphs are used to show trends), or not know what type of graph is best suited for a particular set of data.
- They might have difficulty interpreting graphs in addition to simply reading off data.
- They might not recognize when a graph presents data in misleading ways (e.g., change is exaggerated when the vertical axis is started at a point greater than 0).
- They might not see how bias can affect the reliability of data collected.
- They might create survey questions that are unclear and vague (e.g., wanting to know if the community centre should be open longer hours, but asking instead: Do you want to go to the community centre more often?).
- They might include choices in a survey that are not exclusive or discrete (e.g., asking if you like winter sports, team sports, competitive sports, or other sports).

 Copyright © 2012 by Nelson Education Ltd.

- They might not realize that a circle graph may not provide the actual data values.
- They might not see the connection between the number of degrees as a fraction of 360 and the percent for each sector in a circle graph. Or they might not realize that the percents must add to 100%.
- They might have difficulty choosing an appropriate scale for a line graph (e.g., they might use an inappropriate or irregular scale).
- They might reverse the numbering on the vertical axis of a line graph so that it increases from top to bottom rather than from bottom to top.

Diagnostic Tool: Displaying Data

Use the diagnostic tool to determine the most suitable intervention pathway for displaying data. Provide Diagnostic Tool: Displaying Data, Teacher's Resource pages 312 to 315, and have students complete it in writing or orally.

See solutions on Teacher's Resource pages 316 to 319.

Intervention Pathways

The purpose of the intervention pathways is to help students improve their skills in collecting, displaying, and interpreting data in various ways. This will help them in communicating mathematically and in interpreting more complex graphs in later grades.

There are 3 pathways:

- Pathway 1: Using Circle Graphs and Line Graphs
- Pathway 2: Bias and Sampling
- Pathway 3: Interpreting Graphs

Use the chart below (or the Key to Pathways on Teacher's Resource pages 316 to 319) to determine which pathway is most suitable for each student or group of students.

Diagnostic Tool Results	Intervention Pathway
If students struggle with Questions 7 to 9	use Pathway 1: Using Circle Graphs and Line Graphs *Teacher's Resource pages 320–321* *Student Resource pages 397–402*
If students struggle with Questions 4 to 6	use Pathway 2: Bias and Sampling *Teacher's Resource pages 322–323* *Student Resource pages 403–407*
If students struggle with Questions 1 to 3	use Pathway 3: Interpreting Graphs *Teacher's Resource pages 324–325* *Student Resource pages 408–413*

The 3 pathways are somewhat independent. Some students should be directed to more than one pathway depending on the results of the diagnostic tool.

Copyright © 2012 by Nelson Education Ltd.

Name:________________________________ Date:____________________

Displaying Data

Diagnostic Tool

1. The tables and graphs below show the growth of a tomato plant and a bean plant. Rowan forgets which is which, but he knows that at one point, the bean plant is 5 cm taller than the tomato plant.

   Decide which table and which graph show the growth of the bean plant, and which show the growth of the tomato plant. Explain your thinking.

   ____

   ____

   ____

Growth of __________ Plant

Time (days)	Height (mm)
0	0
1	5
2	7
3	10
4	15
5	18

Growth of __________ Plant

Time (days)	Height (mm)
0	0
1	2
2	5
3	9
4	10
5	12

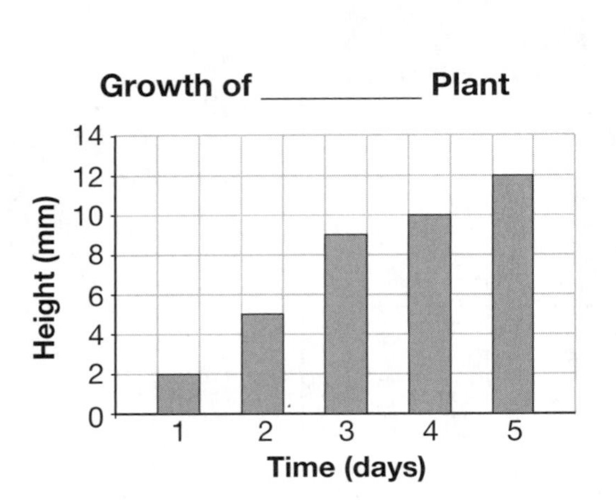

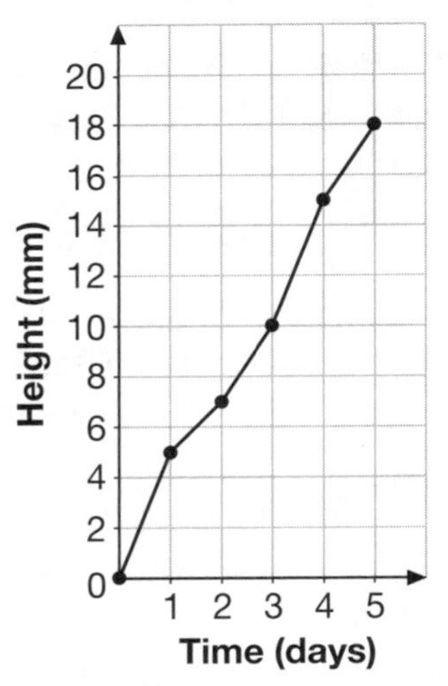

 Copyright © 2012 by Nelson Education Ltd.

Name:______________________________ Date:______________

2. A runner records her time as she completes each kilometre of a 4 km run. Her graph looks like this:

Running Record

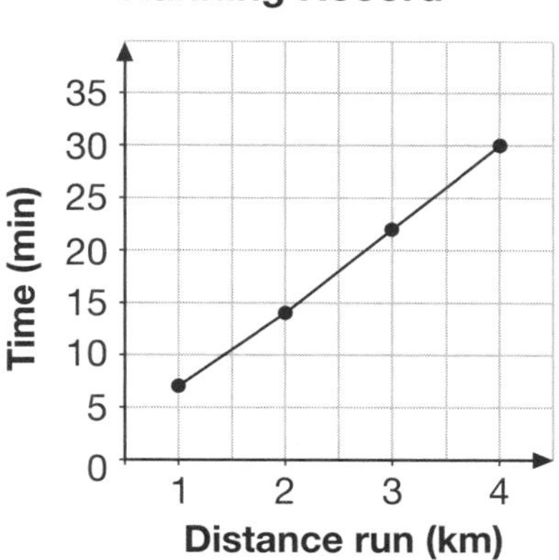

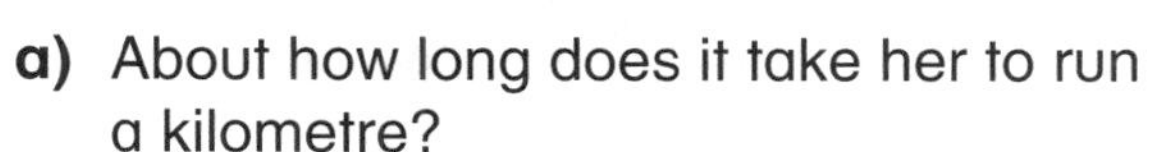

a) About how long does it take her to run a kilometre?

____

b) About how long do you think it will take her to run 7 km?

____

3. Daniel recently opened a fruit store. The table below shows his sales and the fruit sales at the nearby grocery store.

Sales of Fruit at 2 Stores

Month	Daniel's sales ($)	Grocery store sales ($)
1	800	1600
2	1100	1700
3	1000	1500
4	1300	900
5	1400	900
6	1400	900

a) What were each store's total sales in the first month?

______________________ ______________________

b) What does the data tell you about the sales over time in the 2 stores? What trends do you see?

____

____

____

____

c) Do you think that Daniel's store will take over all the fruit sales from the grocery store? Explain your thinking.

____

____

Copyright © 2012 by Nelson Education Ltd.

Name: ______________________ Date: ______________________

4. The survey questions at right were used to determine what people usually eat for breakfast. For each question, tell whether you think it will get an honest response from people without leading them in some way. How might you rewrite the biased questions?

____

____

____

____

____

1. Do you like eating cereal for breakfast?

2. Fruits are good for you. Do you include fruit for breakfast?

3. What do you usually have for breakfast? (Choose all that apply.)
a) eggs
b) cereal
c) yogurt
d) rice
e) other

5. Suppose you survey students in your school to find out common shoe sizes. What might you find out from your survey?

____

____

6. Kristy wants to collect data that is unbiased and reliable to find out what kinds of things students from her school do during the summer. How do you think Kristy should collect the data?

____

____

____

____

Remember

- Data that is unbiased and reliable is data that you think is accurate and also applies to lots of people who are not surveyed.

7. This circle graph shows the types of sandwiches sold in a cafeteria.

a) What is the most popular sandwich?

____

b) Why is this circle graph a good choice for the data?

____

____

____

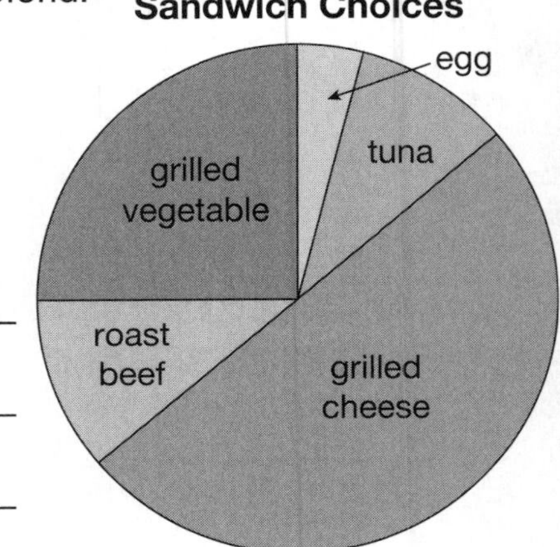

 Copyright © 2012 by Nelson Education Ltd.

Name:________________________ Date:______________

8. a) What could the line graph below be about? (Make sure that the shape of the graph and the numbers make sense for your topic.)

____

____

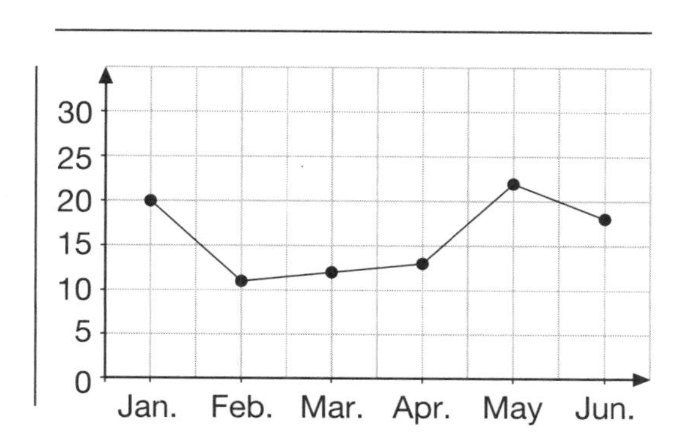

b) Label the axes and give the graph a title.

9. This line graph shows the average temperature each month for Flin Flon, Manitoba.

a) Which are the warmest months in Flin Flon?

____

____

b) How many months have an average temperature above freezing?

____

c) When did the temperature increase the most and decrease the most? Tell how you know.

____

____

____

____

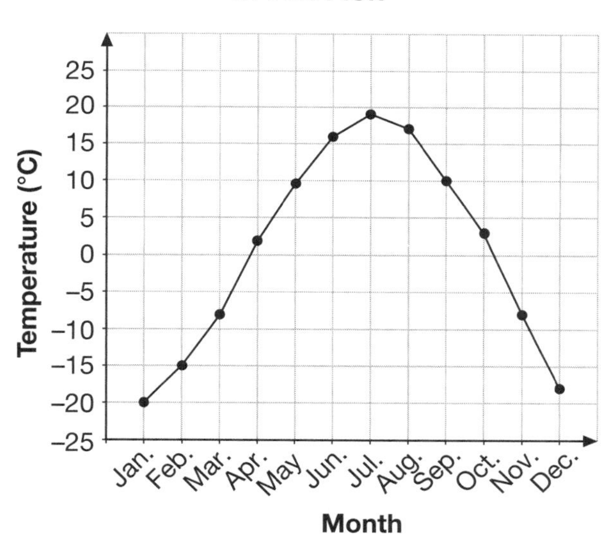

Copyright © 2012 by Nelson Education Ltd.

Solutions and Key to Pathways

Name:______________________ Date:______________

Displaying Data

Diagnostic Tool

Pathway 3

1. The tables and graphs below show the growth of a tomato plant and a bean plant. Rowan forgets which is which, but he knows that at one point, the bean plant is 5 cm taller than the tomato plant.

   Decide which table and which graph show the growth of the bean plant, and which show the growth of the tomato plant. Explain your thinking.

   e.g., The first table goes with the line graph, so the second table must go with the bar graph. Table 1 is the bean plant because on Day 4, its height is 15 mm, while the tomato plant is only 10 mm tall.

Growth of Bean Plant

Time (days)	Height (mm)
0	0
1	5
2	7
3	10
4	15
5	18

Growth of Tomato Plant

Time (days)	Height (mm)
0	0
1	2
2	5
3	9
4	10
5	12

Growth of Bean Plant

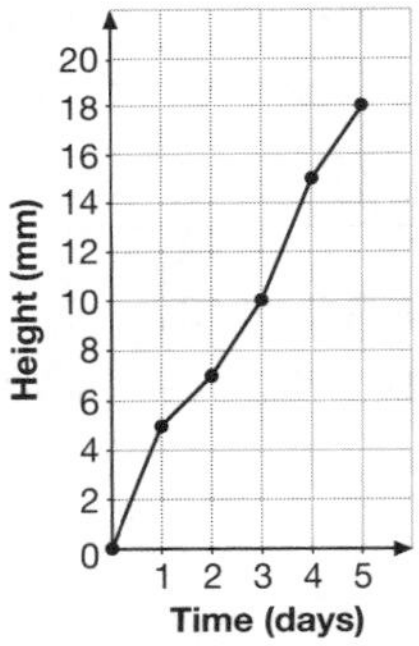

Growth of Tomato Plant

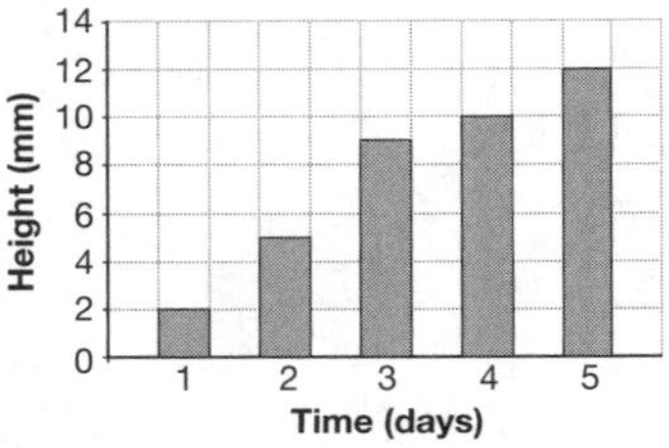

312 Displaying Data, Diagnostic Tool *Leaps and Bounds* Copyright © 2012 by Nelson Education Ltd.

Copyright © 2012 by Nelson Education Ltd.

Name:______________________ Date:______________

2. A runner records her time as she completes each kilometre of a 4 km run. Her graph looks like this:

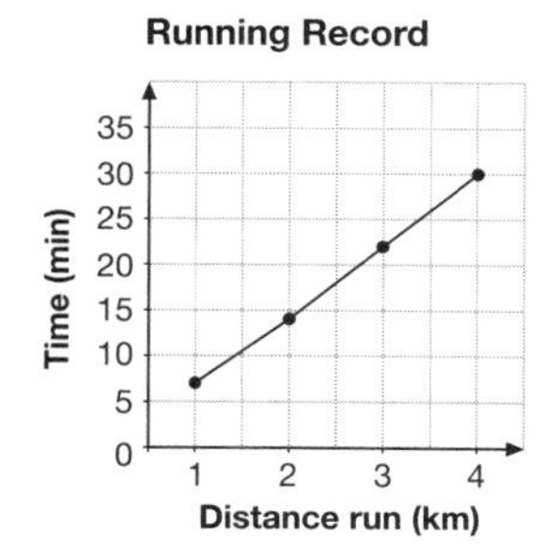

a) About how long does it take her to run a kilometre?

about 7 minutes

b) About how long do you think it will take her to run 7 km?

e.g., about 50 or 55 minutes

3. Daniel recently opened a fruit store. The table below shows his sales and the fruit sales at the nearby grocery store.

Sales of Fruit at 2 Stores

Month	Daniel's sales ($)	Grocery store sales ($)
1	800	1600
2	1100	1700
3	1000	1500
4	1300	900
5	1400	900
6	1400	900

a) What were each store's total sales in the first month?

$800 for Daniel's store $1600 for the grocery store

b) What does the data tell you about the sales over time in the 2 stores? What trends do you see?

e.g., Since Daniel's store opened, fruit sales in the grocery store have gone down and Daniel's sales have gone up. Customers may be buying fruit from Daniel's store rather than the grocery store.

c) Do you think that Daniel's store will take over all the fruit sales from the grocery store? Explain your thinking.

No. e.g., It looks like Daniel's sales and the grocery store sales have stayed about the same for the last 3 months.

Pathway 3

Name:______________________ Date:______________

4. The survey questions at right were used to determine what people usually eat for breakfast. For each question, tell whether you think it will get an honest response from people without leading them in some way. How might you rewrite the biased questions?

1. No, e.g., It promotes eating cereal. "What do you like to eat for breakfast?"

2. No. e.g., You might not want to say you don't eat fruit. "Do you include fruit for breakfast?"

3. Yes. e.g., This question will get an honest response.

1. Do you like eating cereal for breakfast?

2. Fruits are good for you. Do you include fruit for breakfast?

3. What do you usually have for breakfast? (Choose all that apply.)
a) eggs
b) cereal
c) yogurt
d) rice
e) other

Pathway 2

5. Suppose you survey students in your school to find out common shoe sizes. What might you find out from your survey?

e.g., I might find out that boys have bigger shoe sizes than girls, and what sizes are most common.

6. Kristy wants to collect data that is unbiased and reliable to find out what kinds of things students from her school do during the summer. How do you think Kristy should collect the data?

e.g., She could send a questionnaire home with one class in each grade for parents to complete and return to her. She should collect data near the end of the school year when families are making plans.

Remember

- Data that is unbiased and reliable is data that you think is accurate and also applies to lots of people who are not surveyed.

Pathway 1

7. This circle graph shows the types of sandwiches sold in a cafeteria.

Sandwich Choices

a) What is the most popular sandwich?

grilled cheese

b) Why is this circle graph a good choice for the data?

e.g., It shows the fractions of sandwich choices of the whole. If you're interested in the popularity and not the number sold, the circle graph works.

 Copyright © 2012 by Nelson Education Ltd.

 Copyright © 2012 by Nelson Education Ltd.

Name:______________________________ Date:______________

8. **a)** What could the line graph below be about? (Make sure that the shape of the graph and the numbers make sense for your topic.)

e.g., how much was sold in a sporting goods store over the first half of a year

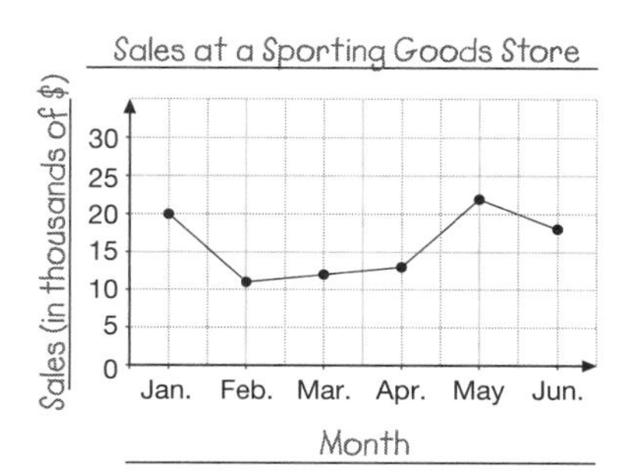

b) Label the axes and give the graph a title.

Pathway 1

9. This line graph shows the average temperature each month for Flin Flon, Manitoba.

Average Monthly Temperature in Flin Flon

Temperature (°C): 25, 20, 15, 10, 5, 0, −5, −10, −15, −20, −25

Jan. Feb. Mar. Apr. May Jun. Jul. Aug. Sep. Oct. Nov. Dec.

Month

a) Which are the warmest months in Flin Flon?

June, July, August

b) How many months have an average temperature above freezing?

7 months

c) When did the temperature increase the most and decrease the most? Tell how you know.

e.g., It increased the most from March to April, because that's the steepest line going up. It decreased the most from October to November, because that's the steepest line going down.

Using Circle Graphs and Line Graphs

You will need

- Types of Graphs (BLM 30)
- straightedges or rulers
- coloured pencils
- materials for graphing (e.g., a compass and protractor, Percent Circles (BLM 31), 1 cm Grid Paper (BLM 12) or 0.5 cm Grid Paper (BLM 13), a spreadsheet or graphing software)
- calculators
- Student Resource pages 397–398

Open-Ended Intervention

Before Using the Open-Ended Intervention

Provide copies of Types of Graphs (BLM 30), and have students locate the table that shows data collected every 5 years. Have them identify the 3 types of graphs—bar graph, line graph, and circle graph. Ask:

- Which graph do you find most useful for understanding the data? (e.g., *the circle graph, because it shows farms with different types of crops*)
- How does each graph show that grains is the most popular crop? (e.g., *In the bar graph, the tallest group of bars is for grains. In the line graph, the highest line is for grains. In the circle graph, the biggest section is for grains.*)
- How are the 3 graphs alike? (e.g., *They show that farms have different types of crops and what these are. The bar graph and line graph show the same data.*)
- What differences are there? (e.g., *The circle graph doesn't show the changes in types of crops over time, from the start to 10 years later.*)

Using the Open-Ended Intervention **Student Resource pages 397–398**

Provide the necessary materials. Read through each part of the task on the student pages together. Ensure students understand how to construct a circle graph and a line graph. You might suggest that they look at the instructions in the guided intervention for this pathway on Student Resource pages 399 and 400.

Give students time to work, ideally in pairs.

Observe whether students

- make up their own data (first-hand data) and construct a circle graph of the data
- use given data (second-hand data) and make a line graph with the proper scale
- explain why a circle graph is appropriate to represent data that is part of a whole
- explain why a line graph is appropriate to represent continuous data over time

Consolidating and Reflecting

Ensure understanding by asking questions such as these based on students' work:

- How does your circle graph show the greatest and least amounts you spent on a charity? (e.g., *The most I spent is in the largest section of the circle—cancer research. I spent the least on the library, and that has the smallest section.*)
- Could you have had the same circle graph if you had been given only $1000? (e.g., *Yes, the sections could be the same size with the same percents.*)
- Could you have had the same line graph if the amounts were different? (e.g., *maybe; the line could be in the same place, but the scale would be different*)
- How could you tell that your circle graph looked right? (e.g., *All the pieces fit together, and the percents added to 100.*)
- Why does a line graph make sense for yearly data? (e.g., *The amount for each year is separate, but you see how those numbers go up and down over time.*)

 Copyright © 2012 by Nelson Education Ltd.

Using Circle Graphs and Line Graphs

Pathway 1 GUIDED

Guided Intervention

Before Using the Guided Intervention

Provide copies of Types of Graphs (BLM 30), and have students locate the table that shows data collected every 5 years. Have them identify the 3 types of graphs—bar graph, line graph, and circle graph. Ask:

- Which graph do you find most useful for understanding the data? (e.g., *the circle graph, because it shows farms with different types of crops*)
- How does each graph show that grains is the most popular crop? (e.g., *In the bar graph, the tallest group of bars is for grains. In the line graph, the highest line is for grains. In the circle graph, the biggest section is for grains.*)
- How are the 3 graphs alike? (e.g., *They show that farms have different types of crops and what these are. The bar graph and line graph show the same data.*)
- What differences are there? (e.g., *The circle graph doesn't show you the changes in types of crops over time, from the start to 10 years later.*)

You will need

- Types of Graphs (BLM 30)
- straightedges or rulers
- coloured pencils
- materials for graphing (e.g., a compass and protractor, Percent Circles (BLM 31), 1 cm Grid Paper (BLM 12) or 0.5 cm Grid Paper (BLM 13), a spreadsheet or graphing software)
- calculators
- Student Resource pages 399–402

Using the Guided Intervention Student Resource pages 399–402

Provide the materials and work through the instructional section with students.

Have students work through the **Try These** questions in pairs or individually.

Observe whether students

- can interpret a circle graph and a line graph (Questions 1, 2, 3, 4, 5)
- can explain why a circle graph or a line graph is appropriate (Questions 2, 4)
- can create a circle graph and a line graph (Questions 3, 4, 5)

Consolidating and Reflecting

Ensure understanding by asking questions such as these based on students' work:

- How does the circle graph in Question 1 show you the most popular way for students to get to school? (e.g., *The largest section is for the bus, so most of the students surveyed take the bus to school.*)
- In Question 3, how could you estimate the size of the sectors of the circle before checking your final graph? (e.g., *40% and 10% make up half the circle, and 28% is about one quarter, so the remaining 6% and 16% represent what's left.*)
- Why does it make sense to use a circle graph in Question 3? (e.g., *I can easily compare the categories by looking at the size of each section.*)
- Why does it *not* make sense to use a circle graph in Question 4? (e.g., *The data doesn't fit into parts of a whole.*)
- Why does a line graph make sense in Question 5c)? (e.g., *You want to show the amount for each day separately, but you also want to see how the numbers go up and down.*)

Pathway 2 OPEN-ENDED

Bias and Sampling

You will need

- Student Resource page 403

Open-Ended Intervention

Before Using the Open-Ended Intervention

Tell students that Statistics Canada did a survey in 2009 about the favourite subject of Canadian students in elementary schools. The report said that the favourite subject of about 32% of all students surveyed was physical education. About 14% chose math, and a bit less than 6% chose computers. Ask:

- Do you think just a few elementary schools were sampled? (e.g., *No, I think they asked every elementary school in their files, but not every school responded.*)
- Do you think 20 elementary schools would be a large enough sample for you to trust the results? Why? (*No*, e.g., *The students in those schools could all be similar and not represent what schools are like across the country.*)

Using the Open-Ended Intervention (Student Resource page 403)

Read through the tasks on the student page together. Discuss how the number of people asked determines the reliability of the results and that a good survey must have unbiased questions. Make sure students understand what makes a good survey question. Suggest that if there is time, they might use their survey with classmates or family to try out their questions.

Give students time to work, ideally in pairs.

Observe whether students

- recognize what data needs to be collected to answer the question
- recognize data collection factors that may influence results, such as bias and sample size
- design good survey questions

Consolidating and Reflecting

Ensure understanding by asking questions like these based on students' work:

- How would you make sure that the data you collected was trustworthy? (e.g., *I would collect data first-hand at the crosswalk at different times of day.*)
- It is important that a survey question not be biased. What might be a biased question in your survey? (e.g., *Would a flashing light at the crosswalk be a good idea since it is such a busy street?*)
- What second-hand data might help your study? (e.g., *I could look up data on how many accidents have occurred at this crosswalk, or use data already collected by the school about how many students walk or ride the bus.*)
- How does the number of people surveyed affect the reliability of the results? (e.g., *The more people you ask, the more reliable the results are. But you can't ask too many people, because that takes too long.*)
- What might be a biased sample? (e.g., *asking 100 crossing guards if a school needs a crossing guard*)

 Copyright © 2012 by Nelson Education Ltd.

Bias and Sampling

Pathway 2 GUIDED

Guided Intervention

You will need

- Student Resource pages 404–407

Before Using the Guided Intervention

Tell students that Statistics Canada did a survey in 2009 about the favourite subject of Canadian students in elementary schools. The report said that the favourite subject of about 32% of all students surveyed was physical education. About 14% chose math, and a bit less than 6% chose computers. Ask:

- Do you think that just a few elementary schools were sampled? (e.g., *No, I think they asked every elementary school in their files, but not every school responded.*)
- Do you think 20 elementary schools would be a large enough sample for you to trust the results? Why? (*No.* e.g., *The students in those schools could all be similar and not represent what schools are like across the country.*)

Using the Guided Intervention **Student Resource pages 404–407**

Work through the instructional section with students. Guide them as they consider reasons why a question may be biased. (Providing choices may affect how someone responds, but you can reduce the bias by including "other.") Make sure students understand the importance of reliable data when collecting samples.

Have students work through the **Try These** questions in pairs or individually.

Observe whether students

- recognize and correct biased survey questions (Questions 1, 2, 3, 5)
- can create good survey questions (Questions 3, 4, 5)
- recognize how sample size may influence results (Questions 3, 5, 6, 7)
- can design a survey to answer a question and know who to ask and when to collect the data (Question 4)

Consolidating and Reflecting

Ensure understanding by asking questions such as these based on students' work:

- In Question 3, how would asking more boys affect the results? (e.g., *It may appear that basketball, soccer, and "other" are the most popular—but if more girls were asked, that could change the results.*)
- In Question 4, who would you ask to do your survey? Why? (e.g., *You could ask people leaving a grocery store where bottled water is sold.*)
- In Question 5b), how did you make up your question that is biased? (e.g., *I chose the shopping mall entrance to survey about increasing bus service to the mall. It is biased because people there would certainly want more buses so they don't have to wait very long. They don't think about costs of having more buses.*)
- How does the number of people surveyed affect the reliability of the results? (e.g., *The more people you ask, the more reliable the results are. But you can't ask too many people, because that takes too long.*)

Copyright © 2012 by Nelson Education Ltd.

Pathway 3 OPEN-ENDED

Interpreting Graphs

You will need

- Types of Graphs (BLM 30, excluding the circle graph)
- materials for graphing (e.g., straightedges or rulers, coloured pencils, 1 cm Grid Paper (BLM 12) or 0.5 cm Grid Paper (BLM 13), or graphing software)
- access to the Internet
- Student Resource pages 408–409

Open-Ended Intervention

Before Using the Open-Ended Intervention

Provide Types of Graphs (BLM 30). Have students locate the table of data, and look at the bar graph and the line graph. Ask:

- How can you tell which type of crop is grown least often?
  (e.g., *The shortest bar is for flowers and trees. In the line graph, the lowest line is for flowers and trees.*)
- Both graphs show the same data. Which do you think is easier to interpret? Why? (e.g., *The bar graph is easier because that's what I'm used to.*)
- In the bar graph, how can you tell which type of crop has changed the least in the amount grown over the 10 years?
  (e.g., *All 3 bars are the same height for flowers and trees. The other crops show bars that increase in height over the years.*)
- How does the line graph show this data?
  (e.g., *The line that shows flowers and trees is almost straight across.*)

Using the Open-Ended Intervention (Student Resource pages 408–409)

Read through the tasks on the student pages together. Discuss the different types of tables and graphs that can be used to show data. Provide grid paper or graphing software for creating graphs. Give students time to work, ideally in pairs.

Observe whether students

- interpret a given line graph correctly
- can create a table and a bar graph to display changes in population over time
- can compare or make predictions using tables and graphs
- recognize why a graph might be misleading

Consolidating and Reflecting

Ensure understanding by asking questions such as these based on students' work:

- Which graph do you think is better for seeing changes over time: a line graph or a bar graph?
  (e.g., *maybe a line graph because you don't have to keep track of heights of the bars; maybe a bar graph because the bars stand out better than dots*)
- How did you predict what would happen in the next 5 years?
  (e.g., *From the trend in the starfish graph, it looks like the population of starfish will increase and the live coral will decrease.*)
- Why do graphs make it easy to predict what will happen?
  (e.g., *You can see how numbers are getting bigger, smaller, or staying the same, and it's easier to notice patterns in the graph.*)
- Why might the graph about minimum wages in 2009 look misleading?
  (e.g., *The vertical axis doesn't show the numbers between 0 and $7.*)

 Copyright © 2012 by Nelson Education Ltd.

Interpreting Graphs

Pathway 3
GUIDED

Guided Intervention

Before Using the Guided Intervention

Provide Types of Graphs (BLM 30—you may want to cover up the circle graph, which is not used in this pathway). Have students locate the table of data, and look at the bar graph and the line graph. Ask:

- How can you tell which type of crop is grown least often? (e.g., *The shortest bar is for flowers and trees. In the line graph, the lowest line is for flowers and trees.*)
- Both graphs show the same data. Which do you think is easier to interpret? Why? (e.g., *The bar graph is easier because the years are shown as different bars.*)
- In the bar graph, how can you tell which type of crop has changed the least in the amount grown over the 10 years? (e.g., *All 3 bars are the same height for flowers and trees. The other crops show bars that increase in height over the years.*)
- How does the line graph show this data? (e.g., *The line that shows flowers and trees is almost straight across.*)

You will need

- Types of Graphs (BLM 30, excluding the circle graph)
- materials for graphing (e.g., straightedges or rulers, coloured pencils, 1 cm Grid Paper (BLM 12) or 0.5 cm Grid Paper (BLM 13), or graphing software)
- Student Resource pages 410–413

Using the Guided Intervention Student Resource pages 410–413

Work through the instructional section with students.

Have students work through the **Try These** questions in pairs or individually. They can use grid paper or graphing software to make their own graphs.

Observe whether students

- can interpret a line graph (Questions 1, 3, 5)
- can compare data in a table with similar data in a bar graph (Question 2)
- can make a table or a bar graph (Questions 2, 6)
- can use a table of data to interpret information (Questions 2, 4)
- recognize why a graph might be misleading (Question 5)

Consolidating and Reflecting

Ensure understanding by asking questions such as these based on students' work:

- What could you see straight away when you looked at the graph in Question 1? (e.g., *The line went up and the number of customers increased after the opening.*)
- How did you compare the 2 cities in Question 2? Did you show the other data in the table or add another set of bars to the bar graph? (e.g., *I used the table to draw the bars for Vancouver on the St. John's bar graph. Then I just compared the 2 groups of bars.*)
- What did you notice in Question 3? (e.g., *The population numbers change as in a cycle. When there are lots of rabbits, the coyotes increase for the next few years until there aren't as many rabbits. Then there aren't as many coyotes, so you get more rabbits.*)
- In Question 5, why might the graph be misleading? (e.g., *The vertical axis doesn't show the numbers between 0 and $9.*)

Copyright © 2012 by Nelson Education Ltd.

Summarizing Data

Planning For This Topic

Materials for assisting students with summarizing data consist of a diagnostic tool and 3 intervention pathways. The pathways differ in their focus on different uses of averages. Pathway 1 addresses the effects of outliers and additional data on the mean, median, and mode. Pathway 2 involves the appropriate uses of mean, median, and mode. Pathway 3 focuses on calculating and using the mean for a set of data.

Note: Throughout this resource, the term *mean* is associated with what is sometimes called the *arithmetic mean*—the value calculated by adding the data values and dividing by the number of data values. Some students will meet other types of means (e.g., geometric mean, harmonic mean) in later grades, but it is not necessary to use the word *arithmetic* at this time.

Each pathway has an open-ended intervention and a guided intervention. Choose the type of intervention more suitable for your students' needs and your particular circumstances.

Curriculum Connections

Grades 5 to 8 curriculum connections for this topic are provided online. See www.nelson.com/leapsandbounds. All pathways are suitable for both the WNCP and Ontario curriculums, although range (mentioned in Pathway 1) is not part of the Ontario curriculum at this level.

Professional Learning Connections

PRIME: Data Management and Probability, Background and Strategies (Nelson Education Ltd., 2006) pages 93–97

Making Math Meaningful to Canadian Students K–8 (Nelson Education Ltd., 2008), pages 532–539

Big Ideas from Dr. Small Grades 4–8 (Nelson Education Ltd., 2009), pages 173–176

Good Questions (dist. by Nelson Education Ltd., 2009), pages 161–162, 175

More Good Questions (dist. by Nelson Education Ltd., 2010), pages 160–161, 176–177

Why might students struggle with summarizing data?

Students might struggle with summarizing data for any of the following reasons:

- They might not realize that sets of data can have more than one mode or no mode.
- They might forget to order the data values before determining the median (middle value).
- They might be able to calculate a mean but not understand what it tells about the data.
- They might find it difficult to know whether the mean, median, or mode is most suitable for summarizing a particular set of data for a particular purpose (e.g., they might not know how to apply mean, median, and mode to realistic situations).
- They might not understand the impact of additional data or outliers on the mean, median, and mode of data sets (e.g., the mean may change, but the median and mode can stay the same).
- They might interpret the range of data as the actual maximum and minimum values, rather than as the spread between them.

 Copyright © 2012 by Nelson Education Ltd.

Diagnostic Tool: Summarizing Data

Use the diagnostic tool to determine the most suitable intervention for summarizing data. Provide Diagnostic Tool: Summarizing Data, Teacher's Resource pages 328 to 330, and have students complete it in writing or orally. Have linking cubes available for students to use for modelling, and provide calculators.

See solutions on Teacher's Resource pages 331 to 333.

Intervention Pathways

The purpose of the intervention pathways is to help students determine mean, median, and mode for sets of data, compare these different types of averages, and understand their purpose. As well, students will gain a better sense of how additional data or outliers can affect the mean, median, and mode of a set of data.

There are 3 pathways:

- Pathway 1: Effects of Changing Data
- Pathway 2: Using Mean, Median, and Mode
- Pathway 3: Calculating the Mean

Use the chart below (or the Key to Pathways on Teacher's Resource pages 331 to 333) to determine which pathway is most suitable for each student or group of students.

Diagnostic Tool Results	Intervention Pathway
If students struggle with Questions 8 to 10	use Pathway 1: Effects of Changing Data *Teacher's Resource pages 334–335* *Student Resource pages 414–419*
If students struggle with Questions 4 to 7	use Pathway 2: Using Mean, Median, and Mode *Teacher's Resource pages 336–337* *Student Resource pages 420–424*
If students struggle with Questions 1 to 3	use Pathway 3: Calculating the Mean *Teacher's Resource pages 338–339* *Student Resource pages 425–429*

Copyright © 2012 by Nelson Education Ltd.

Name: ______________________ Date: ______________

Summarizing Data

Diagnostic Tool

You will need

- a calculator
- linking cubes (optional)

1. Here are the times that Anders spent walking to school each day last week:

23 minutes, 25 minutes, 19 minutes, 26 minutes, 22 minutes

What is the mean (average) time for Anders to walk to school?

mean: ______________________

2. Draw a picture that shows why the mean of 2, 5, and 5 is 4.

3. The mean age of 3 people is 38 years. One person is 46 years old. The other 2 people are the same age. What are the ages of the 3 people?

____

4. Determine the mean, median, and mode for a group of people with these ages:

18, 19, 19, 12, 17, 19, 17, 18, 19, 25

mean: ______________ mode: ______________

median: ______________________

5. Create a set of 5 numbers where the mean is much greater than the median.

____

 Copyright © 2012 by Nelson Education Ltd.

Name: ______________________________ Date: ______________

6. This chart shows the number of drinks served per day in a cafeteria.

Drinks Served per Day

Day of the week	Mon.	Tues.	Wed.	Thurs.	Fri.	Sat.
Number of drinks	240	260	340	320	640	480

a) What is the median number of drinks sold per day?

median: ______________________________

What is the mode? ______________________________

b) Why is the median a good way to describe the average number of drinks sold?

____

____

7. The mean number of hours that Terri works each week is 45. The median number of hours is 42. The mode is 38. If Terri includes her travel to and from work, she would add 5 hours to each work week. What would be the new values of the mean, median, and mode?

mean: ______________ mode: ______________

median: ______________________________

8. **a)** Create a set of 6 numbers where the mean is 8. The numbers should all be different.

____

b) You can increase the mean of your data in part a) by 2 by changing only one number. What could the new set of data be?

____

c) Does the range of your data in parts a) and b) change? Why or why not?

____

____

____

Copyright © 2012 by Nelson Education Ltd.

Name: ______________________ Date: ______________________

9. The chart shows Maddie's and Jaime's quiz marks.

Quiz Marks in Health

Maddie	85%	82%	85%	86%	90%	51%
Jaime	80%	78%	83%	83%	81%	79%

If they each dropped their lowest mark, whose mean mark would gain the most? Explain your choice *without* using calculations.

____

____

____

____

10. The line plot shows the amounts of money spent by students on snacks at a school in one day.

Remember

- You read a line plot by counting the number of Xs in each category. e.g., 2 people each spent $5.

Money Spent on Snacks in One Day

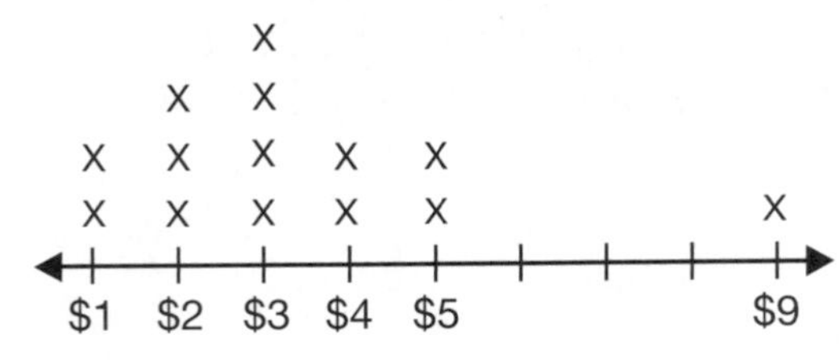

a) What is the mean amount spent on snacks in one day?

____

b) What is the mode? ______________________

c) What is the median? ______________________

d) Identify the outlier (the data that doesn't seem to fit the rest). Describe how the mean, median, and mode would be affected if the outlier was left out.

____

____

 Copyright © 2012 by Nelson Education Ltd.

Solutions and Key to Pathways

Name:________________________ Date:____________

Summarizing Data

Diagnostic Tool

You will need
- a calculator
- linking cubes (optional)

Pathway 3

1. Here are the times that Anders spent walking to school each day last week:

   23 minutes, 25 minutes, 19 minutes, 26 minutes, 22 minutes

   What is the mean (average) time for Anders to walk to school?

   mean: $115 \div 5 = 23$ minutes

2. Draw a picture that shows why the mean of 2, 5, and 5 is 4.

   e.g.,

3. The mean age of 3 people is 38 years. One person is 46 years old. The other 2 people are the same age. What are the ages of the 3 people?

   46, 34, 34

Pathway 2

4. Determine the mean, median, and mode for a group of people with these ages:

   18, 19, 19, 12, 17, 19, 17, 18, 19, 25

   mean: $183 \div 10 = 18.3$ years old   mode: 19 years old

   median: 12, 17, 17, 18, 18, 19, 19, 19, 19, 25; median 18.5 years old

5. Create a set of 5 numbers where the mean is much greater than the median.

   e.g., 1, 2, 3, 4, 100

 Copyright © 2012 by Nelson Education Ltd.

Copyright © 2012 by Nelson Education Ltd.

Name:______________________________ Date:________________

Pathway 2

6. This chart shows the number of drinks served per day in a cafeteria.

Drinks Served per Day

Day of the week	Mon.	Tues.	Wed.	Thurs.	Fri.	Sat.
Number of drinks	240	260	340	320	640	480

a) What is the median number of drinks sold per day?

median: average of 320 and 340 = 330; median: 330 drinks

What is the mode? There is no mode.

b) Why is the median a good way to describe the average number of drinks sold?

e.g., If you didn't want to show all the data, the middle number gives a pretty good idea of a typical day's sales.

7. The mean number of hours that Terri works each week is 45. The median number of hours is 42. The mode is 38. If Terri includes her travel to and from work, she would add 5 hours to each work week. What would be the new values of the mean, median, and mode?

mean: $45 + 5 = 50$ hours mode: $38 + 5 = 43$ hours

median: $42 + 5 = 47$ hours

Pathway 1

8. a) Create a set of 6 numbers where the mean is 8. The numbers should all be different.

e.g., 5, 6, 7, 9, 10, 11 (The sum must be 48.)

b) You can increase the mean of your data in part a) by 2 by changing only one number. What could the new set of data be?

e.g., 5, 6, 7, 9, 10, 23 (Any one number could be increased by 12.)

c) Does the range of your data in parts a) and b) change? Why or why not?

Yes. e.g., The range in part b) is greater because I increased the number at the top end. The range in part a) is $11 - 5 = 6$ and in part b) is $23 - 5 = 18$.

Name:______________________ Date:______________

9. The chart shows Maddie's and Jaime's quiz marks.

Quiz Marks in Health

Maddie	85%	82%	85%	86%	90%	51%
Jaime	80%	78%	83%	83%	81%	79%

If they each dropped their lowest mark, whose mean mark would gain the most? Explain your choice *without* using calculations.

e.g., Maddie's mean mark will increase more because her lowest mark, 51%, is a lot different from her others. Jaime's marks are closer together. Dropping her lowest, 78%, won't affect her mean much.

10. The line plot shows the amounts of money spent by students on snacks at a school in one day.

Remember

- You read a line plot by counting the number of Xs in each category. e.g., 2 people each spent $5.

Pathway 1

Money Spent on Snacks in One Day

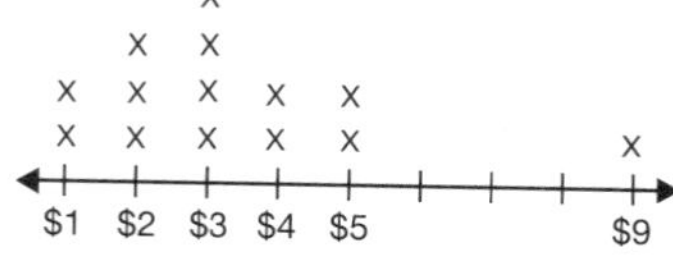

a) What is the mean amount spent on snacks in one day?

47 ÷ 14 is about $3.36.

b) What is the mode? $3 (tallest column)

c) What is the median? $3 (average of 2 middle numbers)

d) Identify the outlier (the data that doesn't seem to fit the rest). Describe how the mean, median, and mode would be affected if the outlier was left out.

The outlier is $9. e.g., The mean drops, but the median and mode do not change.

 Copyright © 2012 by Nelson Education Ltd.

Copyright © 2012 by Nelson Education Ltd.

Effects of Changing Data

You will need

- calculators
- Student Resource pages 414–415

Open-Ended Intervention

Before Using the Open-Ended Intervention

Present a list of heights of 10 students: 158 cm, 162 cm, 159 cm, 156 cm, 155 cm, 170 cm, 158 cm, 150 cm, 155 cm, 155 cm. Ask:

- Which height is the mode? Why? (e.g., *150 cm, since it is there the most*)
- How would you figure out the mean of these heights? (e.g., *I would add the numbers and divide by 10.*)
- Is there one height that seems unlike the rest? (e.g., *170 cm is much taller.*)
- What is the median of the data? How do you figure it out? (*157 cm;* e.g., *You have to put the numbers from smallest to biggest. The middle 2 numbers are 156 and 158. The middle of those 2 numbers is 157.*)

Using the Open-Ended Intervention Student Resource pages 414–415

Read through the tasks on the student pages together. Provide calculators. Give students time to work, ideally in pairs.

Observe whether students

- can calculate mean, median, and mode
- can predict and verify the effects of changing data
- can identify an outlier
- recognize the effect of an outlier on the mean, median, and mode
- realize how the range is related to the spread of the data

Consolidating and Reflecting

Ensure understanding by asking questions such as these based on students' work:

- What did you think about when you were choosing scores so that the mode did or didn't change? (e.g., *I thought about not changing the most frequent number.*)
- What did you have to think about when you were choosing scores so that the median did or didn't change? (e.g., *I thought about whether the new scores would move the median, the middle number, up or down.*)
- What did you have to think about to make sure the mean stayed the same? (e.g., *I would take away the same from some numbers as I added to others.*)
- What did you have to think about to make sure the mean increased? (e.g., *that I increased some numbers more than I decreased others*)
- Suppose you didn't add scores but just increased all the scores by 5. Which of the mean, median, or mode would change? How do you know? (e.g., *The scores all increase by 5 so the mean, median, and mode also increase by 5.*)
- Why might outliers sometimes increase the mean a lot, but sometimes not? (e.g., *If the outliers were both very high, the mean would increase since the total increases. If one was high and one was low, the mean wouldn't change much.*)

 Copyright © 2012 by Nelson Education Ltd.

Effects of Changing Data

Pathway 1 GUIDED

Guided Intervention

You will need

- calculators
- Student Resource pages 416–419

Before Using the Guided Intervention

Present a list of heights of 10 students in a class: 158 cm, 162 cm, 159 cm, 156 cm, 155 cm, 170 cm, 158 cm, 150 cm, 155 cm, 155 cm. Ask:

- ▸ Is there one height that seems unlike the rest? (e.g., *170 cm is much taller.*)
- ▸ What might happen to the mean, median, and mode if you don't include 170 cm? (e.g., *The averages may be different.*)

Using the Guided Intervention **Student Resource pages 416–419**

Work through the instructional section of the student pages together. Provide calculators.

Have students work through the **Try These** questions individually or in pairs.

Observe whether students

- can calculate the mean, median, and mode (Questions 1, 2, 3, 6)
- can predict and verify how a mean changes with specific changes in the data (Questions 1, 3, 7)
- can identify outliers in a set of data and know if they should be included or not (Questions 2, 3, 5, 6)
- recognize the outliers and how they affect the mean, median, and mode, and whether they need to be included (Questions 2, 5, 6, 8)
- can create a set of data with a given mean (Questions 4, 5)
- realize how range is related to the spread of a set of data (Question 8)

Consolidating and Reflecting

Ensure understanding by asking questions such as these based on students' work:

- ▸ In Question 1, how does increasing each hourly rate by the same amount affect the mean, median, and mode? (e.g., *They all increase by the same amount.*)
- ▸ In Question 3c), when the longer time is included, which average is affected the most? Why? (*the mean;* e.g., *Putting in the longer time increases the total, so it increases the mean. Since the median was already in the high end and not near the middle, it didn't change much. The mode didn't change at all.*)
- ▸ In Question 4, can the missing number be less than 10? (*No;* e.g., *The mean is 15, and a larger number has to be added to bring the total up to 75.*)
- ▸ Why might outliers sometimes increase the mean a lot, but sometimes not? (e.g., *If the outliers were both very high, the mean would increase since the total increases. If one was high and one was low, the mean wouldn't change much.*)
- ▸ Why do you need to know the context to decide whether to remove outliers from a set of data? (e.g. *If you make an error and get an outlier, then don't include it. But sometimes, data is really variable and you just haven't collected enough to realize. Then the outlier should be included.*)

Copyright © 2012 by Nelson Education Ltd.

Pathway 2
OPEN-ENDED

Using Mean, Median, and Mode

You will need

- calculators
- Student Resource page 420

Open-Ended Intervention

Before Using the Open-Ended Intervention

Present this data set for hourly rates at a restaurant:
- all waiters and waitresses: $10.00/hour
- all dishwashers, attendants: $9.50/hour
- all short order cooks and bartenders: $10.50/hour
- the restaurant owner: $40.00/hour

- ▸ What do you think the overall average hourly rate would be close to? (e.g., *It would be close to $10/hour.*)
- ▸ A mode is a number that appears most often. What would most likely be the mode? (e.g., *I think the mode would be $10. There are more waiters and waitresses, and they all make $10 per hour.*)
- ▸ If you were the restaurant owner and wanted to tell someone the average salary, would you use the mean, median, or mode? (e.g., *the mean, because it is the highest, since the $40/hour pulls it up*)

Using the Open-Ended Intervention Student Resource page 420

Read through the tasks on the student page together. Provide calculators. Ensure students understand the meaning of mean, median, and mode for this problem. Give students time to work, ideally in pairs.

Observe whether students
- can create a set of data with a specific mean
- can create a set of data with a specific mode
- can create a set of data with a specific median
- can determine mean, median, and mode for a set of data
- recognize that it is important to know which average is being used

Consolidating and Reflecting

Ensure understanding by asking questions such as these based on students' work:
- ▸ What situation can you think of where the mode would be the best average to describe the sample set? (e.g., *I could use mode to describe the most popular size of T-shirt sold in a store.*)
- ▸ Would you use the mean score or median score to predict someone's basketball points for a game? (e.g., *The median score is better to predict the score, especially if they have 1 or 2 scores that are unusually high or low that would affect the mean point scores.*)
- ▸ Why is it important to say the kind of average you are talking about instead of just saying average? (e.g., *If you know something is the mode, it really doesn't tell you much about the data; but if you know it's the median, you know that half the data is above and half below, so you have a better sense of it. If it's the mean, you know the data total, so you have some sense of it.*)

 Copyright © 2012 by Nelson Education Ltd.

Using Mean, Median, and Mode

Pathway 2 GUIDED

Guided Intervention

You will need

- calculators
- Student Resource pages 421–424

Before Using the Guided Intervention

Present this data set for hourly rates at a restaurant to students:

- all waiters and waitresses: $10.00/hour
- all dishwashers, attendants: $9.50/hour
- all short order cooks and bartenders: $10.50/hour
- the restaurant owner: $40.00/hour

- ▸ What do you think the overall average hourly rate would be close to? (e.g., *It would be close to $10/hour.*)
- ▸ A mode is a number that appears most often. What would most likely be the mode? (e.g., *I think the mode would be $10. There are more waiters and waitresses, and they all make $10 per hour.*)
- ▸ If you were the restaurant owner and wanted to tell someone the average salary, would you use the mean, median, or mode? (e.g., *the mean, because it is the highest, since the $40/hour pulls it up*)

Using the Guided Intervention **Student Resource pages 421–424**

Work through the instructional section of the student pages together. Provide calculators.

Have students work through the **Try These** questions individually or in pairs.

Observe whether students

- can determine the mean, median, and mode (Questions 1, 2, 5, 6)
- rationalize whether to use the mean, median, or mode to describe the average of a set of data (Questions 2, 3, 6, 7)
- can create a data set with a given mean (Question 4)

Consolidating and Reflecting

Ensure understanding by asking questions such as these based on students' work:

- ▸ In Question 1, how does knowing the mean number of tickets sold per year help the organizers? (e.g., *It would help them decide which events were most popular over the last 5 years and predict how many people would attend.*)
- ▸ In Question 3, why does the mean not tell you if you are in the top half of the fitness group? (e.g., *The median is the middle score, so it is the only average that will tell you if you are in the top or bottom half of the scores.*)
- ▸ In Question 6, if there was an event for young children at the community centre, would the mean age go down? (e.g., *It depends on how many children come and how many are with their parents or older siblings. If there are more ages under the mean of 18 years old, then the mean would decrease.*)
- ▸ Would that change the average you would use to describe the ages of who goes to the centre? (e.g., *no, since the middle number feels right to use; it is safest*)

Copyright © 2012 by Nelson Education Ltd.

Pathway 3 OPEN-ENDED

Calculating the Mean

You will need

- 3 containers with coloured cubes (e.g., one with 6 white, 6 black, 5 grey; one with 7 white, 3 black, 4 grey; one with 5 white, 3 black, 6 grey)
- square tiles or linking cubes
- calculators (optional)
- Student Resource page 425

Open-Ended Intervention

Before Using the Open-Ended Intervention

Show students 3 containers (or pictures) like the following:

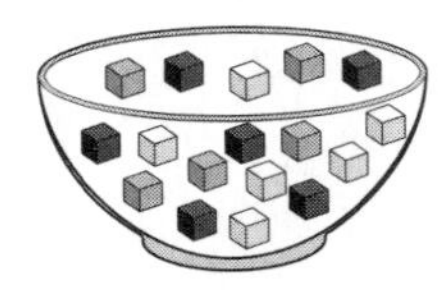 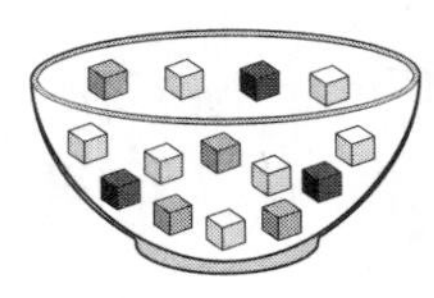 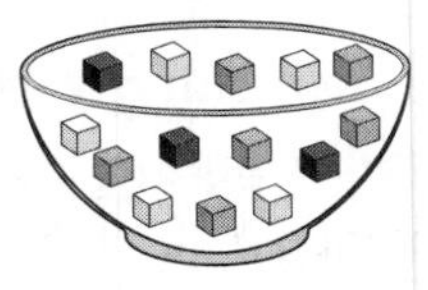

Make sure that students understand the definition of *mean*. Ask:

- ▸ Just looking at the bowls, what is the mean number of white cubes in a bowl? How did you know? (e.g., *The mean number of white cubes is 6. I see 6 white cubes in a bowl, and 1 more than 6 and 1 less than 6 in the other bowls.*)
- ▸ What are the mean numbers of grey cubes and black cubes? (e.g., *I see 5 grey in a bowl, and 1 more than 5 and 1 less than 5 in the other bowls, so the mean is 5. There are fewer black, so I think the mean number of black cubes is less than 5, or 4.*)
- ▸ How can you figure out the mean by stacking the black cubes? (e.g., *Take out all the black cubes and stack them, and then divide them into 3 equal stacks.*)

Using the Open-Ended Intervention (Student Resource page 425)

Read through the tasks on the student page together. Provide the materials and give students time to work, ideally in pairs.

Observe whether students

- can demonstrate concretely what the mean represents and how to calculate it
- can determine the mean for a set of data
- can compare the means of 2 sets of data
- can create a set of data with a specific mean

Consolidating and Reflecting

Ensure understanding by asking questions such as these based on students' work:

- ▸ Why did you try to even out your data rows to calculate the mean? (e.g., *The mean is based on sharing the total data equally, so that's what I did.*)
- ▸ What litter numbers did you pick for Buttercup? (e.g., *6, 9, 7, 10, 8*)
- ▸ Would you have tried to even out your data rows if the last litter number had been 7? (e.g., *I couldn't have made the rows all equal—there would have been 4 rows of 7 and one row of 8. I'd have to imagine splitting the extra 1 tile among 5 rows, and each row would increase by 0.2. The mean would have been 0.2 more.*)
- ▸ Does the mean always increase if 1 or 2 pieces of data increase? (*Yes,* e.g., *if none of the data decreases to make up for it*)
- ▸ What is an easy way to create a set of data with a given mean? (e.g., *You use that number over and over, or you take away from it and add to it by the same amounts.*)

 Copyright © 2012 by Nelson Education Ltd.

Calculating the Mean

Pathway 3
GUIDED

Guided Intervention

Before Using the Guided Intervention

Present the following situation: 4 brothers are pooling their money to buy their mother a present; they have $17, $23, $29, and $31. Ask:

- How would you calculate a fair share for each brother to contribute? (e.g., *I would divide 100 by 4 to get $25.*)
- How much less or more than his share does each boy contribute? (*$8 less, $2 less, $4 more, and $6 more*)
- What do you notice about the amounts? (e.g., *The "less" amounts and the "more" amounts both add to $10.*)

You will need

- small blank cards or pieces of paper
- linking cubes (5 colours)
- calculators (optional)
- 1 cm Grid Paper (BLM 12) or 0.5 cm Grid Paper (BLM 13) (optional)
- Student Resource pages 426–429

Using the Guided Intervention Student Resource pages 426–429

Work through the instructional section of the student pages together. Students can use grid paper or sketch a simple bar graph for Question 10.

Have students work through the **Try These** questions individually or in pairs.

Observe whether students

- demonstrate concretely what the mean represents and how to calculate it (Question 1)
- can determine the mean for a set of data (Questions 2, 3, 4, 5, 9, 10)
- show an understanding of how the mean of a set of numbers relates to the total (Questions 2, 9, 10)
- can create a set of data with a specific mean (Questions 5, 6, 7, 8)

Consolidating and Reflecting

Ensure understanding by asking questions such as these based on students' work:

- In Question 1, you figured out the mean of 3 numbers in a row (consecutive numbers). Would the idea change if there had been 5 numbers in a row? (*No, it would still be the middle number.* e.g., *For 2, 3, 4, 5, 6, the mean is 4.*)
- In Question 5b), how did you decide what the additional speed was to reach a mean of 65 km/h? (e.g., *I looked at how far away each of the 5 given speeds were from 65. I picked a 6th speed so that the amount below 65 matched the amount above 65. Below 65 were 2, 10, and 6 for a total of 18. Above 65, there were already 3, so I needed to be 15 above 65, which is 80.*)
- In Question 9, did you think the new mean mark would go up or down by a lot or a little? Were you right? (e.g., *There are more tests, but the extra test mark is higher than the first mean of 82%. I thought it would go up by a lot because it was so much higher than the mean—but it went up only by 1%.*)
- How did you sketch your answer for Question 10? (e.g., *I sketched a graph with 3 bars. One shows 11 as 9 below 20, and another shows 25 as 5 above 20, so the 3rd bar shows 4 above 20.*)

Copyright © 2012 by Nelson Education Ltd.

Strand: Data and Probability

Probability

Planning For This Topic

Materials for assisting students with probability consist of a diagnostic tool and 3 intervention pathways. The pathways differ in the type of probability or situation involved. Pathway 1 focuses on the combined probability of 2 independent events. Pathway 2 addresses the theoretical probability of a single event. Pathway 3 focuses on experimental probability.

Each pathway has an open-ended intervention and a guided intervention. Choose the type of intervention more suitable for your students' needs and your particular circumstances.

Professional Learning Connections

PRIME: Data Management and Probability, Background and Strategies (Nelson Education Ltd., 2006), pages 101, 104–111, 114–118
Making Math Meaningful to Canadian Students K–8 (Nelson Education Ltd., 2008), pages 544, 546–554, 557–561
Big Ideas from Dr. Small Grades 4–8 (Nelson Education Ltd., 2009), pages 197–206
Good Questions (dist. by Nelson Education Ltd., 2009), pages 162–164, 173–174, 176–179
More Good Questions (dist. by Nelson Education Ltd., 2010), pages 161–162, 178

Curriculum Connections

Grades 5 to 8 curriculum connections for this topic are provided online. See www.nelson.com/leapsandbounds. All 3 pathways are relevant to both the Ontario and WNCP curriculums.

Why might students struggle with probability?

Students might struggle with probability for any of the following reasons:

- They might not know the difference between determining the probability through an experiment and determining the theoretical probability (e.g., theoretical probability does not change, but experimental probability can change with each trial).
- They might not understand that the number of times an experiment is repeated changes the experimental probability of the event, but not the theoretical probability.
- They might think that all outcomes are equally likely even when they are not (e.g., spinning a spinner where the sections are not equal) and might not know how this could affect both the experimental probability and theoretical probability.
- They might not understand that an event such as tossing a coin has the same probability each time and does not depend on previous events.
- They might not be able to use strategies such as tree diagrams or organizational charts to help them figure out combined probabilities.
- They might not be able to tell the difference between dependent events and independent events.

 Copyright © 2012 by Nelson Education Ltd.

Diagnostic Tool: Probability

Use the diagnostic tool to determine the most suitable intervention pathway for probability. Provide Diagnostic Tool: Probability, Teacher's Resource pages 342 to 344, and have students complete it in writing or orally. Provide dice and coins, and also have available the following optional materials: small paper bags and identical coloured objects or pieces of paper (5 blue, 3 purple, 2 white, 2 red, 2 yellow, 1 orange); and Spinners (BLM 18), and pencils and paper clips.

See solutions on Teacher's Resource pages 345 to 347.

Intervention Pathways

The purpose of the intervention pathways is to help students with predicting probabilities using either fractions or percents. Students also conduct experiments to check whether their predictions seem reasonable. Later on, students should be able to relate experimental probabilities to theoretical probabilities and understand that for a greater number of experiments, the probability will be closer to the theoretical value.

There are 3 pathways:

- Pathway 1: Probability: Independent Events
- Pathway 2: Theoretical Probability
- Pathway 3: Experimental Probability

Use the chart below (or the Key to Pathways on Teacher's Resource pages 345 to 347) to determine which pathway is most suitable for each student or group of students.

Diagnostic Tool Results	Intervention Pathway
If students struggle with Questions 6 to 8	use Pathway 1: Probability: Independent Events *Teacher's Resource pages 348–349* *Student Resource pages 430–436*
If students struggle with Questions 4 and 5	use Pathway 2: Theoretical Probability *Teacher's Resource pages 350–351* *Student Resource pages 437–442*
If students struggle with Questions 1 to 3	use Pathway 3: Experimental Probability *Teacher's Resource pages 352–353* *Student Resource pages 443–446*

If students successfully complete Pathway 2, they may or may not need the additional intervention provided by Pathway 1. Either re-administer Pathway 1 questions from the diagnostic or encourage students to do a portion of the open-ended intervention for Pathway 1 to decide if more work in that pathway would be beneficial.

Copyright © 2012 by Nelson Education Ltd.

Name: ______________________________ Date: ____________________

Probability

Diagnostic Tool

1. Daniel used this spinner to see which chore he would do on Saturday. Can you be sure of the number of times you have to spin the spinner before you land on "wash dishes"? Why or why not?

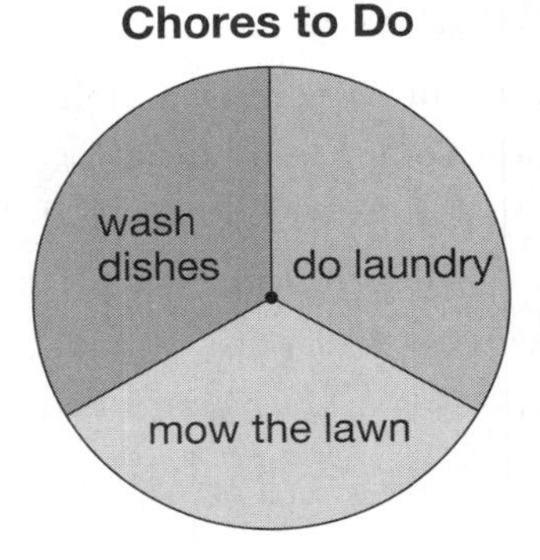

____

____

____

You will need

- 2 dice
- a coin
- a paper bag and identical coloured objects or pieces of paper: 5 blue, 3 purple, 2 white, 2 red, 2 yellow, 1 orange (optional)
- a pencil and a paper clip (optional)
- Spinners (BLM 18) (optional)

2. Suppose that to start a board game, you have to roll doubles.

a) Roll 2 dice 10 times and record whether you roll doubles.

Roll number	1	2	3	4	5	6	7	8	9	10
Doubles?										

b) Determine the experimental probability of rolling a double in 10 tries. Write it as a fraction and as a percent.

________ ________

3. Suppose you tossed a coin 4 times and got heads every time.

a) What was the experimental probability of getting heads? Write it as a fraction and as a percent.

________ ________

b) If you toss the coin 4 more times, will the probability be the same as what you got in part a)? Why or why not?

____

____

c) Toss a coin 20 times. Show your results in the chart.

Heads																					
Tails																					

Write the experimental probability of getting heads as a fraction. ________

 Copyright © 2012 by Nelson Education Ltd.

Name:______________________________ Date:____________________

4. **a)** Carly reached into her drawer and picked a T-shirt, without looking. She has 5 blue, 3 purple, 2 white, and 2 red T-shirts. What is the probability that she picked a white T-shirt?

____

b) Brad reached into his drawer and picked a T-shirt, without looking. The probability of picking a red T-shirt is $\frac{3}{10}$ more than the probability of picking a white T-shirt. How many and what colours of T-shirts might be in his drawer?

____

5. A hat contains 100 cards that show all the counting numbers from 1 to 100. If you pull a card out of the hat without looking, what is each probability below? Express each probability as a fraction and a percent.

a) the probability that the number will be even:

_______ _______

b) the probability that the number will have 2 digits (e.g., 12, 33, or 98):

_______ _______

c) the probability that the number will be a multiple of 10 (e.g., 10, 20, 30,...):

_______ _______

6. Fifi the poodle has 4 sweaters and 2 bows. This tree diagram shows the possible outfits.

pink sweater	pink bow	(pink sweater, pink bow)
	blue bow	(pink sweater, blue bow)
black sweater	pink bow	(black sweater, pink bow)
	blue bow	(black sweater, blue bow)
blue sweater	pink bow	(blue sweater, pink bow)
	blue bow	(blue sweater, blue bow)
red sweater	pink bow	(red sweater, pink bow)
	blue bow	(red sweater, blue bow)

pink sweater

black sweater

pink bow

blue bow

blue sweater

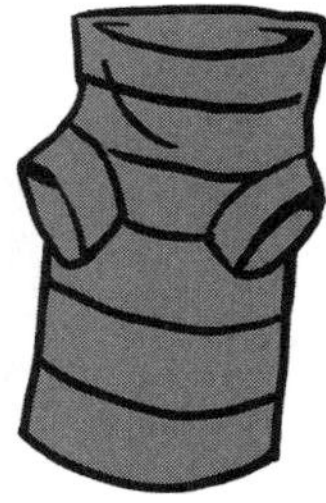
red sweater

Use the tree diagram to figure out the probability that Fifi's outfit will include a pink sweater and pink bow or a blue sweater and blue bow.

____

Copyright © 2012 by Nelson Education Ltd.

Name:______________________________ Date:______________

7. Alyssa and Colton have a paper bag that contains 2 yellow cubes, 3 blue cubes, and 1 orange cube.

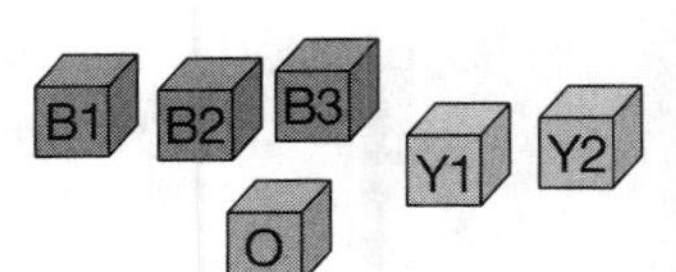

a) Alyssa will pick a cube from the bag without looking and put it back. What is the probability that the cube will be blue? ________

b) Alyssa will pick a cube without looking and put it back. Then she will pick a second cube. The chart below can be used to show the possible results.

Combinations for yellow cube 1 (Y1) are shown.

Complete the chart to figure out the probability that Alyssa will pick a blue cube first and then a yellow cube.

1st cube	Y1	Y1	Y1	Y1	Y1	Y1	Y2	Y2	Y2	Y2	Y2	Y2						
2nd cube	Y1	Y2	B1	B2	B3	O												

1st cube																		
2nd cube																		

The probability of Alyssa picking a blue cube, putting it back, and then picking a yellow cube is ________.

c) Colton will pick a cube from the bag without looking and then put it back. What is the probability that the cube will be orange? ________

d) What is the probability of Colton picking the orange cube, putting it back, and then picking a blue cube? ________

8. You spin a spinner and roll a die. Sketch the spinner for each situation.

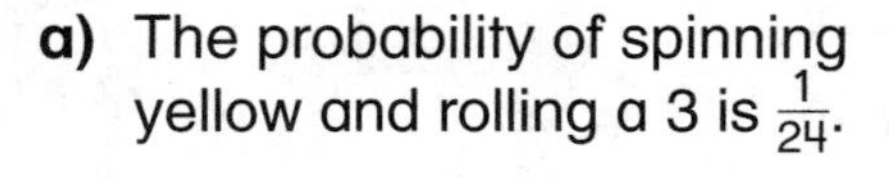

a) The probability of spinning yellow and rolling a 3 is $\frac{1}{24}$.

b) The probability of spinning blue and rolling an even number is $\frac{1}{4}$.

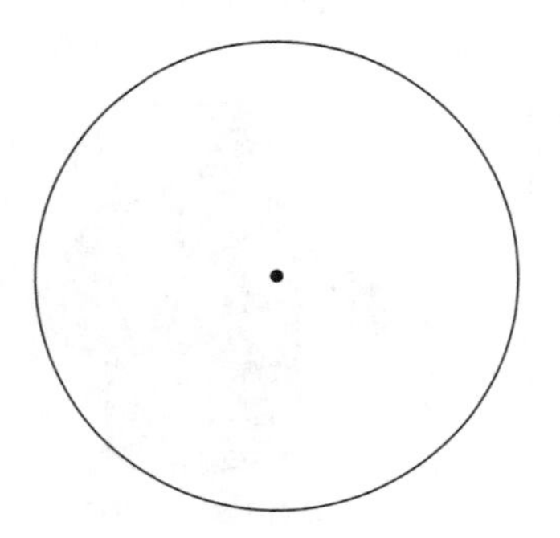

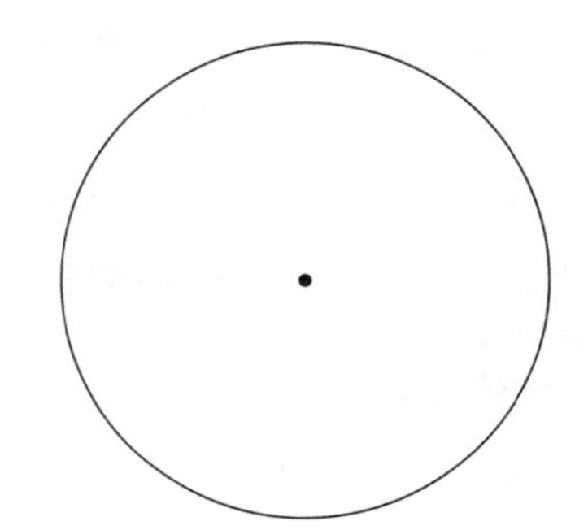

 Copyright © 2012 by Nelson Education Ltd.

Name: ______________________ Date: ______________

Probability

Diagnostic Tool

You will need

- 2 dice
- a coin
- a paper bag and identical coloured objects or pieces of paper: 5 blue, 3 purple, 2 white, 2 red, 2 yellow, 1 orange (optional)
- a pencil and a paper clip (optional)
- Spinners (BLM 18) (optional)

Pathway 3

1. Daniel used this spinner to see which chore he would do on Saturday. Can you be sure of the number of times you have to spin the spinner before you land on "wash dishes"? Why or why not?

Chores to Do

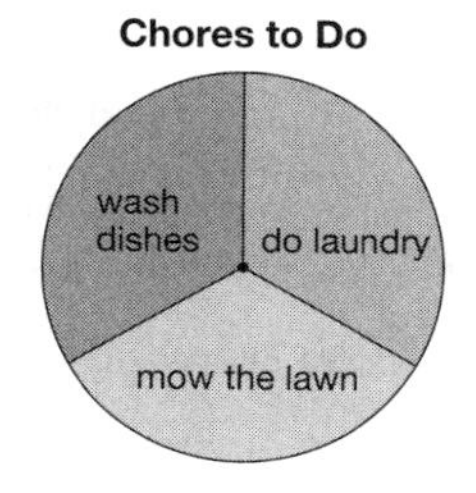

No, you can't be sure. e.g., You could land on it the first time, or you could spin many times and not land on it.

2. Suppose that to start a board game, you have to roll doubles.

a) Roll 2 dice 10 times and record whether you roll doubles.

e.g.,

Roll number	1	2	3	4	5	6	7	8	9	10
Doubles?	✗	✗	✓	✗	✗	✓	✗	✓	✗	✗

b) Determine the experimental probability of rolling a double in 10 tries. Write it as a fraction and as a percent.

e.g., $\frac{3}{10}$ 30%

3. Suppose you tossed a coin 4 times and got heads every time.

a) What was the experimental probability of getting heads? Write it as a fraction and as a percent.

$\frac{4}{4}$ 100%

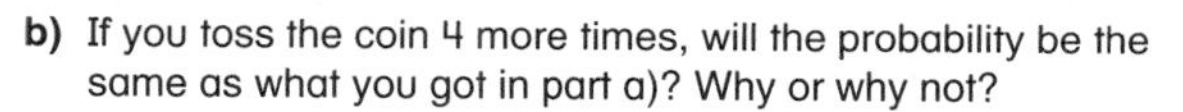

b) If you toss the coin 4 more times, will the probability be the same as what you got in part a)? Why or why not?

e.g., The probability could be different or the same, because it is based on an experiment.

c) Toss a coin 20 times. Show your results in the chart.

Heads	✓	✓			✓		✓	✓	✓					✓	✓		✓			
Tails			✓	✓		✓				✓	✓	✓	✓			✓		✓	✓	✓

Write the experimental probability of getting heads as a fraction. e.g., $\frac{9}{20}$

 Copyright © 2012 by Nelson Education Ltd.

Copyright © 2012 by Nelson Education Ltd.

Name:________________________ Date:____________

Pathway 2

4. a) Carly reached into her drawer and picked a T-shirt, without looking. She has 5 blue, 3 purple, 2 white, and 2 red T-shirts. What is the probability that she picked a white T-shirt?

$\frac{2}{12}$ (or $\frac{1}{6}$)

b) Brad reached into his drawer and picked a T-shirt, without looking. The probability of picking a red T-shirt is $\frac{3}{10}$ more than the probability of picking a white T-shirt. How many and what colours of T-shirts might be in his drawer?

e.g., could be 10 T-shirts: 4 red, 1 white, 5 blue

5. A hat contains 100 cards that show all the counting numbers from 1 to 100. If you pull a card out of the hat without looking, what is each probability below? Express each probability as a fraction and a percent.

a) the probability that the number will be even:

$\frac{1}{2}$ 50%

b) the probability that the number will have 2 digits (e.g., 12, 33, or 98):

$\frac{90}{100}$ 90%

c) the probability that the number will be a multiple of 10 (e.g., 10, 20, 30,...):

$\frac{10}{100}$ 10%

Pathway 1

6. Fifi the poodle has 4 sweaters and 2 bows. This tree diagram shows the possible outfits.

pink sweater	pink bow	(pink sweater, pink bow) (circled)
	blue bow	(pink sweater, blue bow)
black sweater	pink bow	(black sweater, pink bow)
	blue bow	(black sweater, blue bow)
blue sweater	pink bow	(blue sweater, pink bow)
	blue bow	(blue sweater, blue bow) (circled)
red sweater	pink bow	(red sweater, pink bow)
	blue bow	(red sweater, blue bow)

pink sweater black sweater

pink bow blue bow

blue sweater red sweater

Use the tree diagram to figure out the probability that Fifi's outfit will include a pink sweater and pink bow or a blue sweater and blue bow.

$\frac{2}{8}$ or $\frac{1}{4}$

Name:____________________ Date:____________________

7. Alyssa and Colton have a paper bag that contains 2 yellow cubes, 3 blue cubes, and 1 orange cube.

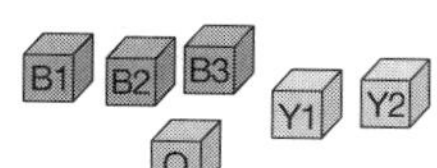

a) Alyssa will pick a cube from the bag without looking and put it back. What is the probability that the cube will be blue? $\frac{3}{6}$ or $\frac{1}{2}$

b) Alyssa will pick a cube without looking and put it back. Then she will pick a second cube. The chart below can be used to show the possible results.

Combinations for yellow cube 1 (Y1) are shown.

Complete the chart to figure out the probability that Alyssa will pick a blue cube first and then a yellow cube.

1st cube	Y1	Y1	Y1	Y1	Y1	Y1	Y2	Y2	Y2	Y2	Y2	Y2	B1	B1	B1	B1	B1	B1
2nd cube	Y1	Y2	B1	B2	B3	O	Y1	Y2	B1	B2	B3	O	Y1	Y2	B1	B2	B3	O

1st cube	B2	B2	B2	B2	B2	B2	B3	B3	B3	B3	B3	B3	O	O	O	O	O	O
2nd cube	Y1	Y2	B1	B2	B3	O	Y1	Y2	B1	B2	B3	O	Y1	Y2	B1	B2	B3	O

Pathway 1

The probability of Alyssa picking a blue cube, putting it back, and then picking a yellow cube is $\frac{6}{36}$ or $\frac{1}{6}$.

c) Colton will pick a cube from the bag without looking and then put it back. What is the probability that the cube will be orange? $\frac{1}{6}$

d) What is the probability of Colton picking the orange cube, putting it back, and then picking a blue cube? $\frac{3}{36}$ or $\frac{1}{12}$

8. You spin a spinner and roll a die. Sketch the spinner for each situation.

a) The probability of spinning yellow and rolling a 3 is $\frac{1}{24}$.

b) The probability of spinning blue and rolling an even number is $\frac{1}{4}$.

e.g.,

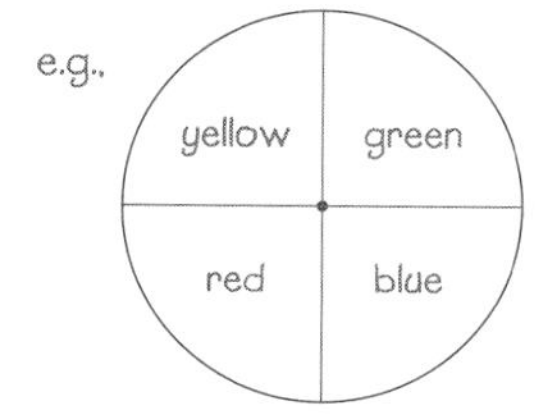

e.g.,

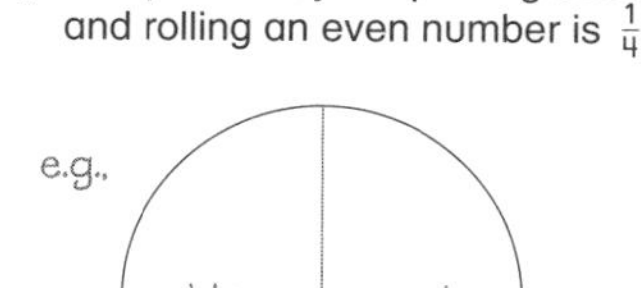

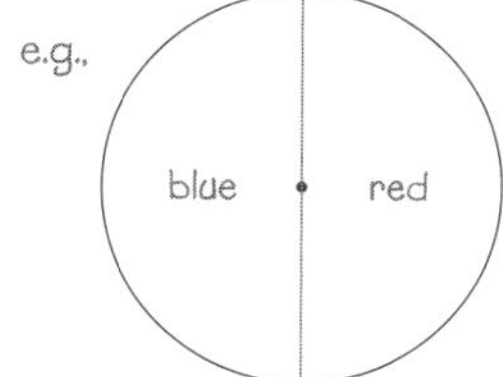

 Copyright © 2012 by Nelson Education Ltd.

Copyright © 2012 by Nelson Education Ltd.

Probability: Independent Events

You will need

- pencils and paper clips
- calculators
- Student Resource pages 430–431

Outcomes of Rock, Paper, Scissors

	P1 R	P1 P	P1 S
P2 R	Tie	P1 wins	P2 wins
P2 P	P2 wins	Tie	P1 wins
P2 S	P1 wins	P2 wins	Tie

P1 means Player 1
P2 means Player 2

Open-Ended Intervention

Before Using the Open-Ended Intervention

Discuss the *rock, paper, scissors* game and how a game is won. (Rock wins over scissors; scissors wins over paper; paper wins over rock.) Have students play the game in pairs 3 times and record the results (P1 means Player 1). Ask:

- ▸ What are all the possible combinations of rock, paper, and scissors for playing the game in pairs? (e.g., *RR, RP, RS, PR, PP, PS, SR, SP, SS*)

Use a chart to help students see all the possible outcomes of a game. Ask:

- ▸ What is the probability that Player 1 will win a game? (e.g., *3 out of 9*)
- ▸ What is the probability of a tied game? (e.g., *3 out of 9*)
- ▸ Do you think one person's choice affects the other person's choice of rock, paper, or scissors? (e.g., *No.*)

Explain that if we think of each person's choice as not affecting the other's, then the events are independent.

Using the Open-Ended Intervention Student Resource pages 430–431

Provide the materials. Read through the tasks on the student pages together. Make sure students realize that they need to decide how many different times there will be (how many sections on a spinner), and what the times will be.

Give students time to work, ideally in pairs, with each student using one spinner.

Observe whether students

- make reasonable predictions about the probability of an event
- understand that spinning the 2 spinners are independent events
- can conduct an experiment to determine experimental probability
- can determine theoretical (or expected) probability

Consolidating and Reflecting

Ensure understanding by asking questions such as these based on students' work:

- ▸ Why did you choose those times for your spinners? (e.g., *I made times so that most of the 2 spins would end up with less than 10 hours of sleep.*)
- ▸ What is the probability of spinning one of the times on your wake-up time spinner? (e.g., *Each time is in one of the 4 equal sections on my spinner, so the probability of spinning any of the times is* $\frac{1}{4}$.)
- ▸ Does the result of spinning the first spinner affect the result you get on the second spinner? (e.g., *No, one spin doesn't affect the other.*)
- ▸ How do you know the theoretical probability of sleeping more than 10 hours? (e.g., *I made a chart of the possible outcomes and then counted the number of outcomes that involve more than 10 hours.*)
- ▸ How did this compare to your experimental results? (e.g., *It was lower.*)

 Copyright © 2012 by Nelson Education Ltd.

Probability: Independent Events

Pathway 1
GUIDED

Guided Intervention

Before Using the Guided Intervention

Show these statements to students, and ask if they think the first outcome affects the next outcome.

1. I tossed a fair coin 4 times and got 4 heads. My next toss will be tails.
2. Four red candies are in a bag of 20 candies. I pull out a red one and eat it. The chance of pulling out a red candy again is less.
   (e.g., *1. No, each toss can be heads or tails. Both are equally likely.*
   2. Yes, the chances of getting a red one the second time are less because only 3 of the 19 candies left are red.)

Using the Guided Intervention Student Resource pages 432–436

Provide the materials and guide students through the instructional section.

Have students work through the **Try These** questions individually or in pairs.

Observe whether students

- can use a chart or tree diagram to determine the theoretical (or expected) probability of 2 independent events (Questions 1, 2, 4, 5)
- can identify when 2 events are independent (Question 3)
- can create a situation likely to lead to a required result (Question 6)

Consolidating and Reflecting

Ensure understanding by asking questions such as these based on students' work:

- How did you figure out the probability of both events happening in Question 1a)? (e.g., *There's only 1 cell in the chart for both heads and 1.*)
- Why isn't the experimental probability the same as the theoretical probability in Question 1b)? (e.g., *The results change every time you do an experiment.*)
- In Question 2, if Jocelyn picked a card and the card was not a star, would it be more likely that Liam would pick a star next?
  (e.g., *No, the probability of Liam picking a star is the same as for Jocelyn.*)
- Choose one of the situations in Question 3. How do you know the events are independent? (e.g., *Rolling a 6 has no effect on tossing a coin to get heads.*)
- For Question 5a), do you think that the probability of the next 3 customers being seniors is big or small? How do you know?
  (*small,* e.g., *since there's only a 1 in 3 chance that a customer will be a senior*)
- How did you decide how many of each colour would go in each bag in Question 6? (e.g., *I tried putting different combinations of white and grey cubes in 2 bags and then made a chart to find the possible outcomes for pulling 1 cube out of each bag. This helped me find the probability of pulling 1 white and 1 grey from each bag and seeing if it is close to* $\frac{1}{2}$.)

You will need

- materials to model games (e.g., Spinners (BLM 18), pencils and paper clips, playing cards, dice, coloured counters, blank cards or paper)
- calculators
- coins (e.g., pennies)
- Student Resource pages 432–436

Copyright © 2012 by Nelson Education Ltd.

Pathway 2 OPEN-ENDED

Theoretical Probability

You will need

- a paper bag (or container that you can't see inside of)
- cubes or counters (2 colours)
- Spinners (BLM 18)
- pencils and paper clips
- dice
- calculators
- Student Resource pages 437–438

Open-Ended Intervention

Before Using the Open-Ended Intervention

Show students a paper bag that contains 2 yellow cubes and 3 red cubes. Ask:

- ▸ How do you know you are more likely to pick a red cube than a yellow cube? (*There are more red cubes.*)
- ▸ Why might you call the probability of picking a yellow cube two fifths? (e.g., *There are 5 cubes and 2 of them are yellow.*)
- ▸ Does it make sense that the probability of $\frac{2}{5}$ for yellow is less than $\frac{1}{2}$? (*Yes.* e.g., *Less than half the cubes are yellow.*)
- ▸ Why is the probability a fraction less than 1? (e.g., *Picking a yellow cube might happen but it might not, and the probability of 1 means it has to happen.*)

Using the Open-Ended Intervention **Student Resource pages 437–438**

Discuss the tasks on the student pages together. Provide the materials.

Give students time to work, ideally in pairs, and try out their games.

Observe whether students

- can calculate the theoretical probability of winning a point in each game
- can determine probabilities and compare likelihoods of winning
- can carry out experiments to see how close they are to their predictions
- can compare experimental probability with theoretical probability

Consolidating and Reflecting

Ensure understanding by asking questions such as these based on students' work:

- ▸ How did you figure out all of the possible outcomes for the original game? (e.g., *I listed all of the possible outcomes for the spinner and the die.*)
- ▸ Which one of your games would you prefer to play? Why? (e.g., *Game 2. It is fair and both players have the same chance of winning a point.*)
- ▸ What makes a game fair? (e.g., *The chance of winning the game is the same as losing the game.*)
- ▸ Did your experimental results match your predictions? (e.g., *Not exactly, since I expected that in Game 2 both players would get the same number of wins, but that's not how it turned out. Player 1 got more wins.*)
- ▸ Why can you write a probability as a fraction or a percent? (e.g., *The denominator is the total number, and the numerator is the number of times you get what you want. A percent is like a fraction out of 100.*)
- ▸ When would the probability of something happening be 0% or 100%? (*when it's impossible or certain to happen*)
- ▸ Why can experimental probability change but theoretical probability can't? (e.g., *Probability can change every time you do an experiment. Theoretical is what you expect to happen. It depends only on the possibilities, which don't change.*)

Copyright © 2012 by Nelson Education Ltd.

Theoretical Probability

Pathway 2 GUIDED

Guided Intervention

You will need

- a paper bag (or a container you can't see inside of)
- coloured cubes
- coloured pencils (red and blue)
- calculators
- coins
- Student Resource pages 439–442

Before Using the Guided Intervention

Show students a paper bag that contains 2 yellow cubes and 3 red cubes. Ask:

- ▸ How do you know you are more likely to pick a red cube than a yellow cube? (*There are more red cubes.*)
- ▸ Why might you call the probability of picking a yellow cube two fifths? (e.g., *There are 5 cubes and 2 of them are yellow.*)
- ▸ Why is the probability a fraction less than 1? (e.g., *Picking a yellow cube might happen but it might not, and the probability of 1 means it has to happen.*)

Using the Guided Intervention Student Resource pages 439–442

Work through the instructional section of the student pages together. Students can use calculators to help them with percent calculations.

Have students work through the **Try These** questions individually or in pairs.

Observe whether students

- can determine the theoretical probability of something happening (Questions 1, 2, 4, 5, 6, 7, 8, 9)
- can create a situation that leads to a required result (Question 3)
- can carry out an experiment to see how close the experimental probability is to the theoretical probability (Question 4)

Consolidating and Reflecting

Ensure understanding by asking questions such as these based on students' work:

- ▸ In Question 1, why does it matter that the sections of the circle are equal? (e.g., *If one section is bigger, then the chances you will land on it are greater.*)
- ▸ Why do the probabilities in Question 5 have a 4 in the denominator? (e.g., *There are 4 coins in the total group.*)
- ▸ In Question 6, does it matter if Sara gets a girl or a boy first? (e.g., *Yes, it matters because if she gets a boy first, then only the boy/girl outcome will work.*)
- ▸ In Question 8, how does the probability change if you don't replace the card? (e.g., *If you don't replace the card, there are fewer cards in total to choose from and one less shape. The probability will be out of 5, not 6.*)
- ▸ Why can you write a probability as a fraction or a percent? (e.g., *The denominator is the total number, and the numerator is the number of times you get what you want. A percent is like a fraction out of 100.*)
- ▸ When would the probability of something happening be 0% or 100%? (*when it's impossible or certain to happen*)
- ▸ Why can experimental probability change but theoretical probability can't? (e.g., *Probability can change every time you do an experiment. Theoretical is what you expect to happen. It depends only on the possibilities, which don't change.*)

Experimental Probability

You will need

- small paper clips
- dice
- centimetre cubes (or other small counters)
- Student Resource page 443

Open-Ended Intervention

Before Using the Open-Ended Intervention

Present the following situation: Suppose you walk to school the same way every day and cross the street at a set of traffic lights. The lights follow this pattern: 60 seconds green, 4 seconds yellow, and 60 seconds red. You leave your house at the same time every morning. You want to know the probability that you will have to wait at a red light before crossing. Ask:

- How would you predict the colour? (e.g., *I would think about what usually happens.*)
- How certain can you be of your prediction about what will happen tomorrow? (e.g., *If I knew what usually happens, I think the same is likely to happen, but not certain.*)
- If you predict there is a 50% chance of waiting at a red light, how many times in a week would you expect to wait at the light? (e.g., *maybe 2 times in 1 week*)

Using the Open-Ended Intervention Student Resource page 443

Provide the materials. Read through the tasks on the student page together. When students design their own game, encourage them to use a different outcome for the winner.

Give students time to work, ideally in pairs, and try out their games.

Observe whether students

- understand that experimental probabilities come from experiments
- make reasonable predictions about likelihood
- can conduct an experiment and keep track of results to determine the experimental probability of a particular outcome
- improve their predictions based on experience

Consolidating and Reflecting

Ensure understanding by asking questions such as these based on students' work:

- When you played the game, which numbered balloon reached the top first? (e.g., *I chose 1, 6, and 10. The balloon number 6 reached the top first.*)
- Is this what you would expect? (e.g., *Yes. There are more ways to get a sum of 6 with 2 number cubes than 1 or 10.*)
- Will this balloon always reach the top first? (e.g., *It is likely but not certain. The next time I play, another number could reach the top faster.*)
- Why is it a good idea to try the game a few times before picking your first "real" 3 numbers? (e.g., *so you know how the game works and you get a sense of how many number pairs there are*)
- When you created your own game, how did you decide how many balloons to use? (e.g., *When you find the difference of the 2 numbers you get, the least is 0 and the most is 5, so you need 6 balloons.*)

 Copyright © 2012 by Nelson Education Ltd.

Experimental Probability

Guided Intervention

You will need

- identical shapes such as tiles, counters, or cubes (2 colours)
- paper bags
- pencils and paper clips
- Spinners (BLM 18)
- Student Resource pages 444–446

Before Using the Guided Intervention

Present the following situation: Suppose you walk to school the same way every day and cross the street at a set of traffic lights. The lights follow this pattern: 60 seconds green, 4 seconds yellow, and 60 seconds red. You leave your house at the same time every morning. You want to know the probability that you will have to wait at a red light before crossing. Ask:

- ▸ How would you predict the colour? (e.g., *I would think about what usually happens.*)
- ▸ How certain can you be of your prediction about what will happen tomorrow? (e.g., *If I knew what usually happens, I think the same is likely to happen, but not certain.*)
- ▸ If you predict there is a 50% chance of waiting at a red light, how many times in a week would you expect to wait at the light? (e.g., *maybe 2 times in 1 week*)

Using the Guided Intervention Student Resource pages 444–446

Provide the materials and guide students through the instructional section.

Have students work through the **Try These** questions individually or in pairs.

Observe whether students

- recognize when outcomes are not equally likely (Question 1)
- can determine the experimental probability from the recorded results (Questions 1, 2, 5)
- can conduct a probability experiment and record results (Questions 1, 4, 5)
- can make reasonable predictions about likelihood (Questions 3, 5)
- can determine numbers to fit a given probability (Question 4)

Consolidating and Reflecting

Ensure understanding by asking questions such as these based on students' work:

- ▸ In Question 1, was there a number you expected to spin to more often? If so, which one and why? If not, why? (e.g., *The circle is divided into 6 equal parts, so the chances of spinning any of the 6 numbers should be the same.*)
- ▸ Why does it matter how many times you toss the coin in Question 3? (e.g., *If you toss 4 times, it is enough to get a probability, but not a very reliable one.*)
- ▸ How many times would you toss the coin to figure out an experimental probability that doesn't change much? (e.g., *a lot more, perhaps 50*)
- ▸ How did you decide how many cupcakes to put in the box in Question 4? (e.g., *I chose about twice as many chocolate as white so the probabilities would be* $\frac{2}{3}$ *and* $\frac{1}{3}$. *I did the experiment 9 times so it would be easy to get thirds.*)
- ▸ In Question 5, would you buy 2 raffle tickets instead of 1? Why? (e.g., *If I want a better chance of winning, I would buy 2 raffle tickets because the probability of winning a prize with 2 tickets is slightly better.*)

Copyright © 2012 by Nelson Education Ltd.

Name: ____________________ Date: ____________

BLM 1: Millions and Billions

World Populations

World	7 090 669 275
Canada	34 278 400
China	1 319 175 334
India	1.21 billion
United States	312.37 million
Brazil	195.34 million
Singapore	5.2 million
Asia	4.1 billion

Populations of Some Canadian Provinces and Cities

Ontario	13.21 million
Quebec	7 907 375
Alberta	3.72 million
Saskatchewan	1 045 622
Toronto	2 503 281
Montreal	1.62 million
British Columbia	4.5 million
Manitoba	1.2 million

Facts about Business

- A computer company bought a software company for $8.5 billion.
- A phone company sold 5.2 million phones in one summer.
- A computer company sold 1.2 million tablets a month.
- A home was recently sold in Vancouver for $17 900 000.
- In 2010, 10.9 billion apps were downloaded.
- In 2009, Canadian farmers produced 4 574 964 240 kg of potatoes.
- A car company recalled 2.2 million cars.
- In 2008, global sales of online games were $6.6 billion.

Solar System Facts

- Jupiter has a diameter of 142.98 million metres at its equator.
- Scientists have found a meteorite that is a remnant of a 4.2-billion-year-old super nova.
- The Sun's core has an average temperature of 13 599 726.8 °C.
- The Milky Way is estimated to be about 13.2 billion years old.
- Earth orbits the Sun at an average distance of about 150 million kilometres every year.
- In 1990, Voyager 1 took a photograph of Earth 6 054 571 520 km away.

Other Facts

- Our ancestors walked on 2 feet about 3.2 million years ago.
- In 2010, Lester B. Pearson International Airport handled 31.9 million passengers.
- There are 3.16 billion seconds in a century.
- The largest exporter of cotton is the United States, with sales of $4.9 billion a year.
- The length of Earth's equator is 400 300 200 000 cm.

Copyright © 2012 by Nelson Education Ltd.

Name:____________________ Date:__________

BLM 2: Place Value Charts (to Billions)

Billions			Millions			Thousands			Ones		
H	T	O	H	T	O	H	T	O	H	T	O

Billions			Millions			Thousands			Ones		
H	T	O	H	T	O	H	T	O	H	T	O

Billions			Millions			Thousands			Ones		
H	T	O	H	T	O	H	T	O	H	T	O

Billions			Millions			Thousands			Ones		
H	T	O	H	T	O	H	T	O	H	T	O

Copyright © 2012 by Nelson Education Ltd.

Name:______________________ Date:______________

BLM 3: Place Value Charts (to Hundred Thousands)

Hundred thousands	Ten thousands	Thousands	Hundreds	Tens	Ones

Hundred thousands	Ten thousands	Thousands	Hundreds	Tens	Ones

Hundred thousands	Ten thousands	Thousands	Hundreds	Tens	Ones

Hundred thousands	Ten thousands	Thousands	Hundreds	Tens	Ones

Copyright © 2012 by Nelson Education Ltd.

ame: _______________ Date: _______________

LM 4: Place Value Charts (to Millionths)

Ones	Tenths	Hundredths	Thousandths	Ten thousandths	Hundred thousandths	Millionths

Ones	Tenths	Hundredths	Thousandths	Ten thousandths	Hundred thousandths	Millionths

Ones	Tenths	Hundredths	Thousandths	Ten thousandths	Hundred thousandths	Millionths

Ones	Tenths	Hundredths	Thousandths	Ten thousandths	Hundred thousandths	Millionths

Copyright © 2012 by Nelson Education Ltd.

Name:______________________ Date:______________

BLM 5: Place Value Charts (to Ten Thousandths)

Hundreds	Tens	Ones	Tenths	Hundredths	Thousandths	Ten thousandths

Hundreds	Tens	Ones	Tenths	Hundredths	Thousandths	Ten thousandths

Hundreds	Tens	Ones	Tenths	Hundredths	Thousandths	Ten thousandths

Hundreds	Tens	Ones	Tenths	Hundredths	Thousandths	Ten thousandths

Copyright © 2012 by Nelson Education Ltd.

Name: ____________________ Date: ____________________

BLM 6: Place Value Charts (to Thousandths)

Hundreds	Tens	Ones	Tenths	Hundredths	Thousandths

Hundreds	Tens	Ones	Tenths	Hundredths	Thousandths

Hundreds	Tens	Ones	Tenths	Hundredths	Thousandths

Hundreds	Tens	Ones	Tenths	Hundredths	Thousandths

Copyright © 2012 by Nelson Education Ltd.

Name:____________________ Date:____________________

BLM 7: Thousandths Grids

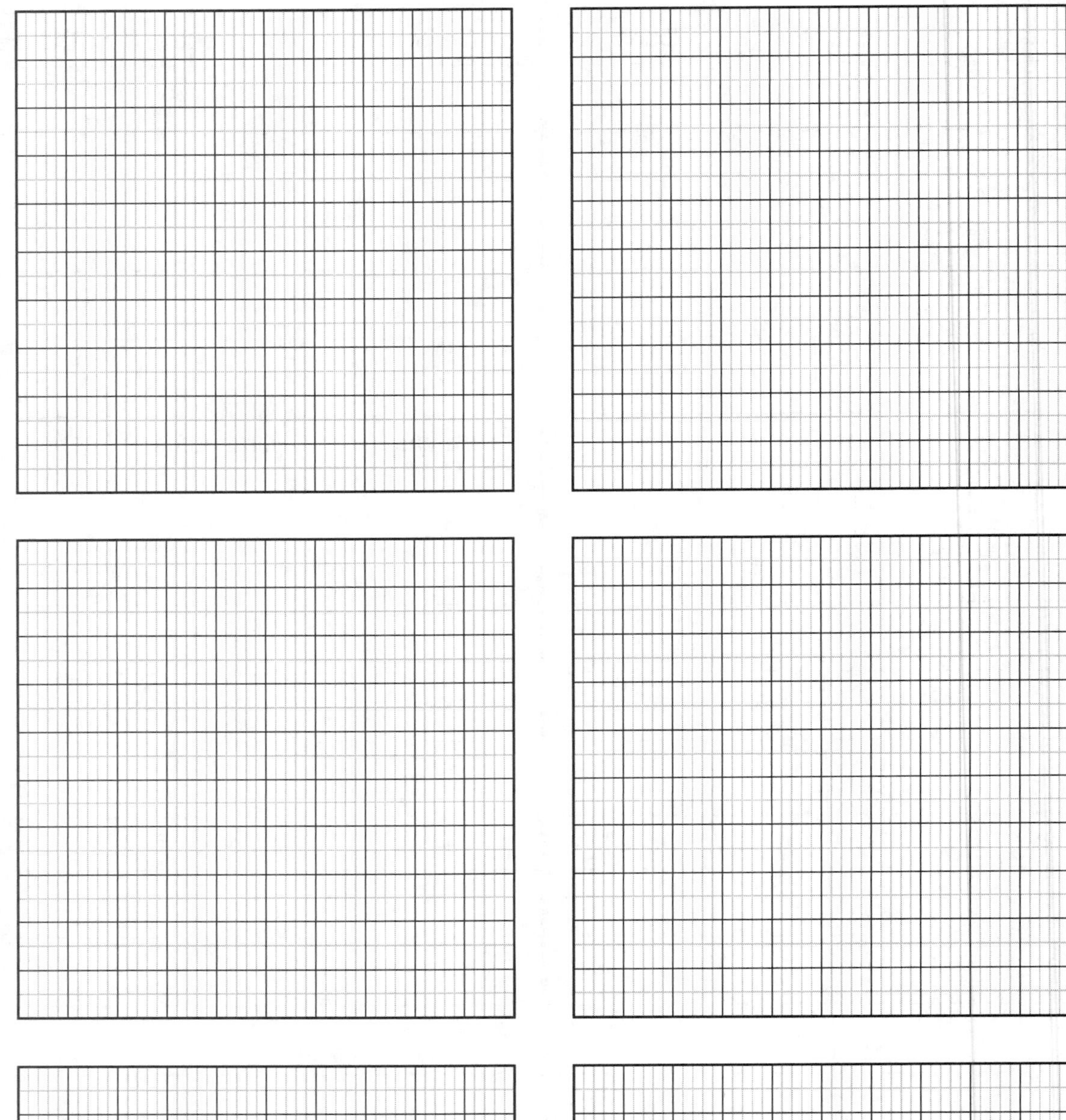

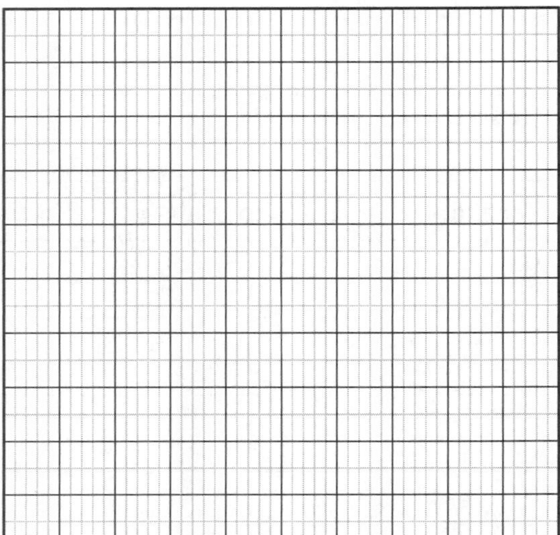

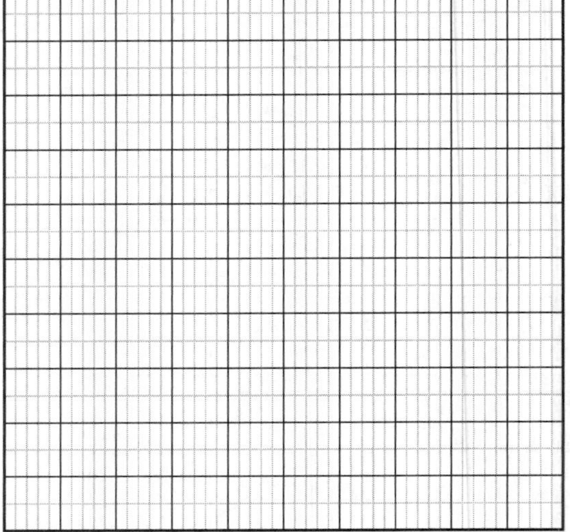

Copyright © 2012 by Nelson Education Ltd.

Name: ______________________________ Date: ______________

BLM 8: Hundredths Grids

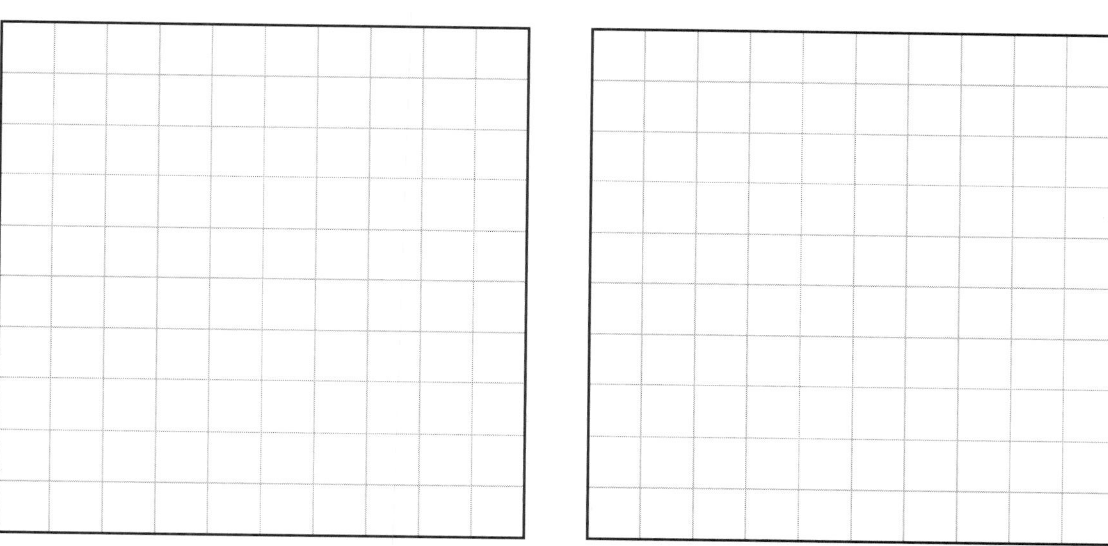

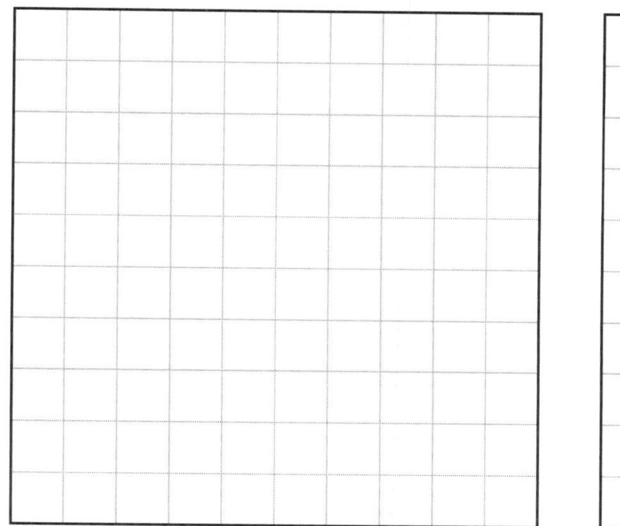

 Copyright © 2012 by Nelson Education Ltd.

Name:________________________ Date:________________

BLM 9: Tenths Grids

Copyright © 2012 by Nelson Education Ltd.

Name:___________________________ Date:_______________

BLM 10: Fraction Strips

 Copyright © 2012 by Nelson Education Ltd.

Name: ______________________________ Date: ______________

BLM 11: 2 cm Grid Paper

Copyright © 2012 by Nelson Education Ltd.

Name: ______________________ Date: ______________

BLM 12: 1 cm Grid Paper

 Copyright © 2012 by Nelson Education Ltd.

Name:______________________ Date:______________________

BLM 13: 0.5 cm Grid Paper

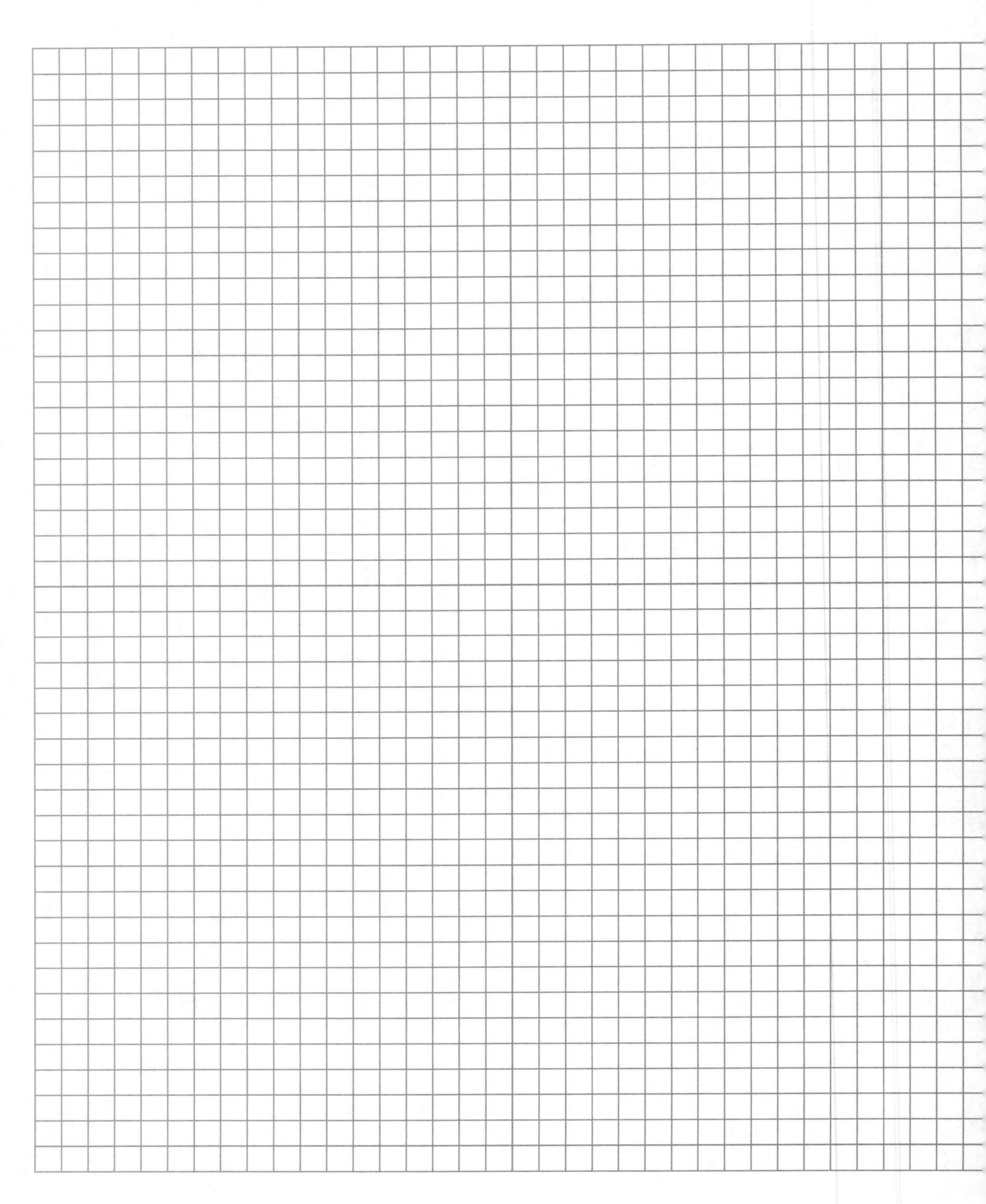

Name:________________________ Date:________________

BLM 14: 2 cm Square Dot Paper

Copyright © 2012 by Nelson Education Ltd.

Name:________________________________ Date:________________

BLM 15: 1 cm Square Dot Paper

Name:______________________________ Date:____________________

BLM 16: Triangle Dot Paper

 Copyright © 2012 by Nelson Education Ltd.

Name: ____________________ Date: ____________________

BLM 17: Circular Geoboard Paper

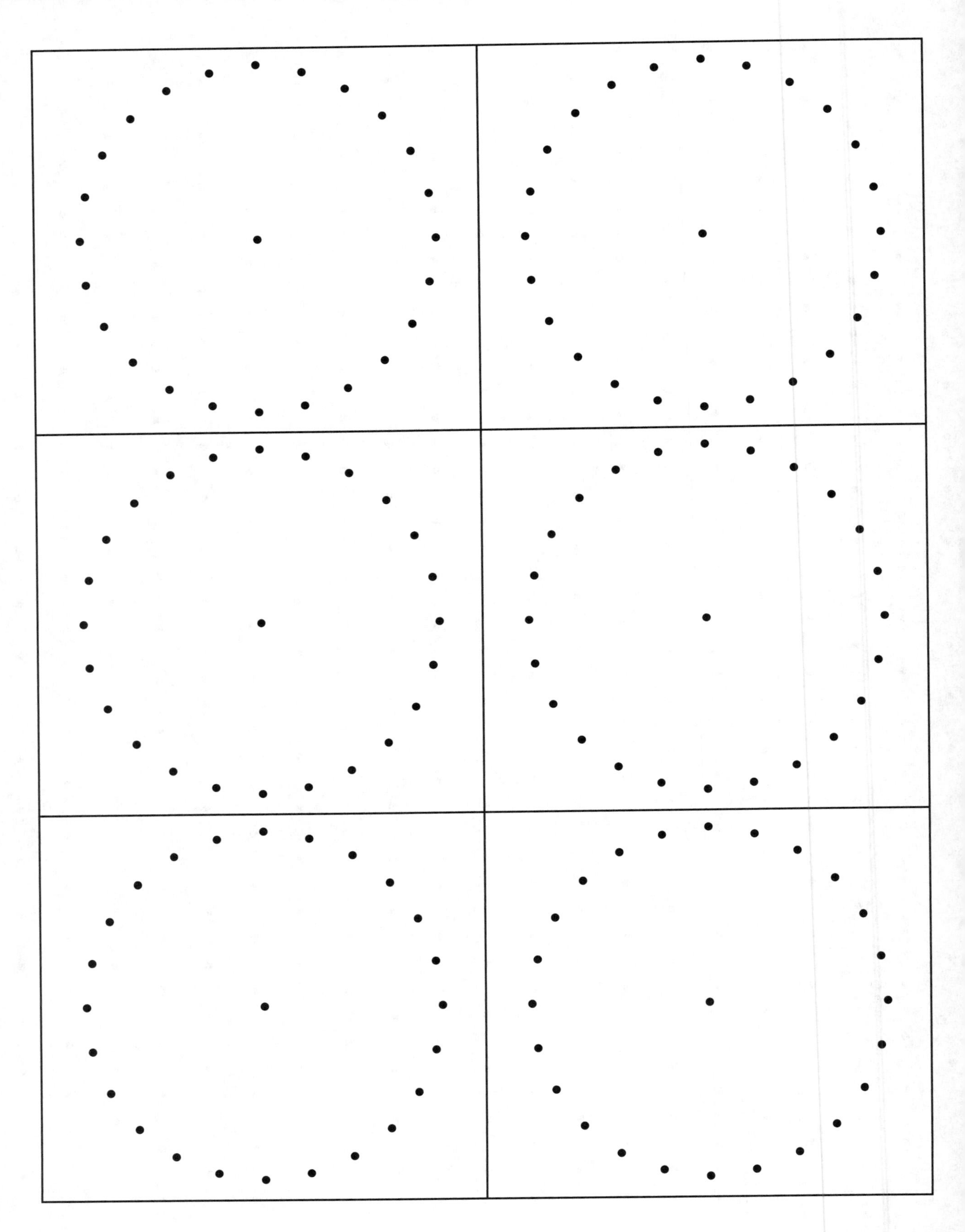

ame:______________________ Date:______________________

3LM 18: Fraction Circles/Spinners

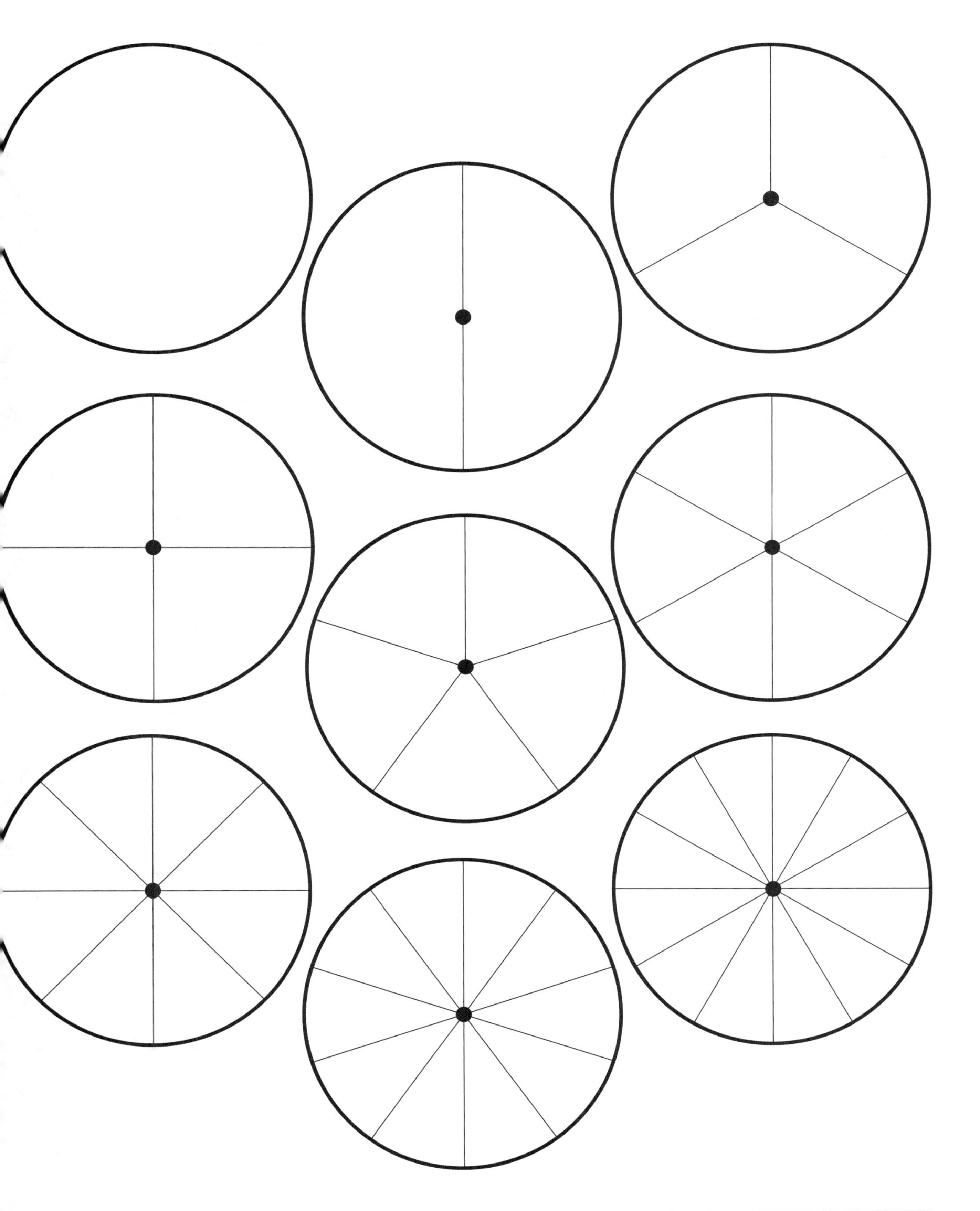

Copyright © 2012 by Nelson Education Ltd.

Name: _______________ Date: _______________

BLM 19: Pattern Blocks: Triangle

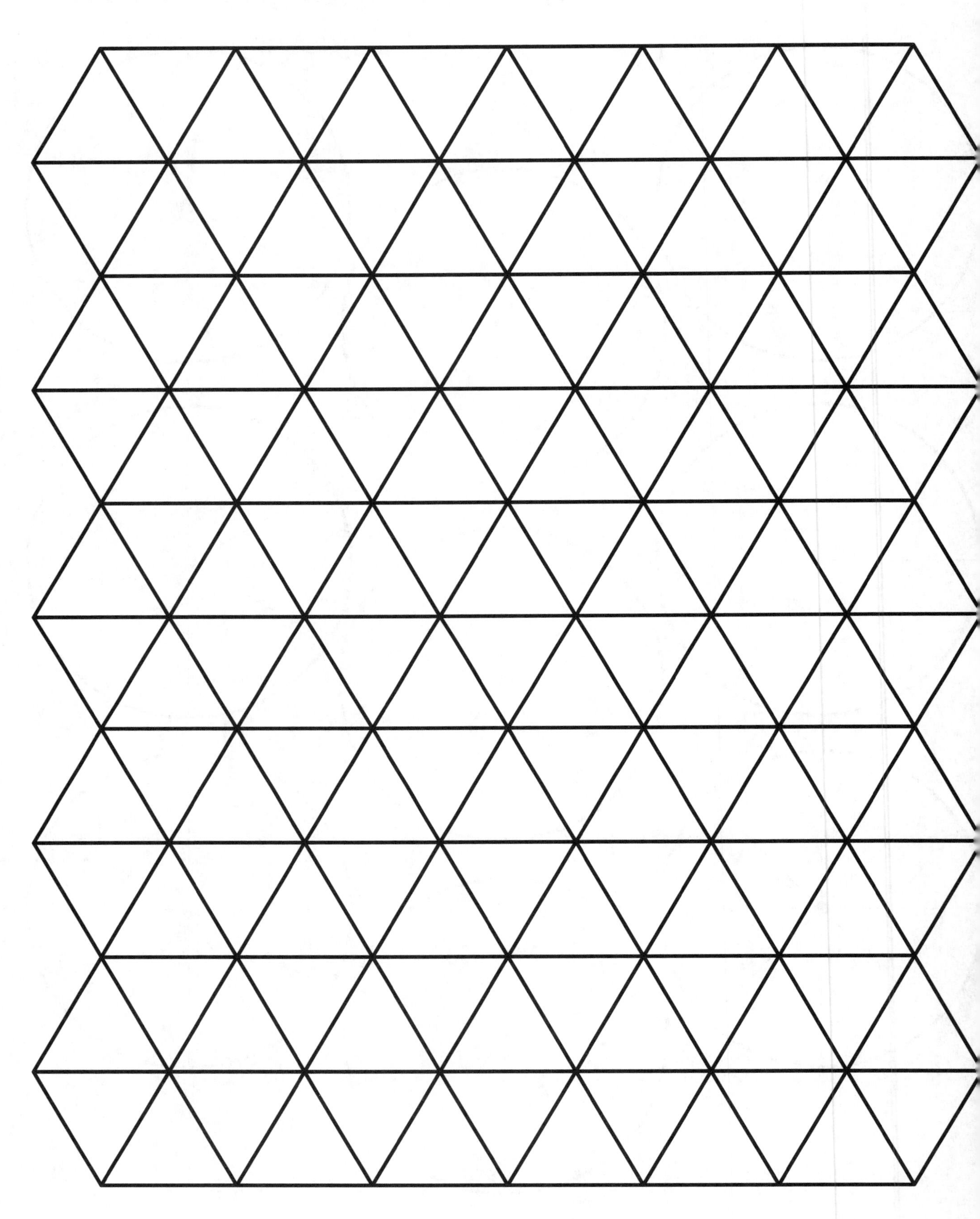

Copyright © 2012 by Nelson Education Ltd.

ame:______________________________ Date:____________________

LM 20: Pattern Blocks: Square

 Copyright © 2012 by Nelson Education Ltd.

Name:______________________ Date:______________________

BLM 21: Pattern Blocks: Rhombus A

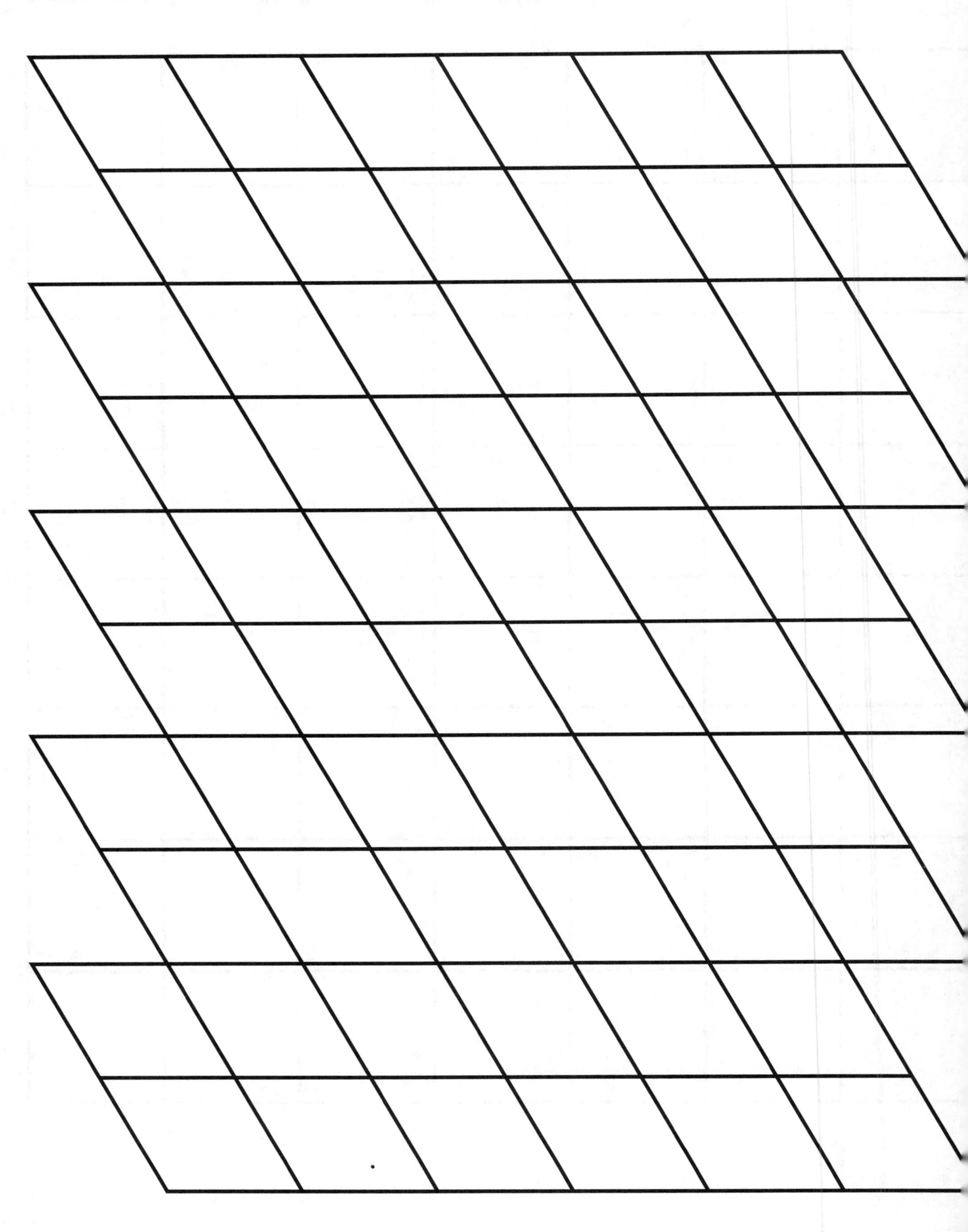

Name: ______________________________ Date: ______________

BLM 22: Pattern Blocks: Rhombus B

 Copyright © 2012 by Nelson Education Ltd.

Name:______________________________ Date:______________________

BLM 23: Pattern Blocks: Trapezoid

Name: ______________________ Date: ______________

LM 24: Pattern Blocks: Hexagon

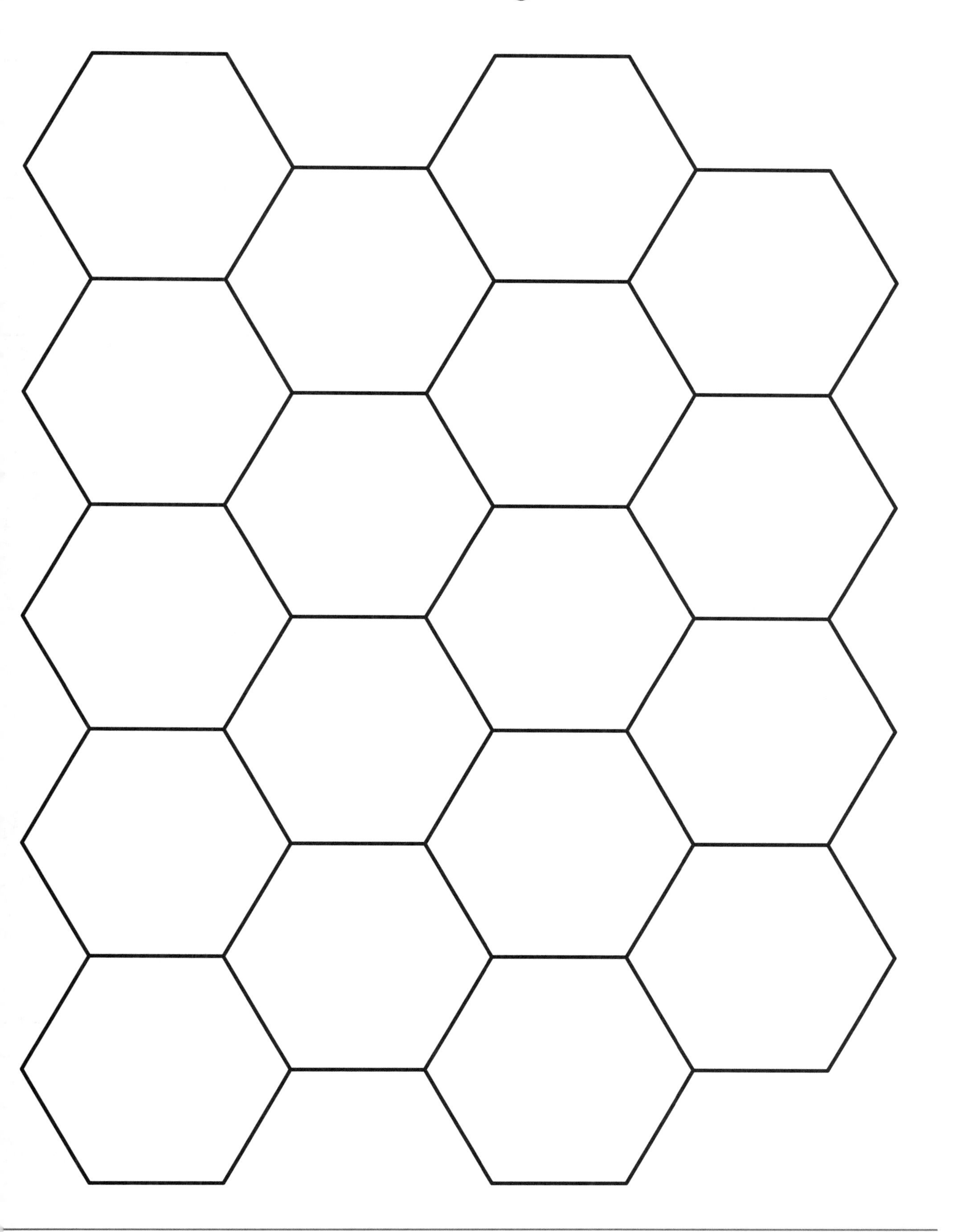

 Copyright © 2012 by Nelson Education Ltd.

Name:______________________________ Date:______________

BLM 25: Number Lines

Copyright © 2012 by Nelson Education Ltd.

ıme: _______________ Date: _______________

LM 26: Regular Polygons

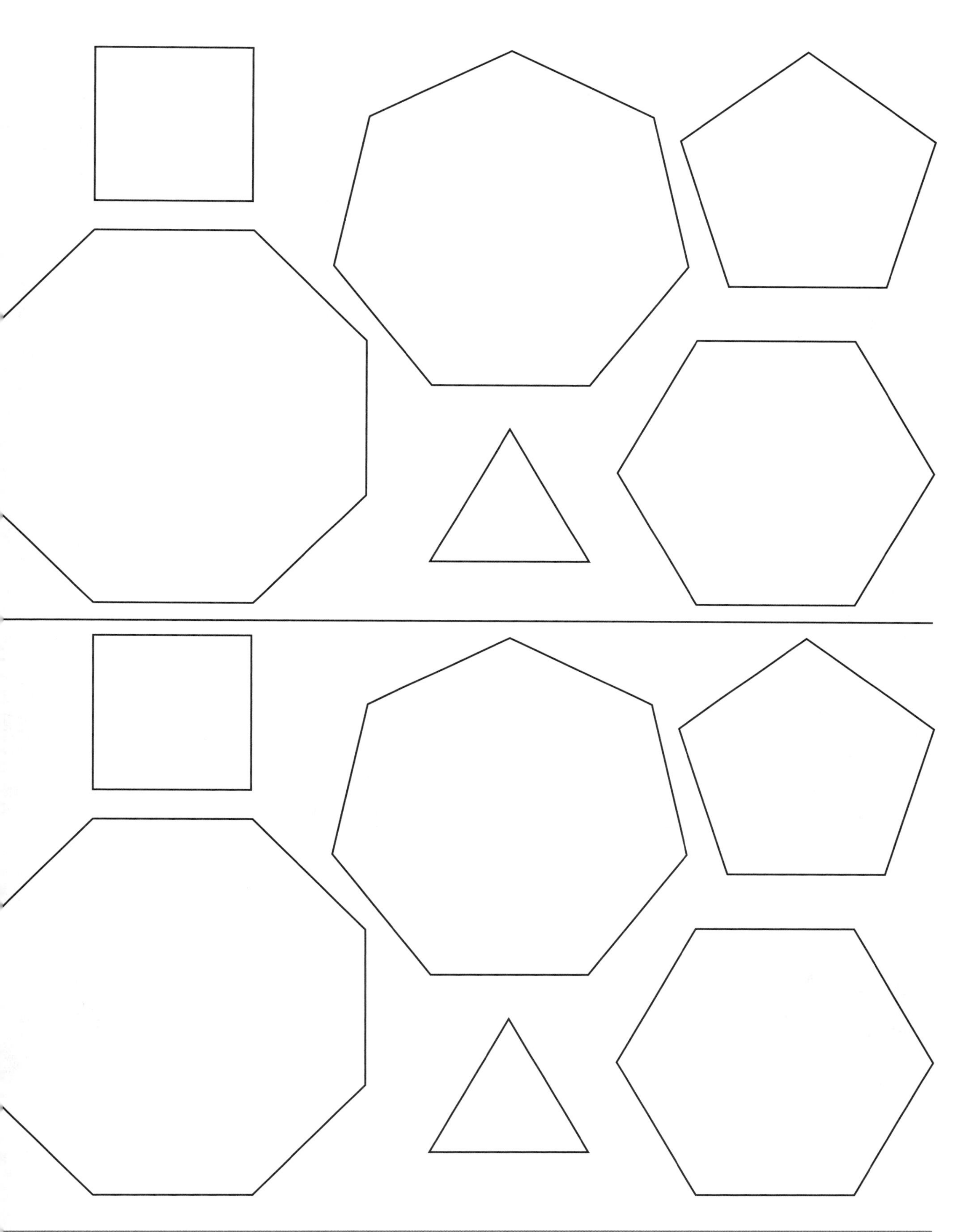

 Copyright © 2012 by Nelson Education Ltd.

Name: ______________________________ Date: ______________

BLM 27: Polygons

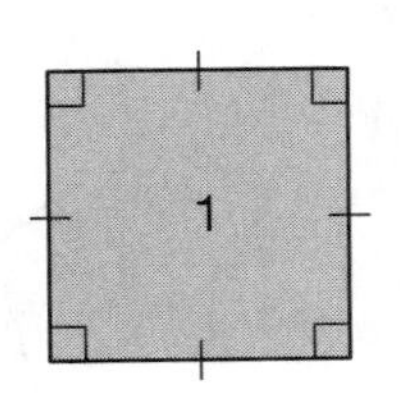

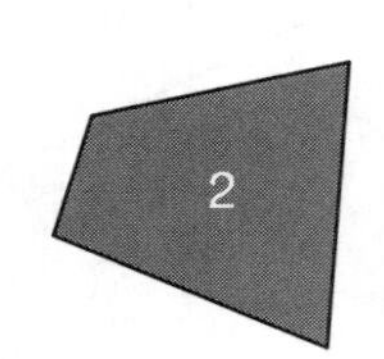

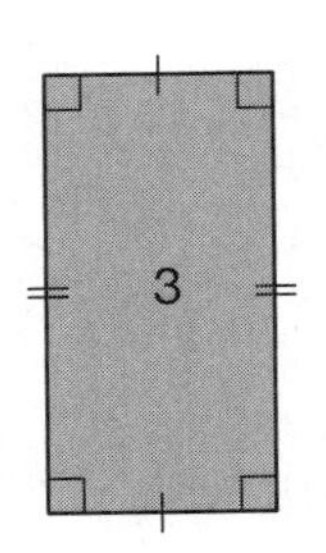

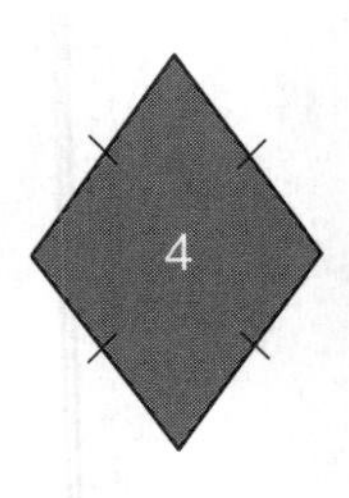

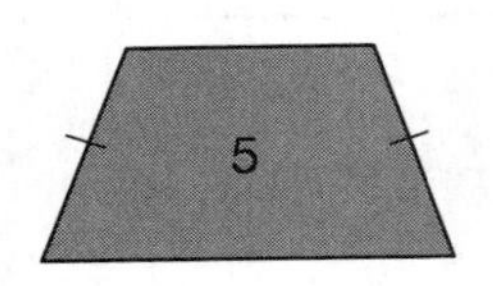

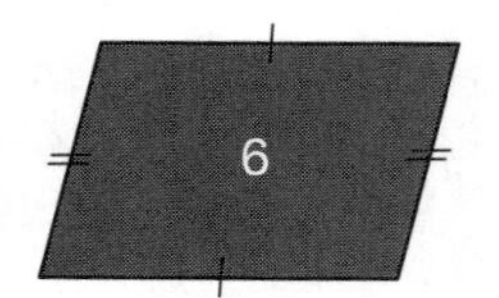

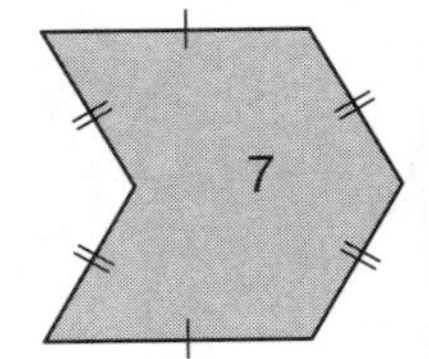

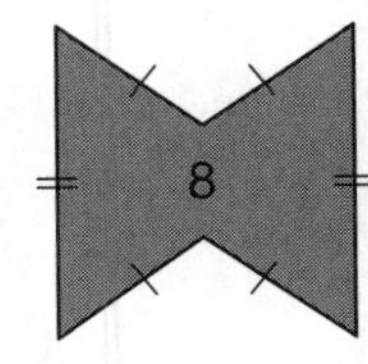

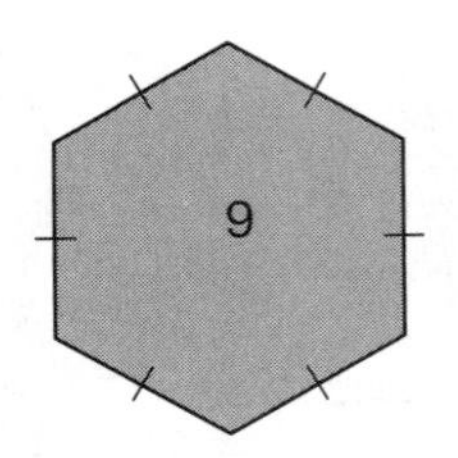

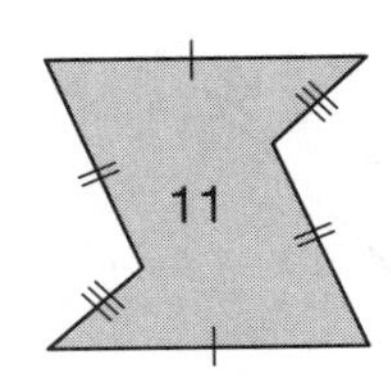

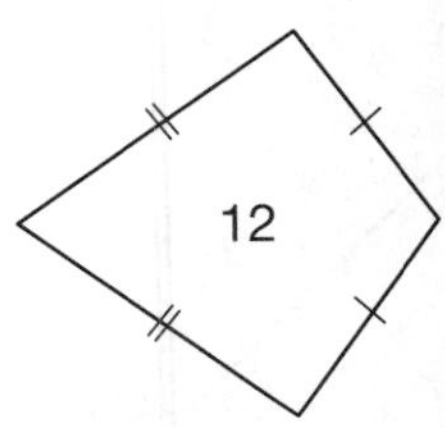

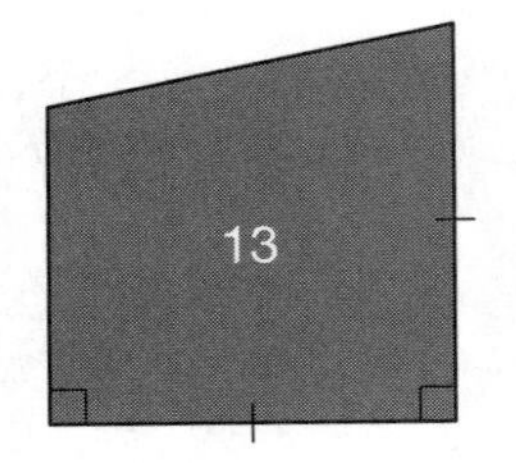

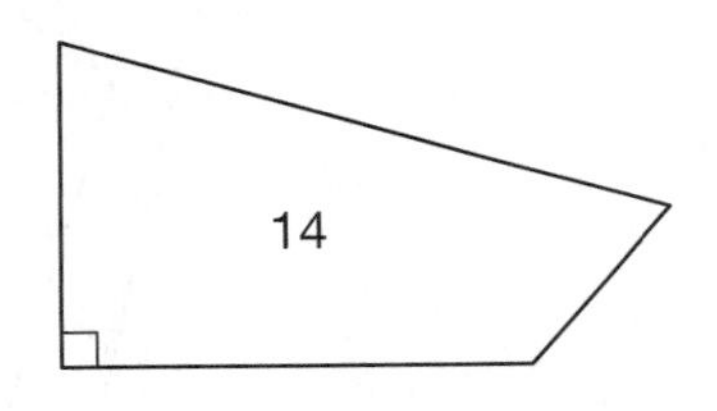

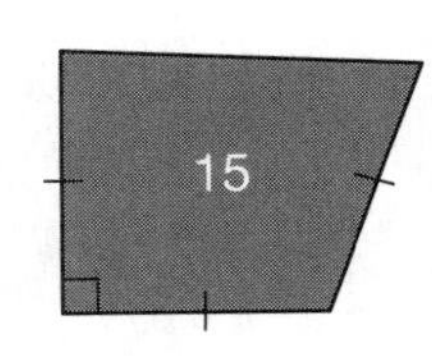

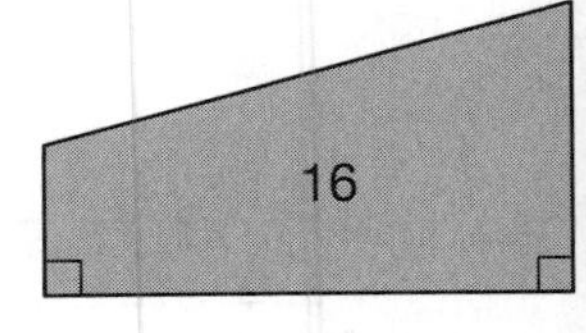

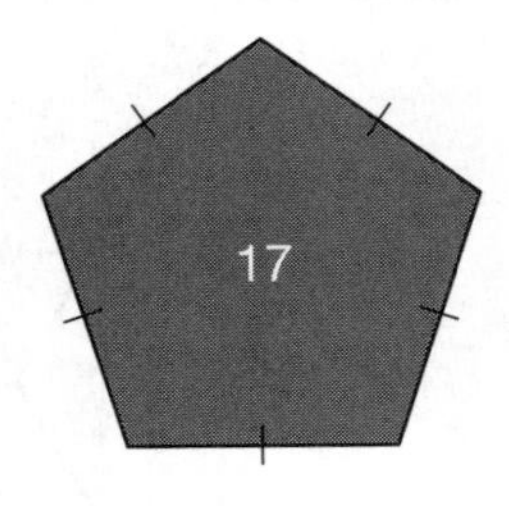

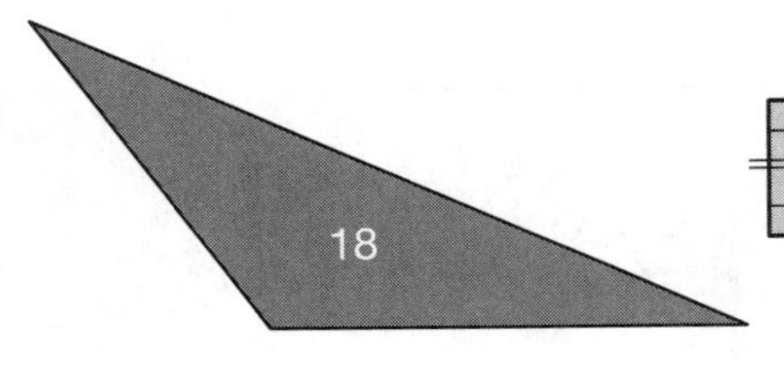

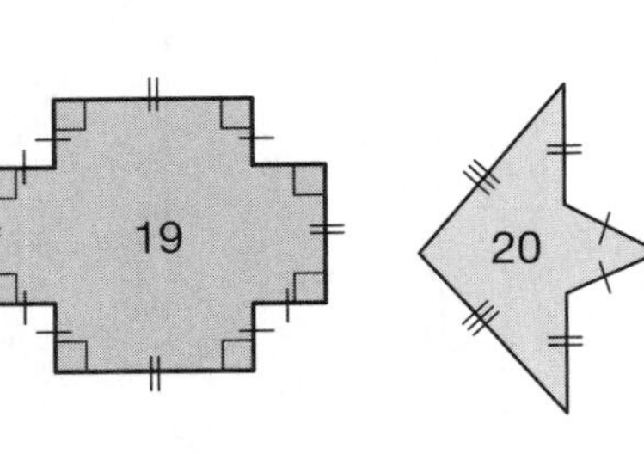

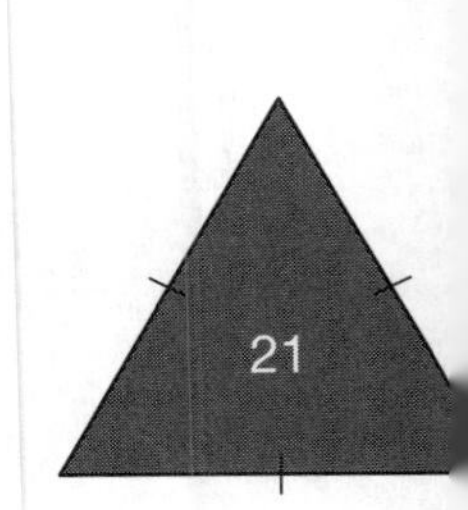

Copyright © 2012 by Nelson Education Ltd.

Name: ______________________ Date: ______________________

BLM 28: Star

 Copyright © 2012 by Nelson Education Ltd.

Name:______________________________ Date:______________

BLM 29: Roof Angles

Name: ______________________ Date: ______________________

BLM 30: Types of Graphs

Types of Crops Over 10 Years

	Year 0	Year 5	Year 10
grains	24	26	28
fruits	6	6	8
cattle food	9	11	12
vegetables	16	18	22
flowers/trees	5	5	5

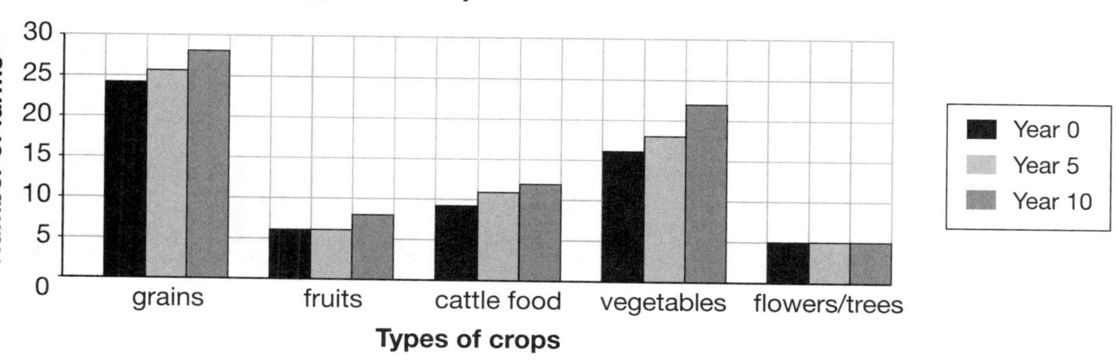

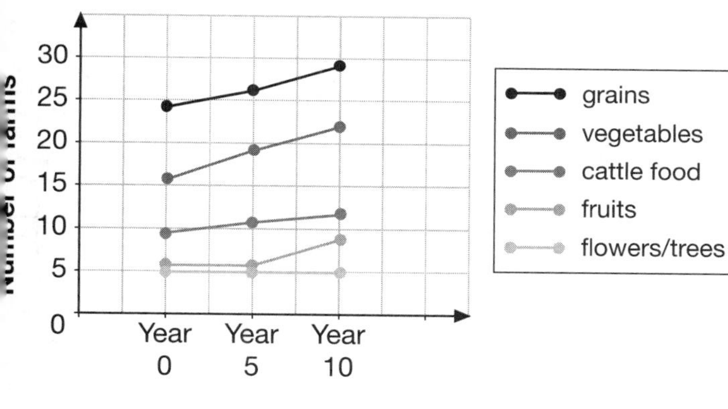

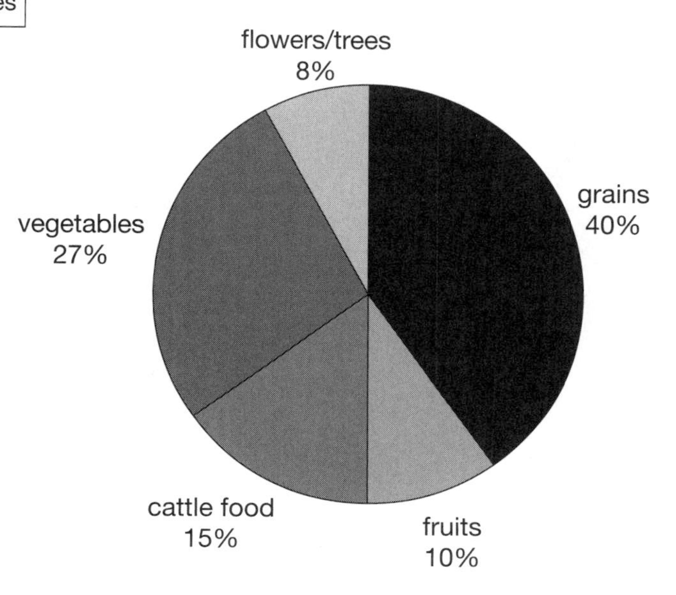

Copyright © 2012 by Nelson Education Ltd.

Name:______________________ Date:______________________

BLM 31: Percent Circles

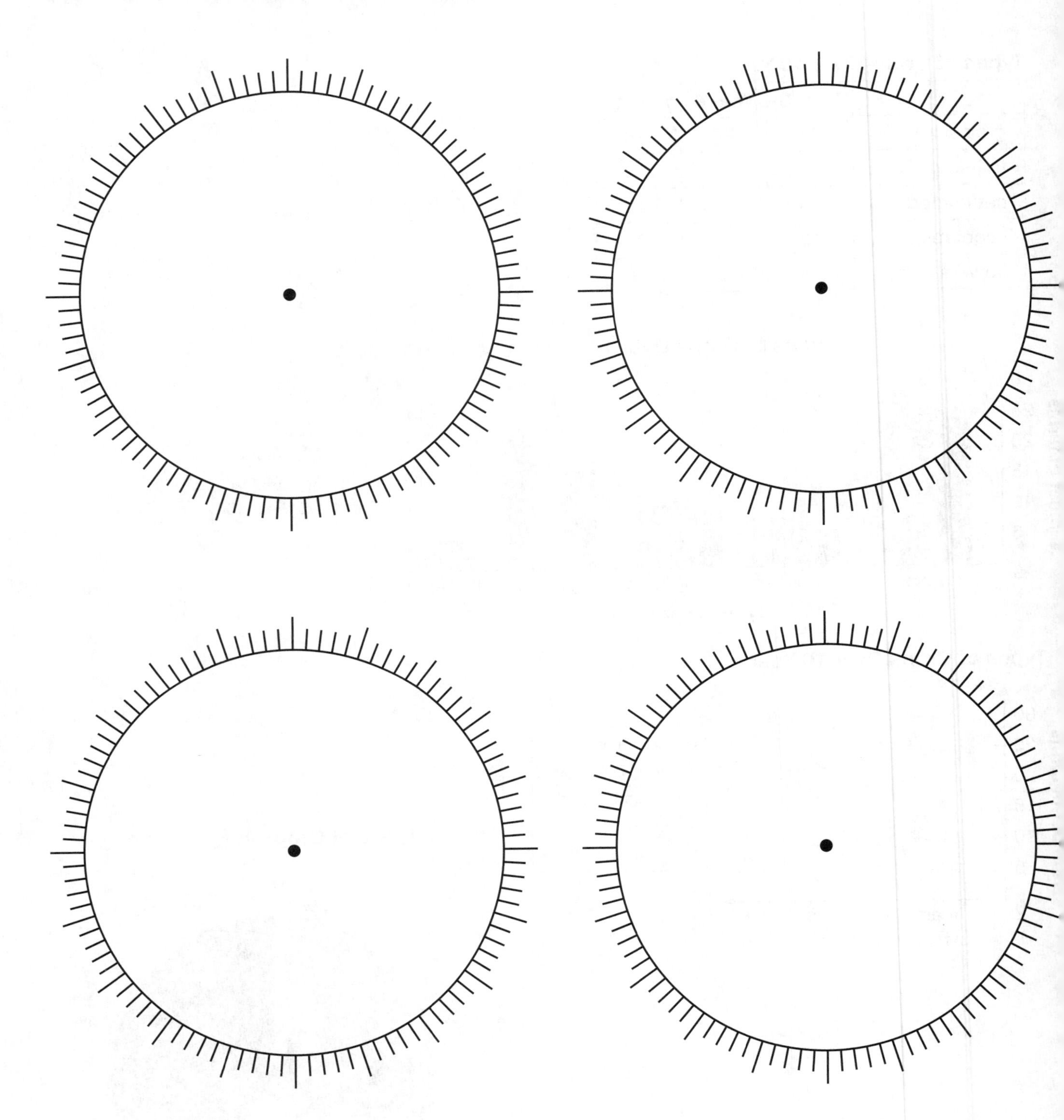

Copyright © 2012 by Nelson Education Ltd.

Index

Copyright © 2012 by Nelson Education Ltd.

E

F

 Copyright © 2012 by Nelson Education Ltd.

Copyright © 2012 by Nelson Education Ltd.

N

O

P

 Copyright © 2012 by Nelson Education Ltd.

R

S

T

 Copyright © 2012 by Nelson Education Ltd.